ROOFING HANDBOOK

Robert Scharff (Deceased)

Terry Kennedy

Second Edition

McGraw-Hill

New York   San Francisco   Washington, D.C.   Auckland   Bogotá
Caracas   Lisbon   London   Madrid   Mexico City   Milan
Montreal   New Delhi   San Juan   Singapore
Sydney   Tokyo   Toronto

Library of Congress Cataloging-in-Publication Data

Scharff, Robert.
    Roofing handbook / Robert Scharff, Terry Kennedy.—2nd ed.
       p.   cm.
    ISBN 0-07-136058-1
    1. Roofing. I. Kennedy, Terry. II. Title.

TH2431.S28   2000
695—dc21   00-062522

McGraw-Hill

A Division of The McGraw·Hill Companies

1 2 3 4 5 6 7 8 9 0 DOC/DOC 0 6 5 4 3 2 1 0

P/N 137056-0
PART OF
ISBN 0-07-136058-1

The sponsoring editor for this book was Zoe G. Foundotos, the editing supervisor was Stephen M. Smith, and the production supervisor was Sherri Souffrance. It was set in Melior following the modified CMS 6x9 design by Joanne Morbit and Michele Pridmore of McGraw-Hill's Hightstown, N.J., Professional Book Group composition unit.

Printed and bound by R. R. Donnelley & Sons Company.

This book is printed on recycled, acid-free paper containing a minimum of 50% recycled, de-inked fiber.

McGraw-Hill books are available at special quantity discounts to use as premiums and sales promotions, or for use in corporate training programs. For more information, please write to the Director of Special Sales, Professional Publishing, McGraw-Hill, Two Penn Plaza, New York, NY 10121-2298. Or contact your local bookstore.

CONTENTS

ACKNOWLEDGMENTS

This work is dedicated to Mr. Robert Scharff. His fine, clear writing style and mastery of the material are the foundation of the book.

Many thanks to Michelle Martineau for her constant help and diligent approach, to Zoe Foundotos for her patience with endless questions and her ability to stay warm and human while being the consummate professional, to Stephen Smith for his grounded and steady hand in the production of this book, and to Emily Rose Kennedy for continual inspiration.

Terry Kennedy

ABOUT THE AUTHORS

Robert Scharff (deceased) was well-known as the author of more than 300 how-to books. He served as a consultant to industry manufacturers and contributed many articles to leading how-to and trade publications.

Terry Kennedy has been a part of the AEC (architecture, engineering, contracting) business for 35 years. He has spent many hours in the field as well as in the office, working with computers since the 1980s. He is currently involved in the settlement of complex construction defect lawsuits, and that work has given him an in-depth knowledge of roofing. Mr. Kennedy has a great deal of respect for the roofing industry and wishes to thank the readers, one and all.

INTRODUCTION

This book was originally written for roofing contractors and crews involved with residential and commercial projects. It has been rewritten to make it an even more valuable companion to the roofing professional's workday. Its purpose is to serve as a reference for both people with broad experience and those who are learning about roofing construction and maintenance. The new companion CD offers live versions of checklists and the resources presented in App. A for immediate use.

The information in the *Roofing Handbook* is suitable for anyone who is involved with roofs in any way: architects, general contractors, building owners, insurance professionals, experts, and equipment manufacturers. The list of people who require knowledge of roofs is long, and this book is a handy reference for any of them.

It has a clear, easy-to-understand format and contains step-by-step construction procedures as well as up-to-date product information and techniques. It also takes the reader through the business of roofing, estimating, and the steps to take to win a contract and launch a job. The techniques related to different types of roofs are covered on a chapter-by-chapter basis, and equipment purchases and safety concerns are also reviewed.

Measures that can protect a business, such as warranties, and contract information to help the roofing contractor maintain a profitable bottom line, including suggestions for new business that is both innovative and complementary to the roofer's present work, are also explored.

Chapter 1 provides a general overview of the roofing industry, including life expectancy and a review of the various types of roofs.

How do you plan for and bid on a roofing job? Do you know your actual cost of doing business? Are you making a profit? How do you coordinate your time on the roof with the other trades? What does customer satisfaction involve? These transitional parts—moving the beginning parts of a job from the office into the field—are covered in Chap. 2, "Planning a Job."

Safety is such an important part of roofing that we dedicate all of Chap. 3 to the subject. From Occupational Safety and Health Administration (OSHA) guidelines to ordinary common-sense decisions, this chapter looks at how to prevent injury. The *Roofing Handbook* even provides a safety plan that is important for a company's workers, increases everyone's sense of well-being, and can actually have a serious effect on bottom-line profits. The safety plan, along with many other lists, is available on the enclosed CD for printing and using as a part of doing business.

Roofing equipment, from hand tools to heavy lifting machinery, is described in Chap. 4. In Chap. 12, the business chapter, the subject of major investments is discussed as a part of the general course of growing a business. In Chap. 4, the subject is examined specifically in relation to how an owner evaluates purchases of expensive equipment. Will the payoff from the purchase be timely for the health of the company or put too much of a crimp in the flow of cash?

Chapter 5 details the techniques of working with built-up roofing. This popular roofing method can be applied successfully to a variety of substrates, and the information in this chapter tells how.

Lightweight and easy to install, single-ply roofing is a popular choice among building owners and roofing contractors. Chapter 6 looks at the available materials and the appropriate installation methods.

Asphalt roofing comes in a variety of forms and can be applied in pleasing designs to new roofs or over existing roofs that have been prepared properly. Chapter 7 details installation methods, as well as roofing accessories such as flashings, gutters, and downspouts.

Chapter 8 covers the special installation techniques required for wooden shingles and shakes. They can now be made fire- and wind-resistant if they are manufactured well and application procedures are followed professionally.

Chapter 9 provides valuable information concerning the installation of slate roofing materials, including repair procedures for existing roofs. With a life expectancy of 75 to 100 years, slate is a fine and stately option for high-quality jobs. Chapter 10 follows up with an in-depth guide to clay and concrete tiles. As everyone involved with roofing knows, concrete tile roofs have spread from Europe to the United States and become more important every year.

Metal roofing, from thin copper to corrugated sheet metal, has been a popular roofing material for many years. Chapter 11 looks at the many benefits offered by metal roofs in both new and reroofing applications.

Chapter 12 guides the roofer through a breakdown of the business of roofing. These basic principles were also noted in other chapters when the material came up. Chapter 13 takes the information reviewed in Chap. 12 and digs into what computers can actually do to enhance efficiency on a nuts-and-bolts basis. The intelligent use of computers is a must in order to be successful at business in the twenty-first century.

Chapter 14 offers the reader a look at some types of work that can augment the cash flow of your roofing business. It also offers a glimpse of the expanding future of the roofing industry. The twenty-first century is ripe with changes and opportunity. Roofing material manufacturers are now marketing their own roof drains and supporting specifications that make roofers responsible for the installation of these features. Nondestructive testing methods are available to roofers as a means of assessing existing roofs for moisture content. The list of possibilities is long and growing.

And to wrap things up, lists of vendors of computer hardware, software, roofing equipment, and supplies, along with a reference to trade groups, magazines, and sources of further information, and a metric conversion table are found in Apps. A and B.

An Overview of Roofing

Throughout history, mankind's need for shelter was second only to the need for food. Prehistoric man took shelter under a roof of stone, arguably the best protection from the elements.

Unfortunately, caves were not available everywhere, so early man looked for a substitute roof. And one of the earliest and still most important principles of roofing was discovered—lapping. When the length of any covering for a building, including the walls, is too short to protect the entire run of a wall or roof, it must be installed in rows, or courses, and each course, beginning at the top, must lap over the next course below. As simple and obvious as this sounds, it is very important for draining water off of a built struture.

In many parts of the world, abundant natural fibers, such as grass, sticks, heather, and straw, were woven into effective shields against the weather. Thatched, pitched roofs, which are still common in parts of Africa, Asia, South America, and Polynesia, proved to be remarkably efficient. Raindrops travel along each reed several inches and then, before they can penetrate the bundle, are conveyed to the roof eave by lapping the courses. From the eave, the raindrops fall harmlessly to the ground.

America's first settlers from Europe continued to use thatched roofs in this country, particularly in the East. As the settlers pushed westward, however, the available hay was needed for animal fodder and other purposes. Settlers searching for an alternative roofing material found it, literally, at their feet. *Kansas bricks*, made of sod, soon became the roofing standard. The sod bricks were heavy and difficult to manage, and, even worse, they often leaked during heavy rainstorms.

Time marched on and so did the search for the perfect roof. Today there are many fine roof coverings available. Asphalt shingles and roll roofing, clay and ceramic tile, metal, slate, wooden shingles and shakes, and cement panels and tiles are used mainly for residential projects. Built-up roofing (BUR) and single-ply membrane roofs are typically used on commercial structures.

Before we take a look at roof classifications and materials, let's consider the three factors that architects, builders, designers, and property owners must address before they choose a roof.

Fire Safety and Protection

Fire safety is a particularly important consideration, since the roof is vulnerable to fire from overhead or airborne sources.

The fire resistance of roofing materials is tested by the Underwriters' Laboratories, Inc. (UL), an independent, not-for-profit public safety testing laboratory. UL established the standard for the testing of roofing materials with the assistance of nationally recognized fire authorities.

Manufacturers voluntarily submit materials for testing. The materials then are classified and labeled according to the classes below. The American Society for Testing and Materials (ASTM) is a voluntary organization concerned with the development of consensus standards, testing procedures, and specifications.

Class A. The highest fire-resistance rating for roofing as per ASTM E-108. This class rating indicates that the roofing material is able to withstand severe exposure to fire that originates from sources outside the building.

Class B. This fire-resistance rating indicates that the roofing material is able to withstand moderate exposure to fire that originates from sources outside the building.

Class C. This fire-resistance rating indicates that the roofing material is able to withstand light exposure to fire that originates from sources outside the building.

Many communities require new roof coverings to meet at least the UL Class C standard. This requirement has the backing of nationally recognized authorities such as the National Fire Protection Association and the International Association of Fire Chiefs.

Some roofing materials, such as slate and clay tile, offer natural fire protection. Asphalt shingles are manufactured to meet the Class C standard or better. Readily combustible materials, such as wooden shingles, do not meet the UL standard unless they have been chemically treated for fire retardancy (see Chap. 9).

Wind Resistance

UL also tests shingle performance against high winds. To qualify for the UL wind-resistant label, shingles must withstand continuous test winds of at least 60 miles per hour for two hours without a shingle tab lifting. Wind-resistant shingles demonstrated their effectiveness under hurricane conditions during the winds brought to Florida in 1992 by Hurricane Andrew, which at the time was the worst hurricane ever recorded in the United States.

Self-sealing asphalt shingles that bear the UL wind-resistant label are manufactured with a factory-applied adhesive. Once the shingles are applied, the sun activates the preapplied thermoplastic sealant and each shingle is bonded to the one below it. Although self-sealers originally were developed specifically for high-wind areas, they are standard in most parts of the country today.

Estimating Life Expectancy

In addition to fire and wind ratings, roof coverings have a life expectancy. Manufacturers offer a guarantee or warranty for their products. For example, most of today's asphalt shingles are designed to provide satisfactory service for 15 to 25 years.

Generally, the longer the life expectancy, the more expensive the shingle material. You might find, however, that a more expensive shingle is the most economical in the long run because the cost of materials and labor is amortized over a longer period of time. Keep in mind that while the labor cost to apply the shingles varies with the product, it is the same whether the life expectancy of the shingle is 15 or 25 years.

Estimate the probable annual cost of a new roof by adding the cost of labor and materials and then dividing the total by the shingle's design life. The formula for determining the annual cost of use is

$$\frac{\text{Total cost (materials and labor)}}{\text{Design life}} = \text{annual cost of use}$$

Classifying Roofs

The National Roofing Contractors Association (NRCA) classifies roofs into two major categories: low slope roofs and steep-slope roofs. Slope is defined as the degree of roof incline expressed as the ratio of the rise, in inches, to the run, in feet (Fig. 1-1). For example, if the span of a roof is 24 feet and the rise is 8 feet, the pitch is 8/24 or 1/3.

Expressed as a slope, the same roof is said to rise 8 inches per 12 inches of horizontal run. If the rise of the same roof span were 6 feet, the pitch would be 1/4 and its slope would be 6 inches per 12 inches of run. Whether a particular roof incline is expressed in pitch or slope, the results of area calculations are the same.

Building Low-Slope Roofs

Low-slope roofs can have slopes as minor as ⅛ inch per 12 inches. These roofs employ a waterproof roofing system and are found primarily on commercial structures.

A low-slope roof system generally consists of a roof membrane, insulation, and one of a number of surfacing options. To control the application and improve the quality of low-slope roofing, a variety of specifications and procedures apply to the assembly of the roofing components. These specifications and procedures are generally accepted and used throughout the United States. Roofing systems that meet these specifications normally can be expected to give satisfactory service for many years.

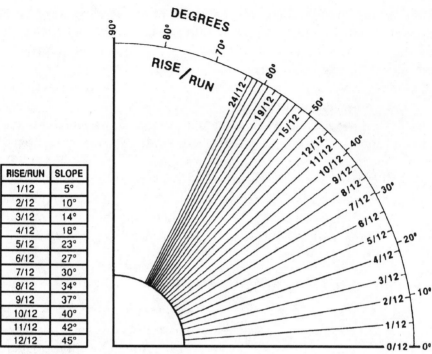

RISE/RUN	SLOPE
1/12	5°
2/12	10°
3/12	14°
4/12	18°
5/12	23°
6/12	27°
7/12	30°
8/12	34°
9/12	37°
10/12	40°
11/12	42°
12/12	45°

FIGURE 1-1 Degrees of slope.

Climatic conditions and available materials dictate regional low-slope procedures, which can vary greatly in different parts of the country. Low-slope roofs are essentially a custom product. They are designed for a specific building, at a specific location, and manufactured on the jobsite.

Membrane Components

Low-slope membranes are composed of at least three elements: waterproofing, reinforcement, and surfacing. Some materials within the membrane might perform more than one function. The waterproofing agent is the most important element within the roof membrane. In BUR and modified bitumen roofing (MBR), the waterproofing agent is bitumen. In single-ply roofing, the waterproofing agent is synthetic rubber or plastic (for more information, see Chaps. 7 and 8).

The reinforcement element provides stability to the roof membrane; it holds the waterproofing agent in place and provides tensile strength. In BUR, reinforcement is typically provided by organic or glass-fiber roofing felts. In MBR, the reinforcement is generally glass-fiber felt or polyester scrim, which is fabricated into the finished sheet by the manufacturer. Polyester and other woven fabrics are used as reinforcements for elastomeric and plastomeric, single-ply membranes. Some single-ply membranes do not require reinforcement because the waterproofing material is inherently stable.

The surfacing materials protect the waterproofing and reinforcement elements from the direct effects of sunlight and weather exposure. They also provide other properties, such as fire resistance, traffic and hail protection, and reflectivity. Some single-ply membranes are self- or factory-surfaced. Aggregate, which is field-applied, and mineral granules, which are usually factory-applied, are the most common types of surfacing materials. Smooth-surfaced coatings, however, are increasing in popularity.

Membrane Classifications

Low-slope roof membranes can usually be grouped, or classified, into the general categories reviewed below. There are, however, hybrid systems that might not fit into a category, or that might be appropriate in several categories. A brief overview of each category is provided here. For specific information, please refer to the referenced chapter.

BUILT-UP ROOFING (BUR)

BUR, which uses asphalt or coal tar products, is by far the oldest of the modern commercial roofing methods. Many commercial buildings in this country have BUR roofs. The large number of 20-, 30-, and even 40-year-old BUR roofs that are still sound attests to the system's durability and popularity. Roofing materials continue to evolve, however, and improvements are continually being made to asphalt and coal tar pitch, the basic bitumen components of BUR. Asphalt tends to be more popular with most roofers than coal tar. (See Chap. 5.)

MODIFIED BITUMEN ROOFING (MBR)

Since the first MBR membranes were manufactured in the United States in the late 1970s, they have become one of the roofing industry's fastest-growing materials. The popularity and specification of MBR membranes has increased steadily for more than two decades. Contractors have found the materials easy to use and easily inspected. MBR systems provide a time-tested, high-performance, reliable roof. See Chaps. 5 and 6.

SINGLE-PLY SYSTEMS

Since they first appeared in the 1950s, single-ply materials have become increasingly popular in the United States. Whether imported from Europe or produced domestically, these high-tech products have proven themselves in a wide variety of climates during more than three decades of use. As described in Chap. 6, there are many different single-ply roofing products.

Constructing Steep-Slope Roofs

Steep-slope roofs have a pitch greater than 2½ inches per 12 inches and are generally found in residential homes. As a part of design, this water-shedding roof system uses the roof's steep slope for water runoff and leakage protection. The slope of the drainage surface dictates the type of roofing material that can be used.

Aesthetic Considerations

Unlike the case with low-slope roofing material, when designing the steep roof, appearance is very important. Let's look at some of the materials used to cover steep roofs. A brief overview of the most common steep-slope materials is provided here. For more information, please refer to the referenced chapter.

ASPHALT SHINGLES

Asphalt shingles are the most commonly used material for residential roofing in the United States today. They are made in a variety of styles. The most popular is the square-butt strip shingle, which has an elongated shape and is available with three, two, or one tab (without cutouts). Less popular today are the hex shingle and individual shingles which are available with interlocking or staple-down tabs. Of the three types, the square-butt strip shingle provides the most attractive roof covering.

Today's generation of dimensional, or architectural, asphalt shingles has elevated the art of roof design, especially in the residential reroofing and new construction markets. Thicker, stronger, and more natural looking than typical three-tab shingles, dimensional shingles add a visual effect of depth and distinction to homes. What is more, with the addition of shadowlines and random, laminated tabs, architectural shingles can capture with uncanny accuracy the old-fashioned warmth and elegance of roofs constructed of wood shakes or shingles and slates. Chapter 7 discusses asphalt products in detail.

WOOD SHINGLES

From a historical perspective, wood shingles and shakes could be considered the most American of all roofing materials. The abundant supply of forested land on the newly settled continent made wooden roofs prevalent in Colonial times, an era in which the most common roof coverings in Europe were slate, tile, and thatch. Wood shingles from trees as diverse as oak, eastern white cedar, pine, hemlock, spruce, and cypress were used, depending on local availability. Today, wood shakes and shingles are used for a variety of architectural styles and effects. (See Chap. 8.)

SLATE ROOFS

As a roofing material, slate is long-lasting and extremely durable. Certain types of slate have a longevity of more than 175 years. No man-made roofing material can make that claim. Like other shingles, slate is bought by the square. Permanence does not come cheap, but maintaining a slate roof and replacing the few slates that might chip or break is relatively inexpensive when compared to replacing other roofs every 20 to 25 years. Chapter 9 discusses the cutting of slate and its application.

CLAY TILE

Their simplicity of form and shape makes traditional tapered mission tiles ideal for funneling and shedding water from pitched roofs. Modern extrusion, pressed-form processes, and high-tech gas-fired kilns have replaced the primitive method of shaping clay tiles over human thighs and then baking them either in the sun or in wood-fired *beehive* kilns. With these advances in manufacturing, tremendous improvements in performance, quality, and product diversity have developed. The features and benefits of clay tiles are discussed in Chap. 10.

FIBER-CEMENT SHINGLES

Because of manufacturing processes and the raw materials used within them, fiber-cement products do not experience the natural decomposition that can occur with wooden roofing: curling, cracking, and splitting. And because of the absence of organic fibers, other problems like dry rot and the attraction of termites are nonexistent. Most fiber-cement shingles and tiles are manufactured with an efflorescence preventative, which inhibits the unsightly leaching of salts associated with other manufactured products. Fiber-cement products are discussed with clay tiles in Chap. 10.

METAL ROOFING

The reasons for selecting metal roofing are both obvious and surprising. Metal roofing allows the owner or architect to make a design statement. When the roof is high-pitched and part of the integral building design, the architectural possibilities of metal roofing are not attainable with any other material. When the roof is low-pitched and serves solely as a functional water barrier, the weatherproofing capabilities of metal roofing exceed those of any other materials available. Chapter 11 discusses the types of metal roofing available and their application methods.

Steep-Slope Roof Styles

While low-slope roofs are generally limited to flat-roof styles and are seldom found on residential structures, steep-roof styles vary greatly (Fig. 1-2).

Of the steep-roof styles, the gable roof is the most common. It has a high point, or ridge, at or near the center of the house or wing that extends from one end wall to the other. The roof slopes downward from the ridge in both directions. This roof style gets its name from the gable, which is the triangular section of end wall between the rafter plate and the roof ridge.

The roof on one side of the ridge is usually the same size and slope as the roof on the other side. The gable roof of the saltbox house is an

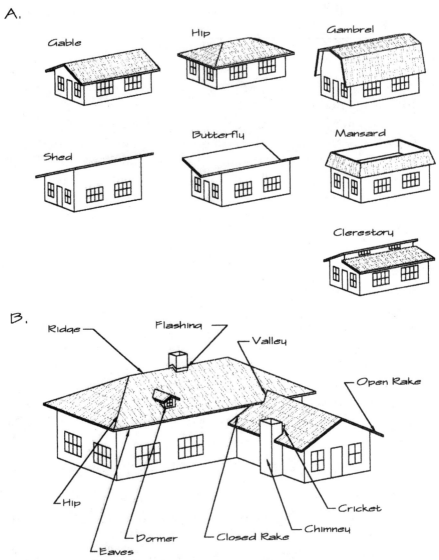

FIGURE 1-2 (A) Typical steep-slope roof styles. (B) Roofing terminology.

exception. An architecture common in New England, the saltbox has different slopes and slopes of different lengths.

A hip roof also has a ridge, but the ridge does not extend from one end of the roof to the other. The lower edge of the roof, or eave, is at a constant height and the roof slopes downward to the eaves on all sides. The point where two roof surfaces meet at an outside corner is called a hip. The junction where two roof surfaces meet at an inside corner is called a valley.

A shed roof slopes in only one direction, like half a gable roof. The roof has no ridge and the walls that support the rafters are different heights. The shed roof has several variations. One is the butterfly roof, where two shed roofs slope toward a low point over the middle of the house.

In another variation, two shed roofs slope upward from the eaves, but do not meet at a ridge. The wall between the two roofs is called a clerestory, and is often filled with windows to let light into the interior of the house.

A gambrel, or barn roof, has double slopes: one pair of gentle slopes and one pair of steep slopes. Like a gable roof, the gambrel roof slopes in both directions from a center ridge. At a point about halfway between ridge and eave, however, the roof slope becomes much steeper. In effect, the lower slope replaces the upper exterior walls of a two-story house. It is common to add projections through the roof, called dormers, for light and ventilation.

Just as a gambrel roof is like a gable roof with two different slopes, a mansard roof is like a hip roof. From a shorter ridge, the roof drops in two distinct slopes to eaves that are the same height all the way around the structure. Up to 40 percent of the building is roof with the mansard roof design. In addition to typical residential applications, mansard roofs are often used for apartment complexes, commercial buildings, and even institutions such as schools.

Planning a Job

The Importance of Planning

With a well-designed computer system, which will be outlined in Chap. 13, the entire roofing business is interlinked: Jobs come in; they are entered into the database; they are estimated; and, if they are won, they become new accounts. If a job is lost at the bid stage, the information about the parties who presented it remains in the marketing database, and announcements about the firm can be sent to the principals of that company on a regular basis.

If the preliminary bid is accepted and set up as a new account, a tentative schedule is (or has been) roughed out. Every step of the process is entered in computers, and they all relate to one another. In effect, planning begins when the request to bid is first entered into the system.

This is extremely important, because planning is probably the most important phase of any roofing job after an estimate is accepted. And in reality, planning always begins while the estimate is being thought through. The more care that is put into the process, the more smoothly and profitably the job goes when it hits the roof deck.

Do not confuse planning with scheduling. Scheduling, which is reviewed later in this chapter, deals with the time frame within which things are done. It is a part of planning but not as comprehensive—planning deals not only with when an action will take place but with how it will happen as well. For example, the slate will be drop-shipped on such and such a date. Planning also covers where the tiles will be ricked on the site for safekeeping. It is this attention to the step-by-step execution of the project that promotes professionalism in roofing.

From the very beginning of estimating, the principles for the job—the estimator, the project manager, and the roofing contractor—should all begin to think through how they will take care of each step of the project. Planning begins with estimating and continues through every stage of putting the project together: scheduling, precontract meetings, preparing contract documents, mobilization, progress tracking, and all the way through to demobilization.

There is an old construction industry cliché that has been repeated in many different formats and applied to many parts of construction but that applies as well to roofing as to any other part of the construction industry: Poor Planning Promotes Poor Performance. It can be stated in many ways and it is given different names, but it is often dubbed the 5 Ps of construction. And it is absolutely true, advance planning can add up to savings in time and materials and result in a better job and future business.

Solid Estimating Practice

The first step in any job that comes in is estimating. And in a sense, estimating—or better, perhaps, the results of estimating—always comes back at the ending of the job, the true end of a job. The final question is, did the estimate work out? Did we make any money?

Estimating can make or break every job the roofer faces. Obviously, if the bid is too high, all the effort put into the negotiations, site visits, take-offs, and preparation of estimate documents is lost. If the figure is too low, it would have been better to have never gotten the job. In this chapter the important aspects of estimating are laid out, and we have provided simple checklists for the bidding process. Value engineering is reviewed as a method for closing bids by discussing the job and alternative approaches to the roofing needs with the principal and the design team.

It is important to remember that each company is located within its own community and that all of the elements involved in laying on a roof which are unique to the company must be considered: terrain, the built environment around the structure that will be roofed, the local labor pool, the crew that will be applying the roof, the local weather, government requirements, street use, special permits, unique access requirements, recycling needs, dump requirements, toxic waste disposal—the process of forecasting what will be required to bring the job in, well done, on time, and at a profit is so complex that it may seem overwhelming to the novice.

However, with time the roofing professional comes to understand the area in which his or her firm operates and knows a great deal about what will be needed in order to perform effectively. Experience is always the best teacher, but everyone makes mistakes, overlooks what may be needed to perform well, or just forgets to add a line item, plain and simple.

In times past, estimates were done manually, and some still do them this way. Estimating on the fly—taking off the square footage, listing any special requirements, and throwing a number at the job—can work and actually lead to a profit. But, except for projects which are very limited in scope, the days of being able to make a list on a brown bag and extend the figures on the butcher paper that your sandwich came in are gone.

Takeoffs

There are any number of ways to estimate a job, but there are certain criteria that should be considered for doing any estimates. Takeoffs are the phase during which the roofer should start serious planning. When the estimator goes to the jobsite for the takeoffs, he or she should play the job out in his or her imagination. Notes should be taken about anything at the site that might slow the job down—imagine that you are doing the job while you are at the site. Figure 2-1 includes a number of the basic areas that should be considered while the estimator is at the jobsite. This checklist can be edited and redone specifically for your company and printed in quantity from the CD that accompanies this book. Even if you are bidding straight off a set of plans, it is a good idea to visit the site and review the logistics you will run into at the location.

Once you are back on the ground, discuss your findings with the homeowner or project manager. If the principal knows about the necessary prep, any proposed supplementary work, and any hidden repair work that may present unforeseen expenses in advance, he or she will be better prepared to accept the initial estimate and any unforeseen tasks that may arise.

✓	Item
	Make all measurements while on the roof.
	Check the condition of the roof and make notes.
	Search carefully for any prep work that needs to be done.
	Check the condition of vents, vent caps, and flashings.
	Check the condition of the chimney, if there is one.
	Check for potential water leakage into the house.
	Look for anything that might cause delays or tie up your work.
	Check for unexpected labor costs.
	Determine if the rafters are secure.
	Determine the number of roofs already on the structure.
	Determine the appropriate shingle if you are adding a roof.
	Is a tear-off necessary for an accurate bid?

FIGURE 2-1 Checklist for estimating site visit.

The more thoroughly you examine the roof for flaws and repairs prior to preparing a bid, the more accurate you can make your estimate. An error in the estimating process can result in lost time and money. And in the end, if the roofer consistently underestimates, it will lead to the loss of word-of-mouth marketing, loss of return clients, and even going belly-up.

It is also important to remember that the roofing contractor is responsible for any damage due to lack of foresight—this is why bringing up unforeseens at the first site visit is very important. If a tear-off is not to be done, make a decision as to what type of shingle is most compatible with those already on the roof. Using the wrong shingle can shorten the expected life of a roof and present an unsatisfactory final appearance.

Calculating Roof Area

The vast majority of roofs are composed of one or more rectangles (Fig. 2-2). The area of the entire roof is the sum of the areas of these rectangles. For example, the area of a shed roof, which has only one rectangle, is found by multiplying the rakeline by the eavesline, or $A \times B$.

The gable roof is made up of two rectangular planes. Its area is found by multiplying the sum of the rakelines by the eavesline, or $A \times (B + C)$. Gambrel roofs have four rakelines. The total area is calculated by multiplying the sum of the rakelines by the eavesline, or $A \times (B + C + D + E)$.

No matter how complicated a roof might be, projecting it onto a horizontal plane easily defines the total horizontal surface. Figure 2-3 illustrates a typical roof complicated by valleys, dormers, and ridges at different elevations. The lower half of the figure shows the horizontal projection of the roof. In the projection, inclined surfaces appear flat and intersecting surfaces appear as lines.

Measure the roof's horizontal projection from the building plans, from the ground, or from inside the attic. Once the measurements are made, draw the roof's horizontal area to scale and calculate the area. Because the actual area is a function of the slope, calculations must be grouped in terms of roof slope. Measurements of different slopes are not combined until the true roof areas have been determined.

The horizontal area under the 9-inch-slope roof is

$$26 \times 30 = 780$$
$$19 \times 30 = 570$$
$$\text{Total} = 1350 \text{ square feet}$$

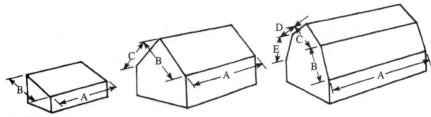

FIGURE 2-2 Simple roof.

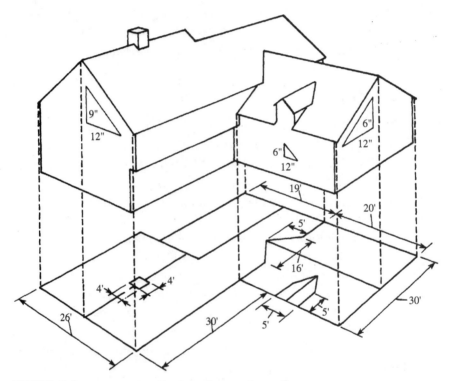

FIGURE 2-3 Horizontal projection of a complex roof.

From this gross figure, deductions can be made for the chimney area and for the triangular area of the ell roof that overlaps and is sloped differently from the main roof.

$$\text{Chimney area} = 4 \times 4 = 16$$
$$\text{Ell roof} = \frac{1}{2} \, (16 \times 5) = 40 \text{ (triangular area)}$$
$$\text{Total} = 56 \text{ square feet}$$

The net projected area of the main roof is calculated as follows:

1350 − 56 = 1294 square feet

The calculation for the horizontal area under the 6-inch-slope roof is

20 × 30 = 600

½ (16 × 5) = 40

Total = 640 square feet

Portions of the higher roof surfaces often project over roof surfaces below them, but the horizontal projections do not show the overlap. Add these duplicated areas to the total horizontal area. Make one final correction to account for these overlapped or duplicated areas before the total projected horizontal area is obtained.

In the example, there are three overlapping areas (ell section):

1. On the 6-inch-slope roof where the dormer eaves overhang 2 feet

2. On the 9-inch-slope roof where the main roof eaves overhang 3 feet

3. Where the main roof eaves overhang the smaller section of the main roof in the rear of the building

In each case, if the eaves extend 4 inches beyond the structure, the duplication calculations are

1. Two eaves overhangs:

$$5 \times {}^4\!/_{12} = {}^{20}\!/_{12} \times 2 = 3\tfrac{1}{3} \text{ square feet}$$

2. Two eaves overhangs:

$$7 \times {}^4\!/_{12} = {}^{28}\!/_{12} \times 2 = 4\tfrac{2}{3} \text{ square feet}$$

3. Overhang covers only half of the 19-foot-wide section:

$$9.5 \times {}^4\!/_{12} = 3\tfrac{1}{6} \text{ square feet}$$

Item 1 is added to the area of the 6-inch-slope roof. Items 2 and 3 are added to the area of the 9-inch-slope roof. Thus, for the 6-inch-slope roof, the adjusted total is

640 + 3 = 643 square feet

and that for the 9-inch-slope roof is

1294 + 8 = 1302 square feet

Fractions are rounded off to the nearest foot. When the total projected horizontal areas for each roof slope are calculated, convert the results to actual areas with the aid of Table 2-1. To use the table, mul-

TABLE 2-1 Estimating Areas Using Roof Pitches and Inclined Areas

| Classification | Incline | |
	Inch per foot horizontal	Angle with horizontal
	⅛	0°-36'
	¼	1°-12'
	⅜	1°-47'
	½	2°-23'
	⅝	2°-59'
Flat roofs	¾	3°-35'
	1	4°-45'
	1⅛	5°-21'
	1¼	5°-57'
	1½	7°- 8'
	1¾	8°-18'
	2	9°-28'
	2¼	10°-37'
	2½	11°-46'
	2¾	12°-54'
	3	14°- 2'
	3¼	15°- 9'
	3½	16°-16'
	3¾	17°-21'
	4	18°-26'
	4¼	19°-30'
Steep roofs	4½	20°-34'
	5	22°-37'
	6	26°-34'
	7	30°-16'
	8	33°-42'
	9	36°-52'
	10	39°-48'
	11	42°-31'
	12	45°- 0'
	14	49°-24'
	16	53°- 8'
Extra steep roofs	18	56°-18'
	20	59°- 2'
	22	61°-23'
	24	63°-26'

Fractional factor	Inclined area per square foot horizontal area	Percentage increase in area over flat roof
	1.000	0.0
	1.000	0.0
	1.000	0.0
$\frac{1}{48}$	1.001	0.1
	1.001	0.1
$\frac{1}{32}$	1.002	0.2
$\frac{1}{24}$	1.003	0.3
	1.004	0.4
	1.005	0.5
$\frac{1}{16}$	1.008	0.8
	1.011	1.1
$\frac{1}{12}$	1.014	1.4
	1.017	1.7
	1.021	2.1
	1.026	2.6
$\frac{1}{8}$	1.031	3.1
	1.036	3.6
	1.042	4.2
	1.048	4.8
$\frac{1}{6}$	1.054	5.4
	1.061	6.1
	1.068	6.8
	1.083	8.3
$\frac{1}{4}$	1.118	11.8
	1.158	15.8
$\frac{1}{3}$	1.202	20.2
	1.250	25.0
	1.302	30.2
	1.356	35.6
$\frac{1}{2}$	1.414	41.4
	1.537	53.7
	1.667	66.7
	1.803	80.3
	1.943	94.3
	2.088	108.8
1	2.235	123.5

tiply the projected horizontal area by the conversion factor for the appropriate roof slope. The result is the actual area of the roof. For example, regarding the ell roof overhang, for the 9-inch-slope roof:

Horizontal × conversion = actual

1302 square feet × 1.250 = 1627.5 square feet

For the 6-inch-slope roof:

Horizontal × conversion = actual

643 square feet × 1.118 = 718.8 square feet

After you convert the horizontal areas to actual areas, add the results to obtain the total area of roof to be covered:

1628 + 719 = 2347 square feet

In the actual job estimate, include an allowance for waste. In this case, assume a 10 percent waste allowance. The total area of roofing material required is

2347 + 235 = 2582 square feet

The same horizontal area projection and roof slope always results in the same actual area, regardless of roof style. In other words, if a shed, gable, or hip roof with or without dormers each covered the same horizontal area and had the same slope, they would each require the same area of roofing to cover them.

Remembering Additional Materials

To complete the estimate, determine the quantity of starter strips, drip edges, hip and ridge shingles, and valley strips needed for the job. Each of these quantities depends on the length or size of the eaves, rakes, hips, ridges, and valleys. Because eaves and ridges are horizontal, their lengths can be measured from the horizontal projection drawing. Rakes, hips, and valleys are sloped. Their lengths must be calculated following a procedure similar to that for calculating sloped roof areas.

RAKES

To determine the true length of a rake, first measure its projected horizontal distance. Table 2-1, used to convert horizontal areas to actual areas, can be used to convert rake lengths. To use the table, multiply the rake's projected horizontal length by the conversion factor for the appropriate roof slope. The result is the actual length of the rake.

For the house shown in Fig. 2-3, the rakes at the ends of the main roof have horizontal distances of 26 feet and 19 feet. There is another rake in the middle of the main house where the higher roof section meets the lower. Its horizontal distance is

$$13 + 3.5 = 16.5 \text{ feet}$$

Adding all these horizontal distances gives a total of 61.5 feet. Using Table 2-1, the actual lengths for the 9-inch-slope roof are calculated as follows:

$$\text{Horizontal} \times \text{conversion} = \text{actual}$$
$$61.5 \text{ square feet} \times 1.250 = 76.9 \text{ square feet}$$

Following the same procedure for the ell section with its 6-inch-slope roof and dormer, the total length of the rakes is found to be 39 feet 1 inch. Add these rake lengths to the total length of the eaves, which is the actual horizontal distance. No conversion is necessary. This provides the quantity of drip edge required for the job. The quantity of ridge shingles required is estimated directly from the drawing, since ridgelines are true horizontal distances.

HIPS AND VALLEYS

Hips and valleys also involve sloped distances. Convert their projected horizontal lengths to actual lengths. Measure the length of the hip or valley on the horizontal projection drawing and multiply it by the conversion factor to get the appropriate roof slope. The result is the actual length of the hip or valley.

On the house illustrated in Fig. 2-3, there is a valley on both sides of the ell roof where it intersects with the main roof. The total measured distance of these valleys on the horizontal projection is 16 feet. The fact that two different slopes are involved complicates the procedure. If there were only one roof slope, the true length could be calculated. Because this house has two slopes, calculations for each slope must be made and then averaged to obtain a close approximation of the true length of the valleys.

$$\text{Horizontal} \times \text{conversion} = \text{actual}$$
$$16 \text{ feet} \times 1.600 = 25.6 \text{ feet for the 9-inch-slope roof}$$
$$16 \text{ feet} \times 1.500 = 24.0 \text{ feet for the 6-inch-slope roof}$$

The approximate length of the two valleys is 24.8 feet, or 12.4 feet each.

The total projected horizontal length of the dormer valleys in Fig. 2-3 is 5 feet. With the ell roof and the dormer both having 6-inch slopes, the actual length of the valleys is calculated to be 7.5 feet. The total length of the valleys for the house is

$$25 + 7 = 32 \text{ feet}$$

Estimate the amount of flashing material needed based on this true length.

Estimating Labor Costs

To estimate labor costs properly, consider two components: (1) work hours required to perform the installation, and (2) taxes and insurance costs associated with the total work hours allocated to the job.

Estimate each segment of the installation separately. Include any and all work required to correct any imperfections. Correcting problems to ensure that the structure, substrates, and projections meet code requirements can account for as many work hours as the installation of the roofing materials, if not even more.

To calculate labor costs, the labor rate per hour is multiplied by the total number of work hours. Once the total number of work hours is calculated, multiply that figure by the labor rate per hour to estimate the labor costs to complete the basic installation for each segment of the job. Combine this formula and the materials estimate to arrive at the costs for the individual job segments.

Sealing around projections, forming and installing flashings and valleys, and installing other finished material all fall into the category of finishing labor. Roofing contractors' familiarity with their crews, along with accurate records of the amount of time each crew requires to perform these tasks, enables them to properly estimate the number of work hours needed for each step.

As mentioned earlier, labor costs are actually composed of two sets of costs. The labor burden must be added to the work hours to calculate total labor costs. On average, the labor burden increases each job's labor costs by 25 to 30 percent.

The labor burden is the percentage of payroll dollars roofers are compelled to pay government agencies in the form of taxes and to insurance companies in the form of premiums. The taxes and insurance mentioned in Chap. 12 are examples of line items that typically make up the labor burden.

Preroofing Conferences

The dynamics of today's roofing industry make preroofing conferences more important than ever before. A preroofing conference is an open discussion between all parties that have anything to do with the roof. Often included in these discussions are the owner, architect, engineers, roofing contractor, deck contractor, siding subcontractor, framer, general contractor, roofing materials manufacturer, insulation manufacturer, and representatives from any trades that have a reason to be on the roof.

The purpose of these meetings is to enable all parties to iron out as many of the logistics required for working together on the project as possible. If differences cannot be worked out or you will not be properly compensated for differences in opinion about the job conditions, it is time to walk away.

The meetings offer everyone a chance to strive to get the best for their own firm and to clarify many items that would be difficult to come to terms on during the heat of construction.

Delegating Responsibility

One important aspect of preconstruction meetings is delegating authority for each task and hammering through what is to happen if the schedule backs up. If the agreements achieved at this meeting are put on paper, they can be a very strong tool for dealing with three major types of problems that can arise after a project hits the roof deck: disputes over a lack of clarity of details and assigned responsibility for tasks at complex areas of the roof; disputes over construction management, the overlapping of the trades, and breakdowns in communication; and cost overruns. Making a firm rule that you will always get a grip on all of these potential situations during precontract negotiations, on a regular basis, can be a big step toward continual profits for your business and preventing construction defect litigation in the future.

For the roofing contractor specifically, these meetings contribute to the precise details of the roofing contract, which stipulates procedures and protections for the roofing firm. All parties involved in the design and construction of the roof system must recognize and meet their responsibilities in order to achieve a proper roofing system. No link in this chain of responsibility must be broken.

Owner Responsibilities

The building owner is the first link in the responsibility chain. The owner's first obligation is to achieve a proper roofing system for the building. The best roof, as determined by the architect, must be accepted and paid for by the owner.

Most owners are not knowledgeable about roofing systems and specifications. Usually they defer to the architect's recommendations. The owner has the additional responsibility of not limiting the architect in the selection of the roofing system and insulation purely because of budgetary concerns. Decisions based solely on price often result in failed roofs.

Once a roof system is accepted, the owner's responsibilities become more limited until the building is accepted and the maintenance program begins. The owner's maintenance responsibilities cannot be overemphasized. An ongoing maintenance program must be developed and followed throughout the life of the building. Ignoring roof maintenance is the same as absolving all other parties from their responsibilities.

Architect Responsibilities

Legal authorities claim that roofs and foundations are the two structural components most likely to lead to a lawsuit involving the architect. So the architect's motive for designing the proper roofing system is twofold: self-preservation and the best interests of and concern for the client.

Get to know the architects who do business in your market area. Speak to other contractors to determine the architects who consistently produce quality work with accurate details, and those whose work is suspect or refused by other contracting firms. Challenge the architect's plans if they do not meet with generally accepted roofing practices. If the plans are sound, the architect will have no trouble clarifying any questions that may arise about the design elements of the structure.

Manufacturer Responsibilities

All roofing materials have limitations. Manufacturers know their products' limitations and are required to accurately represent their line. Manufacturers must be willing to certify their materials. If the materials meet the appropriate standards, certification is not unreasonable. If the materials do not have standards, that too should be verified by the manufacturer. With materials such as bitumens, certification to standards is a must.

Roofing Contractor Responsibilities

The idealized concept of the perfect application of any roofing system has never occurred. With any construction project, there are just too many factors that prevent installations from being completed without any situations that some might consider flaws. Roofing is not performed in a hermetically sealed laboratory—it is vulnerable to dozens of rugged conditions. The responsibility of the roofing contractor is to recognize the many difficult situations specific to the structure, and then to make every attempt to overcome those problems.

The foundation of any high-quality roofing firm is knowledgeable, conscientious supervisory personnel and properly trained work crews. Organized apprenticeship programs can result in work crews that are trained in the contractor's procedures to ensure consistent performance on every job, regardless of which crew performs the work.

Contractors must inform the architect in writing of any recommended changes that they feel are in the best interest of the project. This notification should be made immediately upon discovering that a change is required. Never install a roof that is inadequate—your integrity is always a very important part of your life. And nothing causes business to drop off faster than a failed roofing system and an unhappy client.

Summary of What Should Be Accomplished at Preroofing Conferences

At the preroofing conference, a definite game plan should be pursued. This is a chance to ask all the tough questions that make or break the profit line after the job begins. Never let the opportunity pass you by— avoiding or passing by the opportunity to hammer out the vital questions before hitting the roof deck is probably the most grave error any roofer can make. Figure 2-4 is a roughed-out concept of what should be covered at the conferences. Boot it up from the enclosed CD and adjust it for each meeting.

✓	Item
	All questions about the drawings, details, and specifications have been listed for discussion.
	All questions, restrictive clauses, and contingencies related to the condition of the roof have been listed.
	All prep work that needs to be done has been listed.
	All vent, vent cap, chimneys, and flashing conditions have been listed for the meeting.
	All questions related to specified products have been listed.
	All past and potential water leakage into the structures has been listed.
	All potential delays, including access, weather, materials acquisition, overlapping of trades, site access, framing repair, etc., that may tie up your work have been addressed.
	Questions related to unexpected labor costs have been listed.
	All cleanup issues, including environmental and recycling, have been addressed.
	All opportunities to sell extras and change orders have been addressed.
	Value engineering list is ready for the meeting.

FIGURE 2-4 Checklist for preroofing conferences.

It is important to remember that you should approach the preroofing meeting positively rather than aggressively. Use it as a welcome opportunity to come to an understanding and to sell services like extras that raise the gross value of the project. Below are some notes on value engineering, which should always be addressed at the meeting.

Owners, architects, and designers are becoming more aware of the roof's significance. For this reason, as well as to eliminate callbacks and repair jobs caused by owner neglect, the preroofing conference is the perfect time to establish a maintenance program with the owners or their representatives.

When a preroofing conference is utilized to the maximum, the benefits are widespread. All parties get what they want. The owner and the architect are satisfied that they are getting the best possible roof. The manufacturer is satisfied that the installation will meet warranty requirements. The general contractor, construction manager, and roofing contractor get the roof installed with few problems, and all concerns are coordinated with the other trades.

Closing the Deal with Value Engineering

Value engineering sounds like some kind of complicated system that might come from a structural analysis of the building, but typically the phrase is a marketing expression. In reality, it refers to ideas that the general contractor and subcontractors can offer to the principal and design team that can cut costs for the project.

It should be remembered that the point is to find high-quality methods of attaining cost savings, not questionable methods that may lead to an inferior product—this is very important. Replacing the existing design development concepts with inferior-quality work product can come back in the form of legal ramifications in the future.

Here is a common example of a value engineering offer that might very well overcome the competition and land a contract for your firm. You discover that your suppliers have a line of product that is being discontinued but is of equal quality to that which is specified. The vendor assures you that there is plenty of extra product for repairs that may transpire over the life of the roof. There is a place for the owner to stockpile replacement material, and the savings is considerable. You inform the general contractor. Never go around the general contractor to the building owner or architect; your integrity is very important to long-term business and your own personal sense of well-being.

Writing Real-World Contracts

Most roofing firms have standard contracts that are designed to help the roofer get through each job with few difficulties and maximum profit. These standard contracts allow space for the insertion of details pertinent to the individual job. Review the form at regular intervals to make sure that the language keeps pace with industry changes.

Any matter that is left unsettled by the contract is a potential lawsuit. An accurately drafted roofing contract can reduce or eliminate litigation by providing, in advance, solutions to problems that might arise during the course of the roofing job.

Along with supplying solutions to problems, the contract should put in writing the understandings of all parties concerned with the roofing job (Fig. 2-5). Stipulate all items very carefully; this is a period when taking your time can make a big difference. A preliminary draft of the document should have been presented to the client before the preroofing conference. When the preliminary draft hits the conference table, all parties can voice their conflicting viewpoints and iron out acceptable solutions. Figure 2-6 lists some of the items that must be covered in the contract; use the version on the companion CD to slant it specifically for your firm.

Most contractors who work commercial jobs are familiar with the contract forms published by the American Institute of Architects (AIA). These forms are so widely used that in some instances contractors may sign on the dotted line without reading the contract. A form such as AIA Document A101 (Standard Form of Agreement between Owner and Contractor) might appear to be a fairly straightforward four-page document. Article 1 of the form immediately states, however, that contract Documents consist of this Agreement, the Conditions of the Contract (General, Supplementary, and other Conditions), the Drawings, the Specifications, and all Addenda issued prior to, and all Modifications issued after, execution of this Agreement or repeated in it.

First Meeting

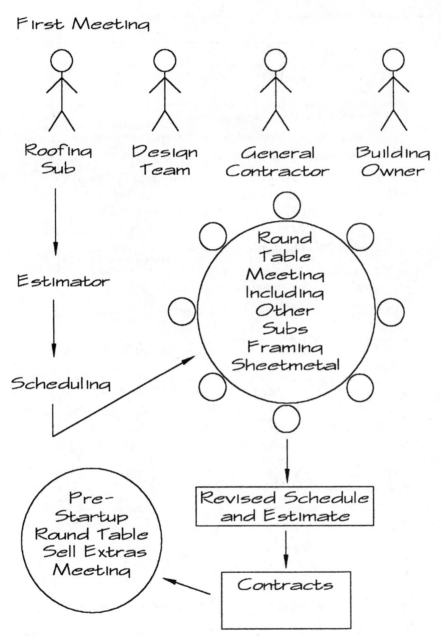

Roofing
Sub

Design
Team

General
Contractor

Building
Owner

Estimator

Scheduling

Round
Table
Meeting
Including
Other
Subs
Framing
Sheetmetal

Pre-
Startup
Round Table
Sell Extras
Meeting

Revised Schedule
and Estimate

Contracts

FIGURE 2-5 Writing real-world contracts.

✓	Item
	Exactly what work is to be done for the stated price
	What, if any, cleanup services and environmental issues are included in the contract price
	Additional work and charges for reroofing
	A statement that any changes to the contract must be made in writing, dated, and signed by all parties
	A statement that the roofing contractor is not responsible for delays in completing the job due to circumstances beyond the roofing contractor's control, including but not limited to diverse weather, labor strikes, accidents, acts of God, site restrictions, etc.
	Terms and conditions under which payment is to be received from the owner or general contractor
	Proof of insurance carried by the owner or general contractor
	A clear understanding of all coverage and liability issues in relation to insurance
	Provision for adequate job specifications to the roofer from the owner or general contractor or their agents
	The provision of various utilities and accessibility to the jobsite by the owner and/or the general contractor
	Any penalties or interest to be assessed as a result of late payment
	A complete understanding of how construction defects and consequential damages will be taken care of

FIGURE 2-6 Items to be included in the roofing contract.

Boiled down, this means that this simple four-page document has just become an unwieldy monster that could be as long as 40 pages, without drawings. Within the entire agreement, the roofing contractor might find language relating to issues of risk, liability, or warranty that changes the general or supplementary conditions. The end result can be the imposition of substantial liability on the roofer.

A simple one-sentence clause that stipulates that the roofer is to verify the accuracy of the architect's plans and designs shifts responsibility for the plans from the architect to the roofing contractor. Another example is the common practice of requiring the contractor to provide the manufacturer's guarantee for a specified period of time. Usually the language for this clause is broad and does not include the same limitations and exclusions contained in the manufacturer's guarantee. Conceivably, this can result in the contractor's being held responsible for conditions from which the manufacturer has been exempted.

Before you bid on a job, it is imperative that you know the risks you are assuming. The following recommendations can help you make an informed business decision about whether or not to go forward with a particular project and the extent to which such factors should be reflected in the final bid price: Examine the documents carefully and note any unusual items that impose a greater burden on the roofing contractor than otherwise would be expected. Have an experienced construction attorney review the documents. The attorney should already be familiar with construction documents, so he or she should require less time to thoroughly examine the documents than a general practice attorney. Assist the construction attorney by highlighting the unusual items noted above. If, after consulting with the construction attorney, there are any areas of significant concern, be sure to raise these issues at the preroofing conference. These changes and clarifications are important if it ultimately becomes necessary to defend your position in court.

However, always remember that a building project is a team effort, and the more you enlist all of the other players—the contractor, the roofer, the architect, the sheetmetal and siding companies—on your team, the better the job will go and the less likely it is that the dead time and financial drain of legal problems will arise. It is very important to remember this from the time the request to bid comes in and to provide a very thorough list of questions for working through at all preconstruction meetings.

Scheduling

Another benefit of preroofing conferences is schedule coordination between trades. It is not efficient to move a trade into a construction area unless the workers can keep at their job. If they are held up waiting for another trade to finish, both time and money are lost. And, as all roofers know, if another sub like the framer omits part of its work—for example, backing on a vertical wall—work is slowed and money is lost because of the downtime.

Scheduling of roofing can be done efficiently with software. For large firms with many big, ongoing projects such as subdivisions or massive condo clusters, a high-end package of critical path software may be needed. Remember, it is simple enough to punch out a beginning timeline of what needs to happen; what gets complicated is rescheduling. The software needs to be able to allow the scheduler to punch in changes as they come about and automatically adjust the entire schedule to take these changes into account.

For a closer look at the complicated uses of the critical path method, see McGraw-Hill's *CPM in Construction Management* by James O'Brien (5th ed., 1999). For companies with more simple and less expensive software, see App. A and contact the vendors for review of the software.

Remember to use preroofing conferences to establish the amount of time each trade needs on the roof and the rate of advance. Naturally, any phase of the construction that requires cooperation of the trades, a projection through the roof, or any other complicated activity should be discussed in detail, prior to the actual roofing. This includes the placement of mechanicals on the roof, air vents, skylights, plumbing vents, ductwork, electrical intrusions, etc.

Investing in Equipment

There will be jobs that come along where large equipment—dewatering, generators, lifting devices, etc.—would cause the work to move at a much faster clip. It is very important to purchase equipment that will reduce labor considerably, and having a job that will show a considerably more handsome profit if a piece of equipment is used is the perfect time to buy.

In Chap. 12 we look at forecasting and leveling, and this is a perfect illustration of how important a tool they can be. In Chap. 13 we review how computers can make a terrific difference when it comes to having an overview of your business. This is an excellent case in point.

If you know from your records that this is a time of year when cash is strong and your payables and receivables are in good shape, it may be time to buy this piece of equipment. It may be best to rent equipment from several different vendors and have your field personnel evaluate it for your company. Then it is wise to discuss the benefits of leasing and of depreciation of an outright purchase with your accountant.

If you have done your homework on marketing, the purchase of a piece of equipment is an excellent time to send out an announcement to all of the contractors, architects, engineers, and building owners who do this type of roof that you are now better equipped to service their needs.

Managing Inventory

Every business needs to keep a certain amount of material on hand. As we all know, these supplies are known as inventory. There was a time when many businesses required large amounts of warehouse space and storage yards, which meant payments and maintenance for large parcels of real estate. But in recent years, companies have changed the way they control inventory.

There are several components of inventory control, beginning with choosing the person who manages the inventory and the staff member who places inventory orders. Depending on the size of the company, the owner might be responsible for both tasks. Larger companies might employ a warehouse manager and purchasing manager to fulfill these responsibilities. Regardless of who holds these positions, accounting and inventory should be handled by different persons. This will promote veracity and timeliness in the control of stockpiles.

One important aspect of inventory management is cost control.

Materials are only one part of the cost picture. Handling costs, storage costs, shipping costs, and losses from theft or obsolescence are all factors that can contribute to shrinking profits.

In today's business environment, roofing contractors have the technology to effectively deal with these cost areas and increase profits by reducing the amount of stored inventory. As you upgrade your existing computer system or put in a new one, bring your vendors into the loop and size up what automated services they offer and how these services work with the software you are going to buy.

A combination of new technology with traditional offsets of the cost of doing business is an excellent means of increasing the profit line. For example, purchasing land and building rental property on the street portions of the warehouse and yard can cover the typical burden of rental space. Purchasing equipment and using it to market your services can help reduce labor costs on an ongoing series of jobs.

By working with suppliers and your software vendors and consultants, an inventory control system based on when you use the product can be worked out with both manufacturers and nonmanufacturing firms. This "just-in-time" system involves ordering materials just in time for use on the job. More and more wholesale roofing suppliers are equipped with crane or conveyor trucks with which they can deliver materials to the jobsite and directly onto the roof. This is particularly true of suppliers delivering slate or tile products.

Naturally, there are some materials that every roofer needs to keep in stock, such as an assortment of fasteners and felts. On almost every roofing job, whether new construction or retro, products like mastics and felt will be needed. Buying these in quantity and having them at the office can save purchasing costs, downtime, and shopping time on payroll. You can also use materials from the warehouse to start jobs before the materials are delivered by the supplier.

In Chap. 12 we discuss using computer software when estimating. The new, just-in-time materials management is a direct benefit of electronics. And it should be remembered that it can loop right into your system at the beginning of a job, during estimating. Once the job is awarded, it is a simple matter for the software to translate the estimate sheet into a materials list. This list can then be e-mailed directly from the computer to vendors for bidding.

Using the computer's integrated software, you can create a record—or file of materials on order—that includes the job name, job number, date ordered, supplier name, vendor number, and other required information.

Furthermore, the felts and fasteners used on a particular job can be automatically removed from inventory by the computer and reordered or placed on an order sheet. These software programs can, with one instruction from the system operator, order materials, update accounts payable, and calculate the vendor's discount schedule, thus condensing several time-consuming tasks into a one-step process.

The just-in-time inventory philosophy keeps obsolete products off warehouse shelves. With so many colors, styles, and types of roofing materials available, one shingle might be popular today and out of style tomorrow. The longer material stays on your shelves, the more money you lose. The just-in-time policy eliminates the need to invest in large quantities of stock that might not immediately turn into profits.

By reducing the volume of materials kept in physical inventory and having products delivered directly to the jobsite when needed, roofing contractors can reduce insurance, handling, labor, and carrying costs, which translates into an improved bottom line.

Disposing of Old Material

Waste disposal is included here because it has become a very important part of planning any roofing job. The site must be examined closely. If this is new work and the roof is a part of a larger general contract, then waste disposal must be discussed with the principal and the general contractor. If it is a reroof, then it is very important to plan how the material will be evacuated from the roof deck and to know the nature of the material and how it will be evacuated from the site.

Environmental concerns are of increasing importance to the roofing industry. What used to be standard procedure needs to be analyzed and updated to comply with today's evolving social and legal climate.

For years, disposing of old roofing and packaging material was a matter of trucking it to the nearest landfill. Today, many landfills are close to capacity, and some municipalities and cities allow only specific materials to be dumped in their landfills. This means that you might end up sorting debris for disposal at two or more landfill sites, then driving to a landfill that takes each type of material.

The cost of disposal, which used to be of little consequence, is rapidly becoming a significant expense for any roofing project. Consider, too, the additional costs involved if certain types of debris are not allowed in landfills anywhere near the vicinity of the job—perhaps the entire state or even the area of the country—or the cost of hauling debris to one or more landfills if the local landfill will not accommodate the volume of debris generated by a major reroofing job. The cost of either hauling this debris to another site or arranging to have it buried by itself can devastate your profit margin.

In order to limit expenses, roofing contractors are adopting ecological policies. Any reroofing procedures that can be altered in order to reduce costs are being scrutinized. The list in Fig. 2-7 can be expanded as you bring your firm to the forefront in environmental practices, as many large firms have been doing for years. This checklist can be printed and used as a part of the planning and bidding process.

Environmental issues are an area of increasing importance to the roofing industry. What used to be standard procedures need to be analyzed and changed to comply with today's evolving social and legal climate. Do not adopt an attitude of foot dragging and complaining—as in all the rest of your business, become proactive. Use your company's deep environmental knowledge and policies as marketing tools and to segue into more business. Environmental services such as offering products like power-generating shingles can enhance profits (see Chap. 14 for more detail). Remember, markets like environmentally friendly roofing, which aren't well known by everyone, can produce good, strong cash flow in time.

With landfills close to capacity and some municipalities and cities allowing only specific materials to be dumped in their landfills, you might end up sorting debris for disposal at two or more landfill sites and then driving to the landfills that take the material. Use this as a proactive concept rather than complaining; make everyone aware that your company is completely abreast of this information. It is very important—not only can serious troubles for your crew arise from the disposal of hazardous materials, but the principal may end up being responsible for the material even after it is deposited in the disposal site. Adapt Figs. 2-8 to 2-10 from the CD to your firm.

✓	Item
	The cost of disposing of the waste in a local land-fill, if possible
	Will debris from the job need to be sorted for disposal?
	If sorting is required, how many work hours are required?
	Can all debris be disposed of in normal fashion?
	If not, how is disposal accomplished?
	At what cost?

FIGURE 2-7 Checklist for waste disposal.

✓	Item
	Recycle existing roofing materials such as aluminum, copper, etc.
	Reuse existing materials, such as insulation or grave surfacing, on the same roof.
	Reassess the existing roof system to determine if it can accommodate the new roof on top of it.
	Can the materials packaging be incorporated into the roofing system?

FIGURE 2-8 Ecological policies.

✓	Item
	The cost of disposing of the waste in a local landfill, if possible
	Will debris from the job need to be sorted for disposal?
	If sorting is required, how many work hours are required?
	Can all debris be disposed of in normal fashion?
	If not, how is disposal accomplished?
	At what cost?

FIGURE 2-9 Environmental factors to consider when bidding a job.

✓	Item
	Recycle existing roofing materials, such as aluminum, copper, etc.
	Reuse existing materials, such as insulation or gravel surfacing, on the same roof.
	Reassess the existing roof system to determine if it can accommodate the new roof on top of it.
	Can the materials packaging be incorporated into the roofing system?
	What recycling rebates are available from the state and local governments?

FIGURE 2-10 Procedures that can be altered to reduce job costs.

One way to improve profitability while becoming more green is to study, look for, and implement innovative methods for disposing of old roofs. Roofing materials such as tile that cannot be used on the existing roof might be of use to contractors in other parts of the country. Salvaging this material and selling it through a broker or to another contractor eliminates disposal costs, increases revenues, and allows a fellow contractor to purchase materials at a discount. If treated in a proactive manner, recycling and environmental matters can become an important part of your business, as well they should.

Roof Safety

Safety does not come to the jobsite in the worker's toolbox. It has to be carefully cultivated. Whether the roof is to be covered in tiles or shingles, a good roofer knows that business overview, management, workmanship, skill, the right tools, proper loading equipment, and a watchful attitude go hand in hand to create a quality roofing job and a safe roofer.

Protecting the Worker

Considerable research has been directed toward improving personal protective equipment in the construction industry. No matter what the degree of sophistication of the safety equipment, however, it is effective only when it is used properly.

All workers should wear proper protective equipment that is in good repair. Roofers working with flammable substances, climbing ladders, and handling materials should wear clothes that are comfortably snug, particularly around the neck, wrists, and ankles. There should be no loose cuffs, flaps, or strings. Machinery operators should not wear neckties, loose sleeves, rings, watches, or long hair, all of which can be caught in equipment and lead to injury. Other personal protective equipment should include the following:

- Safety glasses, goggles, and plastic face shields when working with material that might become airborne

- Hard hats approved for potential exposures when there is any possible hazard from above

- Closed-cuff chrome-tanned leather, or insulated gloves in good condition and suited to the handling of roofing materials

- Safety shoes that meet nationally recognized standards

- Approved respiratory protective devices proper for the existing hazard (dust, fumes, smoke, vapors, mist, etc.)

- Safety belts when working at elevated levels that are not protected by handrails or when working from suspended scaffolds

Print out the safety checklist (Fig. 3-1) on the CD included with this book to take to the jobsite.

First aid kits and a Red Cross first aid manual should always be handy. When working with roofing chemicals, have these additional items available:

- Gauze for burns and eye injuries

- Baking soda to neutralize acid

- Mineral oil, olive oil, or baby oil to help rinse burning materials from the eye

- Blankets for shock

Training for Physical Safety

Every year, roofers suffer back injuries from incorrectly lifting heavy materials. Besides being extremely painful, back injuries can be the start of other recurrent back problems. Lifting improperly can cause muscles to tear and/or ligaments to rupture. A muscle tear is exactly what it says: a tear in a muscle from excess strain or stretching. Ligament rupture occurs when the tissue breaks from the bone.

To avoid back strain, stop any lifting if you cannot keep your hips under your upper body when you begin to stand up. The American Chiropractic Association (ACA) offers five lifting rules that a roofer

✓	Item
	Safety glasses, goggles, and plastic face shields when working with material that might become airborne
	Hard hats approved for potential exposures when there is any possible hazard from above
	Closed-cuff, chrome-tanned leather, or insulated gloves in good condition and suited to the handling of roofing materials
	Safety shoes that meet nationally recognized standards
	Approved respiratory protective devices proper for the existing hazard (dust, fumes, smoke, vapors, mist, etc.)
	Safety belts when working at elevated levels that are not protected by handrails or when working from suspended scaffolds
	First aid kit and Red Cross first aid manual easily accessible
	Ample gauze for burns and eye injury readily available
	Baking soda available to neutralize acids
	Mineral oil, olive oil, or baby oil to help rinse burning materials from the eye easy to access
	Blankets for shock available

FIGURE 3-1 Job safety checklist.

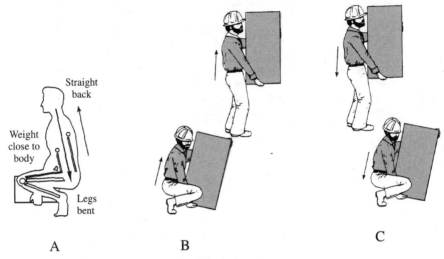

FIGURE 3-2 (*A*) Use leg muscles when lifting any size load, never the back. (*B*) The proper way to lift a shoulder-high load. (*C*) Put down the load as carefully as it was lifted.

should remember and use in order to help avoid back strain (Fig. 3-2).

- Plant feet 12 to 18 inches apart on solid ground in front of the object.

- Squat down in front of the object and keep your back straight when lifting.

- Lift with the muscles of your legs, thighs, arms, and shoulders, not the back.

- Keep your arms close to the object and have a clear view of the path to be taken.

- Never try to lift more than you can handle. Use a mechanical lifting device if the load is too heavy.

Guarding against Falls

Most roofers love the thrill of working high above the ground, but the thrill can be lost with a single slip. The risk of falling can be significantly lessened if federal regulations developed by the Occupational Safety and Health Administration (OSHA) are followed.

Climbing Ladders

The ladder, the indispensable mainstay of every construction site, is often taken for granted. Perhaps the complacent feeling toward the ladder comes from its familiarity. It is, after all, the most widely used roofing tool. But it is probably safe to say that most roofing-related OSHA citations involve the improper care and use of ladders (Fig. 3-3). It is also safe to surmise that nearly all those citations, and the injuries that result from falls off ladders, could have been prevented with a healthy dose of common sense.

An analysis by OSHA of accidents involving ladders reveals that the four principal causes of such accidents are

- Structural failure of the ladder itself

- Failure to secure the ladder at the top and/or bottom

- Ascending or descending improperly

- Carrying objects in hands while ascending or descending

Using Scaffolding

Scaffolding requires frequent safety checks. Any damaged scaffolding accessories, such as braces, brackets, trusses, screw legs, etc., must be immediately removed for repair or replacement. All scaffolding

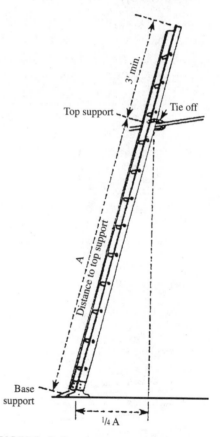

FIGURE 3-3 Make sure the ladder is about 1 foot away from the vertical support for every 4 feet of ladder height between the base support and the top support.

must have solid footing and should be capable of carrying the maximum intended load. OSHA requires that scaffolding be capable of carrying at least four times the maximum intended load. Never use unstable objects, such as barrels, loose bricks, or boxes, to support scaffolds or planks.

OSHA regulations require that guardrails and toeboards be used on all open platform sides and ends that are more than 10 feet above the ground. Scaffolds that are 4 to 10 feet high with a width less than 45 inches must also have guardrails.

Safeguarding Roof Edges

Roof edges should be guarded. Standard guardrails proposed by OSHA should be constructed from 2 × 4 boards and installed to a height of 36 to 42 inches. According to OSHA, guardrails can be counterweighted, mounted on the roof at joists, or attached to the wall or roof of the structure (Figs. 3-4 through 3-6).

The warning line system offers several advantages over guardrail systems. The warning line system consists of stanchions and rope or wire that is rigged and supported so that it complies with the OSHA standard (Fig. 3-7). This standard says that flags or pennants are to be 34 to 39 inches from roof surfaces. Use the warning line system on all flat roofs, regardless of size, when mechanized roof application equipment is used.

Barricading Open Roof Holes

Not only can a worker fall over the edge of a rooftop, there is also the danger of falling through the rooftop. During roof construction, many

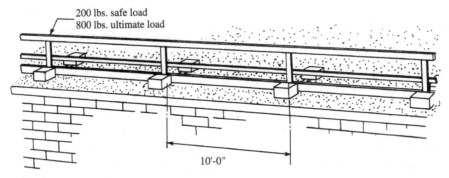

200 lbs. safe load
800 lbs. ultimate load

10'-0"

FIGURE 3-4 Guardrail counterweight system.

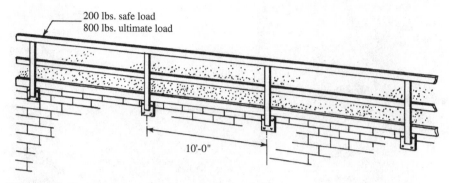

FIGURE 3-5 Guardrail wall-attached system.

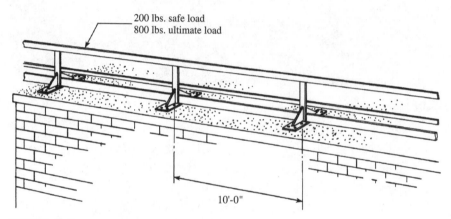

FIGURE 3-6 Guardrail roof-attached system.

situations require openings. Obviously, holes through the deck are hazardous unless they are protected by some type of barricade.

It is essential to guard all openings in roof surfaces with a standard barricade and toeboard on all exposed sides. A toeboard is a strip of wood or metal, 3 to 6 inches in height, that is placed along all exposed edges of any deck opening. Besides helping to keep a worker from falling, toeboards keep tools from being kicked off and injuring a worker below.

Any opening that has a maximum gap of 3 feet or less can be covered with plywood or material that can bear all intended stress. An opening that encounters heavy traffic needs extra reinforcement. As the size of the opening increases, add extra support to the covering. If

FIGURE 3-7 Warning line system.

the opening is too big to cover, erect OSHA standard guard rails. Never place a board or any makeshift bridge over the opening. This only encourages people to walk over the hazardous area.

Specialized barricade options for floor openings and hatchways include

- A hinged cover, with someone in constant attendance when the cover is open
- Portable railings on all exposed sides except at entrances to stairways
- Standard railings for temporary floor openings

Installing Catch Platforms

Catch platforms are installed at or near the roof perimeter. When attached below the roof (Fig. 3-8), the catch platform allows complete freedom for roofing operations. The platform can also serve as a work surface for perimeter work. The system is technically feasible for some buildings, but masonry veneers invite uncertain and possibly danger-

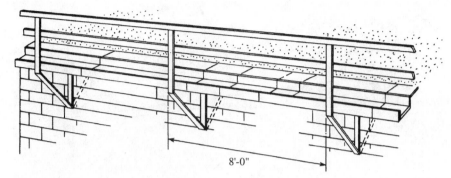

FIGURE 3-8 Perimeter catch platform.

ous horizontal anchorage. The system is useless in situations where walls have high windows or are constructed with lightweight panels.

The roof-attached catch platform system shown in Fig. 3-9 is installed remote from the roof edge. This eliminates the use of lifelines. It can also serve as a work surface for perimeter operations. The catch platform shown in Fig. 3-10 is hung over the roof edge and attached only at the roof deck level. This, in turn, eliminates the need to fasten it to the wall system. This catch platform scheme protects personnel in continuous roofing operations, but limits perimeter work. Lifelines or other safety systems must be used when installing, removing, and patching holes in the roofing system left by safety devices.

Curbed openings pose a threat too, though not as much as flush openings. Rail or cover these openings if they do not extend more than 36 inches. A roofer could accidentally back into a curb edge and fall over it and into the opening if these areas are not properly protected.

Falls also occur when workers are setting panels. Keep walking on girts and purlins to a minimum when this work is being performed. Assure worker control over hoisted panels by using taglines. Also, check roof decking when it is placed to ensure that the panels are supported on all four sides.

Caring for Injuries

A minor slip on a roof can cause a serious accident. Eliminating the causes of slips is the main method a roofing contractor can employ to significantly reduce falls at the jobsite.

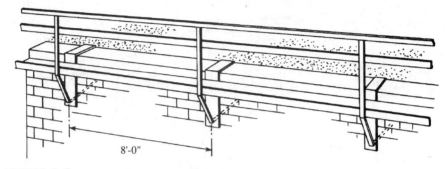

8'-0"

FIGURE 3-9 Roof-attached catch platform.

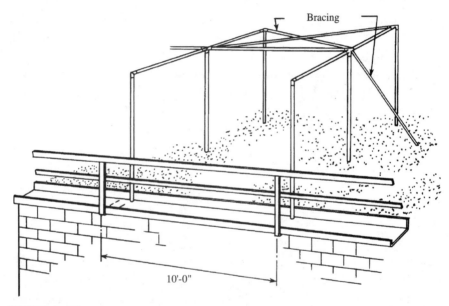

Bracing

10'-0"

FIGURE 3-10 Another type of roof-attached catch platform.

Keep anyone who falls from one level to another still. Have him or her lie flat, even if he or she does not seem to be injured. Do not move the victim. Back injuries are often hidden, and any manipulation of the person can aggravate the problem and cause serious damage.

Ideally, only trained medical professionals should move any worker who has fallen. If a worker falls into an awkward position, however, it might be necessary to slowly and gently straighten his or

✓	Item
	All Jobs, Trucks, Warehouse, and Office
	Fire extinguisher—with *service inspection* up to date
	Safety glasses—in stock and signage posted
	Safety cans for flammable fuels—recently inspected
	Industrial-grade first aid kits—on site and in trucks, recently inspected and stocked
	Flat Roofing
	No mechanical equipment or material within 6 ft of any edges of roofs or large openings
	Warning lines or fall monitoring systems in place, checked, and working
	Tear-Off
	Chute used for disposal at distances greater than 20 ft
	Inspected safety glasses and dust masks available and worn by crews
	Tear-off box roped off
	Special Hazards
	Electrical lines shut off or roped off
	Roof openings covered or roped off
	Unsafe decking properly covered
	Flammable vapors discharged
	Radiation hazard on roof—see owner
	Kettle
	Inspected
	Face shields and proper clothing worn by crew
	Fire extinguisher—inspected and easily reached
	Placement reviewed for safety
	Propane storage secure and 15 ft from kettle
	Guardrails at outlet area
	Dangerous areas roped off
	Proper signage in place

FIGURE 3-11 Jobsite safety checklist (on roof).

✓	Item
	At All Jobs
	Ground fault box at power source
	Emergency phone numbers readily available
	Hard hats worn by all crews
	Steep Roofing—3½ in 12
	Scaffolds, guardrails, fences, safety lines—in place and checked
	Hoist
	Inspected
	Counterbalance with pan weights inspected
	Line secured at night
	Guardrails at hoist area
	Hard hats worn on ground
	Ladders
	Tied off securely and inspected
	Safety feet inspected and in place
	3 ft of ladder above roof edge
	All ladder accesses inspected regularly
	Flammable vapors discharged
	Inaccessible at night
	Scaffolding
	All staging and scaffolds secured and inspected
	Leveled and inspected
	Planks and plywood inspected
	Guardrails inspected
	All scaffold installations inspected regularly

FIGURE 3-12 Jobsite safety checklist (on ground).

her back. Do this only if there are other hazards present, such as falling further or trouble breathing.

The victim should always see a doctor after a fall for a full examination.

Initiating a Safety Program

Safety cannot be accomplished simply by writing a safety policy, designating a safety officer, or performing safety training. There must be follow-through. The only thing that a company can do to enhance its employees' natural desire not to get hurt is to give safety as much presence as possible (Figs. 3-11 and 3-12). The first prerequisite for any safety program is a commitment from top management. Without this commitment, the safety program is going to be handicapped at best and doomed to failure at worst.

Management must understand that safety can enhance profitability. Safety can lower insurance premiums; reduce OSHA, Department of Transportation (DOT), and Environmental Protection Agency (EPA) citations and fines; improve morale and productivity; reduce substance abuse; avoid negative publicity; and even be used as a marketing tool to separate your company from your less safe competition.

Look closely at the type of work you do and the equipment you use. Different operations have different hazards. In addition to the information in this chapter, the National Roofing Contractors Association and OSHA can also help answer questions about hazards and how to protect your company and your employees.

Tools and Equipment

o gain an understanding of where the roofing industry has been, one need only study the equipment catalogs from past decades. In the 1960s, for instance, large, heavy equipment that moves material from one point on the roof to another was advertised.

The emphasis during this period was on new construction. The equipment, therefore, was manufactured to help the roofer construct the roof system. Felt layers, gravel spreaders, hot-stuff spreaders, and dispensers were popular pieces of equipment during this period.

Safety also started playing a larger part in the roofing trade as mechanized equipment began to pose new problems for roofing contractors. Mechanized equipment made the physical work easier, but also exposed workers to new risks.

The powered equipment fell off the edge of the roof, ran into and over other workers, and caused a great deal of concern.

In response, brakes, kill switches, and belt guards were introduced, engines were made smaller, and less riding equipment came to be available.

The need for lighter-weight equipment grew as the building industry moved toward lighter-weight materials for roof deck construction, such as steel, gypsum, and lightweight concrete. The heavy, bulky equipment was redesigned using new and improved materials. The resultant lighter equipment did not sacrifice strength, durability, or application.

During this same period, buildings were going from multistory to single-story, but at a cost that required the continued use of powered equipment. Because of safety and weight requirements, more walk-behind equipment than the rider-mounted variety came into use.

The 1970s and 1980s, while strong in new construction, also saw a greater emphasis on reroofing. Many of the roofs built during the post-World War II era reached the end of their life expectancy. To accommodate reroofing requirements, roof membrane tear-off equipment was developed, along with large powered sweepers and gravel-removal equipment. Truck-mounted cranes used to move material and equipment on and off the roof safely also gained in popularity.

The introduction of single-ply membranes during this same period also had a great effect on the equipment industry. The single-ply system does not need the melters or other equipment associated with traditional heat application systems. In some cases, the need for tear-off equipment was eliminated when membranes were installed over existing substrates.

In spite of all the changes in building design and construction, some equipment, such as the melter kettle, has changed little since its inception. Kettles or tankers are basically the same melting pots today that they were when first introduced.

We can expect to see continued changes in equipment design and safety as new materials and application methods become popular. The modern roofer should be aware of the new and existing equipment available in order to take advantage of the many benefits offered. See the products and equipment in App. A for more information. As explained in Chap. 12, carefully purchased equipment can quickly pay for itself in reduced worker hours, quicker completion times, and an increased number of jobs.

When introducing new gear, follow the manufacturer's operating instructions and warranty information at all times. Use approved safety precautions, safety equipment, and worker education to keep the jobsite a safe and productive work area.

Hand Tools

A roofer's hand tools are perhaps his or her most important equipment. Professional roofing mechanics keep their hand tools clean and ready to use whenever they are on the job. The basic tools are

- A nail bag with proper belt and hatchet holder
- A double-knife sheath attached to belt
- A utility knife with replaceable blades
- A roofer's hatchet, depending on the type of roofing
- A roofer's all-purpose utility bar
- A set of tin snips
- A chalkline and chalk
- A 12-foot measuring tape
- A roofer's toolbelt or container for carrying tools

Knee and hip pads are also important. Wearing knee pads lessens job stress and makes working much easier. For protection on very steep roofs, use hip pads.

Some types of roof coatings require special hand tools. For instance, a nail stripper can increase the production rate up to 50 percent on jobs that require the installation of strip asphalt shingles and three-tab asphalt shingles.

A note of caution: When using the nail stripper (Fig. 4-1), remember that it hangs from your chest on a roof and at belt height on a side wall. Be sure to take the

FIGURE 4-1 Typical nail stripper.

stripper off when you roll out felt. It can get in the way and pose a safety hazard.

Utility knives are a must for roofers installing built-up roofing (BUR) systems, but they are considered a "don't" by many manufacturers of sheet membrane systems. A utility knife can cut through the second piece of a membrane and cause an instant leak. Because of this, shears or scissors are the preferred tool for single-ply systems.

Power Tools

Before deciding what power equipment to use on a job, consider several important criteria. The power supplied must be sufficient to get the job done (and then some), must be reasonably efficient in terms of fuel consumption and overall operating costs, and must be portable.

The gas engine is the number one power source on roofing jobs because of its low purchase price, reasonable operating efficiency, and portability.

In addition, tools such as drills, screw guns, circular saws, and other small hand tools, require an electrical power source. Gas- or diesel-powered generators are a must if you encounter problems with electrical supply at the jobsite frequently. Study how much labor is wasted searching for power supplies, running cords, throwing breakers, etc. If your crews need electric tools on the average jobsite, generators can pay for themselves quickly.

Generators

Before you buy a generator (Fig. 4-2), figure out the maximum power you need. Look at which tools or combinations of tools are used at any given time. Add up the wattage of the tools to arrive at your wattage requirements. Most generators are rated in watts (1000 watts = 1 kilowatt). If an electric tool is not rated in watts, wattage can be figured by multiplying volts times amperes (watts = volts × amperes).

Another way to decide which generator to buy is to rent several types and try them out. Generators are available where other types of construction and industrial machinery are offered for rent.

Pneumatic Tools

Roofers also use pneumatic power, especially those involved in asphalt shingle applications. While pneumatic power can be used for many tasks performed by electric tools, air power is used primarily for nailing and stapling (Fig. 4-3).

FIGURE 4-2 MGH 10,000 generator; Master Distributors (www.mastergenerators.com).

FIGURE 4-3 Max CN450R pneumatic nailer; maxusajdominice@cs.com.

Economic conditions play a key role in any roofer's choice of fastening methods. A modern, lightweight portable compressor and pneumatic nailer or stapler outfit are fairly inexpensive. Air equipment can triple production compared to typical hand application. Installation speed quickly translates into labor and the ability to handle more projects. Savings and more completed jobs enable the roofer to recoup the initial investment in pneumatic equipment quickly.

Special Equipment

BUR and ethylene propylene diene terpolymer (EPDM) applications require special machinery and equipment, such as hot luggers (Fig. 4-4), kettles (Fig. 4-5), roofing torches, gravel spreaders (Fig. 4-6), tankers, and so on. Single-ply installation and roof tiles also require specialized tools, and some tools have general uses, such as wet-roof equipment. These are discussed in their appropriate chapters.

The decision to buy gear should be made with care. Throwing money at tools randomly can lead to a rapid drain of capital. Study the

FIGURE 4-4 55-gallon hi-lo lugger; Reeves Roofing Equipment Co. (www.reevesequipment.com /EquipCatalog/index.htm).

FIGURE 4-5 450-gallon pump kettle with afterburner/safety loader system; Reeves Roofing Equipment Co. (www.reevesequipment.com/EquipCatalog/index.htm).

FIGURE 4-6 Hot process (left) and cold process (right) powered gravel spreaders; Reeves Roofing Equipment Co. (www.reevesequipment.com/EquipCatalog/index.htm).

market and make sure there is potential for an on-going cash flow before pulling out your checkbook.

Dewatering Devices

Sometimes the only practical way to get rid of the wet stuff is with the old-fashioned broom and mop routine. It is not easy to do, but it is a well-known fact that it works. On small jobs or in tight spaces, using a broom and mop is the only way short of evaporation to get rid of water. If you encounter this situation, schedule one or two men at the jobsite ahead of the rest of the crew so everyone's time is not wasted waiting for the deck to dry.

When possible, it makes sense to mechanize water removal. The types of water-removal equipment currently available can be placed into three basic categories: dryers, blowers, and pumps.

DRYERS

Roof dryers are relative newcomers. The principle is simple and effective for drying up dew or shallow spots of water. The dryer rapidly evaporates the water. Obviously, great care should be used with an open-flame device on the rooftop to assure that nothing or no one gets burned.

A later version of a roof dryer is a cart-mounted blower that uses a small engine to spread the heat produced by a light propane (LP) gas-fired burner. The blower can move large volumes of water, and the superheated air serves to evaporate the dampness that remains. Care must be used to keep the blower moving while the torch is lit because the heat is intense enough to possibly ignite materials.

BLOWERS

Blowers, also called air brooms, are useful for getting rid of shallow ponded water. Available in backpack and cart-mounted configurations, air brooms do a more effective job than hand brooms with more speed and less labor.

Backpack models are the most popular because of their light weight, low cost, and portability. In the recent past, the small two-cycle engines and the blowers themselves have been greatly improved while the purchase price has actually gone down.

Look for a commercial-industrial model with a long blow pipe that easily reaches the deck surface. If you anticipate that you will use the air broom more than two hours per day, consider a cart-mounted model. Cart-mounted units are easier to operate over extended periods of time and come with more horsepower and air-handling volume, since weight is not a consideration.

Do not use a blower for small jobs, or when the blown water and debris might fall onto parked cars or any areas where people might be present.

PUMPS

The fastest and most productive way to get rid of large amounts of water is to pump it off. Several dewatering pump packages have been on the market for many years.

The most popular package utilizes a marine type, gas-operated one-inch pump with an automatic shut-off. These pumps can be operated unattended and can move a respectable volume of water at approximately 2000 gallons per hour (gph).

When buying, include a strainer screen at the suction end of these and other pumps. Without the strainer, debris can enter and damage the pump.

Several pump packages offer specific advantages. At the low end of the cost scale is a hand-operated diaphragm pump that is used to start a siphon. The volume and speed are much lower than those of powered pumps (around 200 to 600 gph), but the cost of maintenance and upkeep is low.

A small two-cycle engine, similar to the engine on the backpack blower, also powers a centrifugal pump. The price tag is low and the volume of water moved is respectable at approximately 1750 gph.

Hydraulically powered roof-dewatering pumps are available. The cost of the pump itself is quite nominal, given its production capabilities (approximately 6000 gph).

Reroofing Equipment

The roofing business can be very solid because of the immense number of retrofit opportunities. Reroofing is a mainstay for a large portion of the roofing industry, with virtually all roofers involved to some extent. With good management, reroofing can become a cash flow.

FIGURE 4-7 Powered roof ripper; Reeves Roofing Equipment Co. (www.reevesequipment.com/Equip-Catalog/index.htm).

POWERED ROOF REMOVERS

Relatively easy to operate, powered roof removers (Fig. 4-7) have a small gasoline engine that propels the unit and operates the flat blade on the front, which is from 12 to 24 inches wide. The blade oscillates front to back, slicing through the membrane as the machine moves forward. Generally, it takes a crew of four to eight persons to keep up with one person operating the roof remover.

POWERED ROOF CUTTERS

Designed for use on BUR roofs, the powered roof cutter (Figs. 4-8 and 4-9) can speed production on virtually all flat decks. Weighing 200 pounds or less, the powered roof cutter cuts up the old membrane into sections that are easy to tear off and to handle. It can be used ahead of a powered roof remover and makes a nice even cut for tie-ins at the end of the day. This popular machine has been around for many years and is available in models with 12- and 14-inch blades.

POWERED GRAVEL REMOVERS

There is no doubt that a quality gravel remover (Fig. 4-10) saves labor and dollars every time out. The small *patch-scratcher* class gravel removers cut a path from 4½ to 8 inches wide and are useful for small

FIGURE 4-8 Single-blade roof cutter; Reeves Roofing Equipment Co. (www.reevesequipment.com/Equip-Catalog/index.htm).

FIGURE 4-9 Dustmaster Ultra Cutter; Garlock Equipment Company (info@garlockequip.com).

FIGURE 4-10 Rocker powered gravel remover; Garlock Equipment Company (info@garlockequip.com).

patches and for scraping narrow paths on long runs such as expansion joints or wall flashings.

The full-size gravel removers cut a path wide enough to do entire roof surfaces and are available in a wide range of sizes from 12 to 24 inches. They also come with chimney shrouds and snorkel-type filters similar to those on roof cutters.

Powered gravel removers come in a variety of power and blade configurations. To determine the best model for your needs, ask roofers in your area which one works best for them. The effectiveness of a blade design or cutting head depends on the type of aggregate used and the climate.

POWERED SWEEPERS

Today's powered sweepers (Fig. 4-11) are available in several configurations. For example, there are models that feature 5-horsepower (hp) engines and 36-inch brooms. One model might utilize a belt drive to operate the wheels and broom, while another model might have a gear drive to the wheels and a belt drive to the broom. Most gas-powered sweepers are self-propelled and simple to operate.

The right broom for the job is directly related to the task at hand.

VACUUM SYSTEMS

Introduced in the mid-1970s, vacuum systems were designed to efficiently, quickly, and cleanly remove roofing gravel from the roofing sur-

FIGURE 4-11 Sweeper 86; Garlock Equipment Company (info@garlockequip.com).

face and convey it off the building to a container on the ground. A full-sized roof vacuum takes only a two- or three-person crew and, without dust, can remove gravel at a rate of 150 squares a day to the dump truck.

Roof vacs are at their best when the jobsite conditions are at their worst. When gravel has to be removed from tall buildings that are cut up and inaccessible, a roof vac reduces logistical problems tremendously. The system utilizes a power unit, usually a 60- to 100-hp diesel engine, which drives a blower that creates a vacuum in a container or hopper. The 4-inch hoses extend to the rooftop, where the operator uses a pickup wand to vacuum up the loose gravel.

Most vacs handle water as well, and most full-size vacs feature a reverse blower that conveys gravel back up to the roof. This method is not as productive or as efficient as some other hoisting/conveying methods, but it can return gravel to the rooftop efficiently at some jobsites.

Trash Chutes

Reusable trash chutes are made of high-density polyethylene or other durable materials and come in 4-foot sections that are 30 inches in diameter. Trash chutes are shipped and stored flat, then bolted together to make a cylinder.

Trash chutes are used primarily for high-rise work, but they are also convenient at lower levels. The use of trash chutes can eliminate costly grounds cleanup and avoid damage to trucks, cabs, and windshields.

Lifting and Climbing Gear

Ladders

Ladders have multiple uses in roofing operations. Roofers commonly use various types of ladders made from both aluminum and wood. Designs include single, straight, extension, and pushup ladders. A stepladder is a self-supporting, portable ladder that opens out, like a sawhorse. The maximum permissible height for a stepladder is 16 feet. Always use stepladders only when they are fully opened and do not use them as regular working platforms.

ROOF LADDER

A roof or *chicken* ladder (Fig. 4-12) is used on steeply pitched roofs. Basically, it is a plank that has evenly spaced cleats and is designed to accommodate the roofer's movements, but not materials. The crawl-board plank should be at least 10 inches wide and 1 inch thick. The cleats should be a minimum of 1 inch by 1½ inches and equal in length to the width of the crawl board. Space the cleats equally and not more than 2 feet apart. Drive the nails through the cleats and clinch them on the underside of the crawl board. Secure the crawl board to the roof by a ridge hook or other effective means.

ROOF BRACKETS

Adjust roof brackets to fit the pitch of the roof. Space the brackets a maximum of 10 feet (Fig. 4-13) apart and secure them with multiple nails capable of supporting twice the maximum anticipated load. All planks should be a minimum of 2 × 10, scaffold-grade lumber. Extend the planks a minimum of 6 inches and a maximum of 12 inches past the bracket, or overlap a minimum of 12 inches with both planks supported by the bracket. Either the roof edge should be protected with a 36-inch parapet, standard guardrail, and a scaffold platform, or the crew should use safety belts, lanyards, and lifelines.

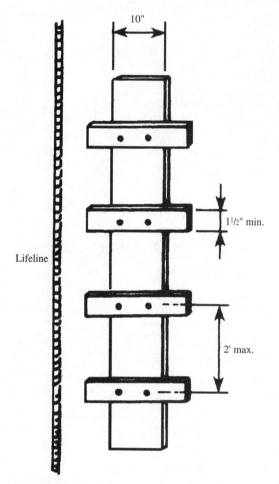

FIGURE 4-12 A roof or chicken ladder.

An adjustable roof seat resting on self-gripping cleats can be useful on steeply pitched roofs.

Scaffolds

Scaffolding consists of elevated platforms that rest on rigid supports; it must be strong and stiff enough to support workers, tools, and materials safely. Scaffolds can be made of wood or metal. Wood staging is normally tacked together on the job, while metal scaffolds are manufactured in a variety of styles.

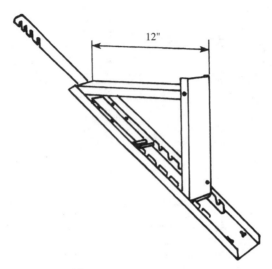

FIGURE 4-13 Roof bracket used on steep-slope roofs.

Most of the scaffolding used today is manufactured metal. The scaffolds are easily assembled and dismantled. Many designs are available, including set tubular metal and rolling mobile. Figure 4-14 shows a tubular stationary scaffold unit.

All scaffolds and lifts must be used in accordance with all local, state, and federal Occupational Health and Safety Administration (OSHA) laws and regulatory codes.

Because of the high cost of warehousing and labor, it is cost-effective to subcontract scaffolding. This allows crews to move quickly on the roofing, making the job profitable. Bid the scaffold subcontractor into the job with any miscellaneous costs you will accrue, and add a profit to the line item.

Hoists

A hoist of some sort is usually used to lift the roofing material onto the roof. A variety of roof-mounted hoists, truck hoists, telescopic hoists, trolley hoists, scissor-lift trucks, and cranes are available, one of which is appropriate for any situation.

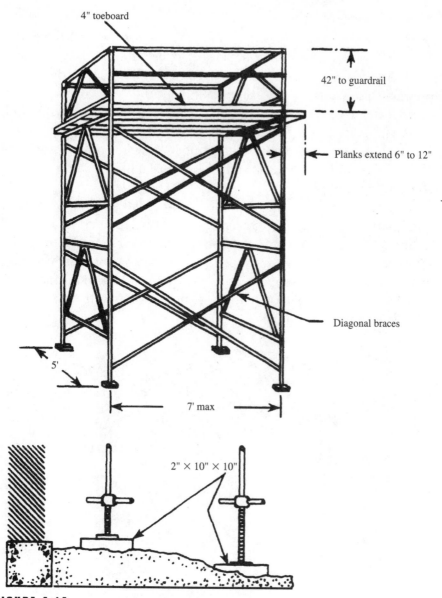

4" toeboard

42" to guardrail

Planks extend 6" to 12"

Diagonal braces

5'

7' max

2" × 10" × 10"

FIGURE 4-14 Typical tubular stationary unit and how it is secured.

ROOF-MOUNTED HOISTS

A roof-mounted hoist is, as the name implies, set up on and operated from the roof. The hoist framework supports the winch and power drive, and a wire cable extends from the roof to lift the materials off the ground. Accessories at the cable end adapt the hoist to the different roofing materials. These accessories include gravel buckets with dump gates for discharging gravel, hoisting forks for lifting full pallets of insulation or rolls, hoist slings for lifting long, heavy rolls of single-ply sheet goods, and platforms for miscellaneous materials or for lowering tear-off (Fig. 4-15).

Roof-mounted hoists come in three basic styles: beam, swing-beam, and platform. Typically, the beam hoist can lift loads up to 300 pounds, while monorail or swing-beam hoists are capable of lifting loads of up to 1 ton in weight (Fig. 4-16). Jobsite factors, such as setup space, material flow, parapet walls, and the sheer bulk of the materials, determine the best hoist. As with all major purchases, try renting first, track labor savings, and choose carefully.

FIGURE 4-15 Typical platform hoist.

FIGURE 4-16   A swing beam hoist in use.

TRACK HOISTS

Track or platform hoists come in a variety of sizes with several power and attachment options. There are many distinct advantages to using track hoists, not the least of which is the low initial cost. These rigs are lightweight and portable enough to be carried easily, in a knocked-down state, in a pickup truck. They can be moved around the building as work progresses, a feature that is especially helpful when using scaffolding. Steep-slope and low-slope roofs are no problem for track hoists.

The basic rig consists of a *track*, which is an aluminum frame that resembles, but should not be confused with, a ladder. Tracks come in several sizes and can be joined with other track sections to assemble the proper amount of reach. Track sections are joined by splice plates that slide into the sides of the tracks. This assures a snug fit and straight track.

TELESCOPIC HOISTS

Another ground-to-roof hoist is the trailer-mounted, towable, telescoping hoist. This type of hoist is capable of extending up to 130 feet, and can carry loads that weigh from 400 to more than 1000 pounds. This type of hoist has actually been around for many years, but advances in technology have added greater height capabilities and lifting capacities. This versatile, easily transported hoist requires minimal setup space. Because of these features, telescoping hoists are gaining in popularity.

TROLLEY HOISTS

The trolley track hoist was developed specifically for the BUR contractor. It generally has a capacity of approximately 400 pounds, though some can move more weight. The hoist is counterbalanced with iron or concrete weights set on a mechanically attached metal pan or tray. The counterbalance weight is at least half that of the weight of the material being hoisted. Do not exceed the manufacturer's recommended weight limit. The aerial platform or high lift provides a means of working on a roof or providing transportation to the rooftop.

Trolley hoists are driven by gasoline, diesel, or electric-powered motors. Most hoists of this type have a dual braking system. Another safety feature, required by OSHA, is 4-foot gates on either side of the trolley hoist.

Scissor-Lift Trucks

Also known as *high lifts*, this type of hoist incorporates the truck used to haul materials and the lifting equipment. The body of the truck uses hydraulics to lift to the rooftop and then the roofing materials are walked or carried by wheelbarrow directly onto the roof.

Scissor-lift trucks are versatile tools for roofers, particularly because they can be used for both tear-offs and BURs. Most scissor-lift trucks have the capability of dumping, making this a very nice piece of equipment. Drawbacks to a high lift are that the reach is limited to approximately 18 feet and the truck has to be driven all the way to the building, which can be a problem on some jobsites.

Set up the unit on level ground with its outriggers down and unloaded evenly. For the residential shingler who does reroofing and mostly one story work, this rig is hard to beat. The scissor-lift truck is also designed to work with a belt conveyor.

Cranes

Cranes are another popular piece of lifting equipment. If you frequently have jobs that require a crane rental, it might be more economical for you to purchase or lease this piece of equipment. A variety of attachments are available, from tear-off boxes and gravel buckets, to hoisting forks and single-ply slings. Keep in mind that of all the equipment available for moving materials, the crane option is probably the most expensive initial investment. Be sure to try rentals and explore leasing before buying.

Conveyors

Conveyors can be used to transport all types of material to the construction worker on the roof. Conveyors are often misused like the ladder and hoist, however, and should not be used to transport equipment or workers.

Belt-flight and chain-driven conveyors provide more protection on large, flat, BUR roofing projects. They have evolved to meet the needs of most new single-ply roofing systems. One advantage to the conveyor is its ability to continuously move material to the top without having to stop and return or even slow down. Another advantage is the conveyor's ability to move gravel at a rate that outpaces the best of crews.

TRUCK-MOUNTED CONVEYORS

Truck-mounted conveyors offer several advantages over some other methods of getting materials to the job and onto the roof. When the conveyor unit is mounted on the truck, there is essentially one piece of equipment that moves the material from warehouse to jobsite and then onto the roof. This eliminates the need for other tow-behind equipment and saves setup time. A continuous-drive conveyor, even moving at relatively slow speeds, can deliver materials at a rate faster than most crews can load and unload them, often up to 720 bundles an hour.

Maintaining Tools and Equipment

Preventive maintenance is recognized by most roofing contractors as being less expensive and more sensible than crisis maintenance, but preventive maintenance lacks the urgency of purpose. Preventive maintenance is not urgent unless it is made urgent.

A sound maintenance program is not always the easiest thing to get started, but once going it can pay some nice dividends. As with safety, put one person in charge of your maintenance program and work with him or her to get the job done. Then follow up to see that the system works correctly. Involve all mechanics and operators and encourage them to report any malfunctioning equipment immediately.

Establish a reporting method for everyone to use. The best way to accomplish this is to enforce a policy that all equipment be kept clean and looking good. When equipment is kept clean and nice looking, employees take the attitude that the equipment is valuable to the owner and care must be taken when it is used. Cleaning also helps certain pieces of equipment perform better technically. Equipment that appears cared for reflects the entire attitude of the company—hit-and-miss or professional.

Built-Up Roofing

oday's low-slope roofing technology offers several basic choices to the contractor, architect, and building owner: single-ply; overlapped and heat-sealed; modified bitumen roofing (MBR) applied as one-, two-, or three-ply systems; and built-up roofing (BUR) utilizing bitumen asphalt or coal tar as the waterproofing and adhesive agents. Each system offers advantages for specific applications.

As stated in Chap. 1, BUR is by far the oldest of the modern low-slope methods and is still a popular commercial application (Fig. 5-1). BUR consists of layers of reinforcing felt sandwiched between bitumens. Asphalt and coal tar are the bitumens of choice. These thermoplastics become more fluid with heat and revert to a more solid state as they cool. They are versatile waterproofing materials.

Welding the Plies

The plies are welded when heated, mopped bitumen melts, and then fuses with, the saturant bitumen in the roofing felts. Therefore, correct application temperatures are vital to the creation of a quality roof membrane. A high bitumen temperature must be maintained to create the welding process.

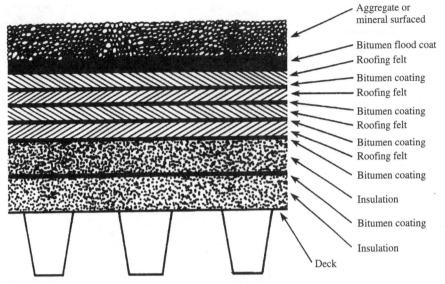

Aggregate or mineral surfaced
Bitumen flood coat
Roofing felt
Bitumen coating
Roofing felt
Bitumen coating
Roofing felt
Bitumen coating
Roofing felt
Bitumen coating
Insulation
Bitumen coating
Insulation
Deck

FIGURE 5-1 Typical four-ply BUR with flood coat.

The properties of bitumen are such that bringing it to a high temperature for a long period of time can reduce the softening point of asphalt and raise the softening point of coal-tar bitumen. Bitumens can be kept at high temperatures for short periods of time without damage. In fact, they must be heated to high temperatures in order to achieve complete fusion and strongly bond the plies.

Manufacturing Asphalt

Asphalt is made up of three elements: saturants, asphaltenes, and resins. Saturants are light oils that make the asphalt soft and flexible so that it expands and contracts with the application surface. Asphaltenes provide the body, rigidity, and strength. Resins bond the saturants and asphaltenes and furnish the asphalt with its resilience.

The crude oil source contributes to the quality of the finished asphalt product just as much as the manufacturing processes do. Different crudes impart distinct characteristics. For example, some California crudes have relatively low saturants and flash points, typically 460° to 525°F.

The flash point is the temperature at which the asphalt's light oils begin to evaporate and can be ignited. As the flash point increases, the chance of combustion and the vaporization of light oils both decrease. The result is a harder flux that is better suited for paving applications than for roofing applications. The best roofing-grade asphalts are produced from the softer crude sources, or from blends that incorporate a sufficient percentage of soft material to meet industry standards and provide the anticipated service life.

After the crudes are selected, the process of blending and blowing the crude-oil fluxes determines the characteristics of the finished asphalt product. Manipulating these two processes produces the following four grades of asphalt.

Type I (dead level). This grade of asphalt is relatively susceptible to flow at roof temperatures and has good adhesive and self-healing properties. It generally is used in aggregate-surfaced roofs on slopes up to ½ inch per 12 inches.

Type II (flat). Type II asphalt is moderately susceptible to flow at roof temperatures. It generally is used on roofs with slopes between ½ inch per 12 inches and 1½ inches per 12 inches.

Type III (steep). Steep asphalt is relatively nonsusceptible to flow at roof temperatures. It generally is used on roofs with a slope between 1 and 3 inches per 12 inches.

Type IV (special steep). This asphalt is relatively nonsusceptible to flow at roof temperatures. It is useful in areas where   high year-round temperatures are experienced. It generally is used on roofs with a slope between 2 and 6 inches per 12 inches.

In 1977, the National Roofing Contractors Association (NRCA) adopted the Equiviscous Temperature (EVT) concept. This followed the determination that the flow characteristic, or viscosity, of roofing bitumens at the point of application to the substrate is an important factor in the construction of BUR membranes. EVT was then defined as the temperature at which the viscosity of a bitumen is sufficient to support the proper adhesion, waterproofing, and application rate.

If the EVT information is not furnished by the manufacturer, use the temperature given in Table 5-1 as a guide.

TABLE 5-1 Heating Temperatures of Asphalt

ASTM D-312 type no.	Asphalt type	Maximum heating temperature (°F)
I	Dead level	475
II	Flat	500
III	Steep	525
IV	Special steep	525

Manufacturing Coal Tar

Coal tar is a byproduct of the coking of bituminous coal during steel making. Coal tar offers superior resistance to the elements. Its chemical structure consists of closed, ring-like molecules that make it highly stable and resistant to aging, chemical breakdown, and moisture. Coal tar is known for its resistance to ponded water. Another advantage of coal tar is its cold flow property, or the ability to self-heal cracks and fissures.

Traditionally, coal tar's *flowability* has kept it from being a choice product for roofs with more than minimal slope. As a *liquid solid*, it just does not stay put. BUR technology developed partly in response to this fact. Several thin layers of bitumen, alternating with felt plies, form a strong, flexible, waterproof membrane that stays in place, at least on gently sloping roofs.

Three types of coal tar are presently used for BUR, dampproofing, and waterproofing systems:

- Type I coal-tar pitch, used in BUR systems with coal-tar saturated felts.

- Type II coal-tar pitch, used in dampproofing and membrane waterproofing systems for below-grade and plaza deck waterproofing. In these types of constructions, the bituminous materials are combined with plies of coal-tar-saturated felts and fabrics.

- Type III coal-tar bitumen, used in BUR systems, has a slightly higher softening point than Type I coal-tar pitch.

Coal-tar products are described by the American Society for Testing and Materials (ASTM) Standard D-450, Types I and III, Coal Tar Pitch

Used in Roofing, Dampproofing and Waterproofing. Types I and III are suitable for use in the construction of coal-tar BUR, the difference being that Type III has less volatile components than Type I. The EVT concept is equally applicable to coal-tar products, although the recommended viscosity base is considerably lower for the coal-tar products than for roofing asphalts.

If the information is not furnished by the supplier, the following temperature ranges are appropriate for both mechanical spreader and mopping application techniques:

- Coal-tar pitch (Type I), approximately 360°F, plus or minus 25°F

- Coal-tar bitumen (Type III), approximately 375°F, plus or minus 25°F

Selecting Felts and Base Sheets

Several materials are used as reinforcing sheets and base sheets in BUR construction.

SATURATED ROOFING FELTS

Either asphalt or coal-tar bitumen can be used as a saturant for organic felts. Organic felts are manufactured from the fiber of paper, wood, rags, or a combination of the three. Prior to saturation, organic felts are described by a number that indicates the weight of the felt. For example, an unsaturated organic felt might be designated as 27-gauge. This means that the unsaturated product weighs approximately 27 pounds per 480 square feet of paper measure and is a nominal 27 mils in thickness.

The saturated product is generally referred to as a nominal number. For example, No. 15 in a No. 15 asphalt-saturated organic felt describes the nominal, or approximate, weight in pounds of one ply of No. 15 asphalt-saturated organic felt needed to cover one roof square, or 100 square feet.

SATURA5TED AND COATED FELTS

Saturated and coated felts can be manufactured from either organic or inorganic materials. Generally, they are factory-coated on both sides and are surfaced on one or both sides with very fine mineral sand or other release agents to prevent adhesion inside the roll prior to application.

IMPREGNATED FELTS

These felts are generally lighter in weight than organic felts. They are termed impregnated, rather than coated, because the surface is not completely covered with asphalt. These felts usually provide an open, porous sheet through which vapors can be vented during membrane application.

PREPARED MATERIALS

Prepared roofing materials are saturated and coated felts with talc, mica, sand, or ceramic granules incorporated into their weather surfaces. This provides both weather protection and decoration. Prepared roofing materials can be manufactured from organic or glass-fiber base felts or a combination thereof. A commonly used prepared roofing material is the reinforced flashing membrane, which consists of a glass-fiber base felt that is laminated with cotton or glass-fiber fabric and coated with asphalt. Prepared roofing materials are usually packaged in rolls that are one roof square, or 100 square feet, in size.

Basic sheets include the following.

No. 30 Asphalt-Saturated Organic Felts: These medium-duty, unperforated felts are used as the underlayment for shingle, slate, and tile work and, in some cases, as the base ply in BUR systems.

No. 40 Asphalt-Saturated and Coated Organic-Felt Base Sheets: Generally, these materials are used as the first ply or base layer for some BUR membranes. They are also used for temporary roofing or temporary flashings and as vapor retarders beneath insulation. These sheets are designed to prevent the intrusion of moisture from below the roof and to minimize the occurrence of ridges and wrinkle cracking.

They also are incorporated where the designer requires a tougher or thicker base to help span joints in the deck or substrate immediately below the roof membrane, and where mechanical fastening of a base ply is required.

Asphalt Glass-Fiber Mat Sheet: This material is composed of a glass-fiber mat that is impregnated and coated on both sides with asphalt. It is available in perforated or unperforated forms and is designed for use as a base ply in the construction of BUR membranes.

Venting Asphalt-Saturated and Coated Inorganic Sheets: Several manufacturers make a special, heavy-gauge felt to vent certain roof conditions. The venting characteristic of the felt is assisted by mineral granules on the underside of the felt. These venting sheets are, in some cases, specified for use over damp or concrete substrates as an underlayment or base ply.

The sheets can be nailed in place and/or spot-mopped with hot bitumen. If a vented base ply does not have a buttonhole attachment and is to be mopped in place, spot mopping is recommended, preferably with a mechanical spot mopper. This prevents solid mopping or strip mopping of the asphalt under the base sheet, which restricts the venting properties of the base sheet. If employed in a BUR system, these sheets provide for the venting of moisture vapor pressure by lateral, or horizontal, movement.

The following are the ply sheets used between coats of bitumens.

No. 15 asphalt-saturated organic felts are manufactured as:

- Perforated felts, which are ply marked for use in two-, three-, or four-layer applications.

- Unperforated felts, which are generally used as the underlayment of shingles and for general building paper usage.

No. 15 coal-tar saturated organic felts are ply marked for two-, three-, or four-ply applications and are commonly used in waterproofing specifications and in coal-tar BUR specifications.

Asphalt-impregnated glass-fiber felts are used in the construction of BUR membranes.

Asphalt-impregnated glass-fiber felts (Type IV) are slightly heavier than the Type III asphalt-impregnated glass-fiber felts. They have a breaking strength in both longitudinal and transverse directions that is nearly twice that of Type III asphalt-impregnated glass-fiber felts.

CAPS AND FLASHINGS

Several caps and flashings are available for use in BUR applications.

Wide-Selvage Asphalt Cap: This roll roofing material is composed of an asphalt-saturated felt that is coated on approximately one-half of the weather side. The asphalt-coated surface is then treated with colored mineral granules. It is generally used for two-layer applications over a base ply or as succeeding sheets over the initial plies of a BUR system.

No. 90 Mineral-Surfaced Roll Material: This product is manufactured either with no selvage or with a 2-inch selvage, depending on the manufacturer. It is supplied with various ceramic-colored granule surfacings, and is used as an organic roll roofing material and specified for a variety of flashing applications.

No. 50 Smooth-Surfaced Material: This roll asphalt product is used in some roof membrane specifications as a heavy-duty base sheet. It is also useful in the construction of temporary roofs and for flashing applications.

No. 60 Smooth-Surfaced Material: This roll roofing product is an organic felt that is saturated with a mineral-stabilized asphalt compound and surfaced with talc or mica. It is used when inexpensive roll roofing is desired, as well as for flashings and temporary roofs.

Inorganic Cap Sheet: This roll material is composed of a heavyweight inorganic felt that is saturated and coated on both sides with asphalt and then surfaced on the weather side with various colored mineral granules. It is used as a top, or cap, surface over other BUR plies, for flashings, for granular-surfaced lightweight roof system constructions, and for decorative purposes.

Reinforced Bituminous Flashing: This material is a heavy saturated felt that is reinforced with glass fiber. The material is surfaced with fine mineral matter.

Surfacing BUR Systems

BUR systems consist of multiple layers of bitumen, which are the waterproofing medium, and various forms of roofing felts, which are the reinforcements. Each type of base, cap, and ply felt is designed to perform a specific function in the roofing system. These various forms permit the installation of BUR systems over many different substrates.

Bitumen is available in various grades to accommodate conditions that might be encountered, i.e., slopes, products, and weather conditions.

There are three basic types of BUR systems.

Surfacing Smooth Systems

Smooth-surfaced inorganic BURs offer several advantages. For example, they are lightweight, generally less than ⅓ the approximate 400 to 700 pounds per square of gravel-surfaced roofs. They are easy to inspect and repair. The absence of gravel facilitates fast visual inspection of the smooth surface. If they are damaged, patching takes little time and there is no need to scrape away gravel. Because of these advantages, maintenance of smooth-surfaced roofs is simpler than that for other BUR roofs.

When reroofing is desired, or necessary, the job can often be done without removing the old smooth-surfaced membrane. If removal is necessary, there is less material to remove. In addition to all these advantages, a smooth-surfaced roof often costs less than a gravel-surfaced roof with the same design life.

Surfacing Gravel Systems

Gravel-surfaced, inorganic BUR systems are similar in construction to smooth-surfaced roofs, except in the final surfacing. Instead of the light mopping of asphalt or a light application of one of the other roof coatings used on smooth-surfaced roofs, a flood coat of approximately 60 pounds per square of hot bitumen is applied, followed by the appropriate aggregate.

Gravel-surfaced systems are typically more durable than smooth-surfaced BUR systems. The aggregate covering helps the asphalt flood coat resist the aging effects of the elements. The gravel also helps stabilize the flood-coat bitumen and permits heavier pours than typically are used in systems that do not employ gravel. This results in additional waterproofing material in the assembly.

Other, more obvious advantages are the improved fire resistance offered by the aggregate and added protection from penetrating forces like hail. Gravel-surfaced roofs are generally limited to slopes of 3 inches or less to minimize gravel loss and membrane slippage.

Surfacing Mineral Systems

Mineral-surfaced cap sheets form the last visible ply in this type of BUR membrane. The inorganic mat is coated with weather-grade asphalt into which opaque, noncombustible, ceramic-coated granules are embedded. The resulting sheet yields a roof that enhances the appearance of the building it protects.

Mineral-surfaced cap sheets provide BURs with several unique properties. The ceramic-coated granules on these products offer a uniform, factory-applied surfacing that helps the underlying bitumen resist weathering and aging. Unlike coatings, these granules are not typically a maintenance item. Many smooth-surfaced, coated roofs require periodic recoating. The cap sheet system also has improved fire resistance and reflective properties.

Surfacing with Reflective/Protective Coatings

This surfacing option for BUR membranes involves the application of a protective or reflective coating to a smooth-surfaced BUR system. The makeup of these coatings is based on the properties they impart to the BUR membrane. Some are light or aluminum-colored for reflectivity. Others are asphalt or coal-tar based and serve as an adhesive into which loose mineral granules are broadcast. Various coatings are used to enhance protective values of the roof, for example, fire resistance or some other physical property of the membrane.

Selecting Accessories

Sheathing paper and asphalt roof cement are two important accessories. Rosin-sized sheathing paper is a rosin-coated building paper that is usually used in BUR systems to separate felts from wooden-plank roof decks.

The sheathing paper prevents the first ply of felt from adhering to the wooden-plank decking and prevents the bitumen from dripping through the roof deck. If rosin-sized sheathing paper is not available, a heavy felt kraft paper or reinforced sisal paper can be used instead.

Asphalt roof cement is usually composed of petroleum asphalt or coal tar and solvents. It is sometimes referred to as plastic cement or roofing mastic. Asphalt cement is suitable for trowel application to roofing and flashing materials. One of the most frequently used product formulas is designed for special wall flashing. This formula contains a substantial amount of mineral ingredients that minimize sagging when the material is used on vertical surfaces.

Plain plastic cement is used in general roof repairs and to set metal. It contains less filler than flashing cement, which makes it more flexible and workable with roofing felts. Flashing cement is trowelable and can be obtained with varying amounts of solvent to make it workable under different temperature conditions.

Handling, Protecting, and Storing Materials

BUR materials, like all roofing materials, are intended to be installed in a dry and undamaged condition. The use of wet or damaged materials can contribute to the failure of the roofing system. It is recommended that the following guidelines for the handling, protection, and storage of all material be strictly followed. (See also Fig. 5-2.)

- Store roll goods on end and on a clean, raised platform to keep the ends of the rolls free from foreign matter.

- Store all roofing and flashing materials in a dry place. Plastic covers, stretchwrap, and shrinkwrap are not intended for job storage since moisture can condense within and on the material inside.

- Store all cartons, insulations, drums or cartons of asphalt, and cans of cement on raised, level platforms.

- Protect materials from the elements with waterproof tarpaulins. Improper storage of packaged asphalt can result in asphalt

✓	Item
	Store roll goods on end and on a clean, raised platform to keep the ends of the rolls free from foreign matter.
	Store all roofing and flashing materials in a dry place. Plastic covers, stretchwrap, and shrinkwrap are not intended for job storage, since moisture can condense within and on the material inside.
	Store all cartons, insulations, drums or cartons of asphalt, and cans of cement on raised, level platforms.
	Protect materials from the elements with waterproof tarpaulins. Improper storage of packaged asphalt can result in asphalt spilling from the containers, contamination, and/or moisture absorption. The latter causes foaming when heated in the kettle, which can result in injury to the roofing mechanic.
	Store solvents and solvent-containing material in a cool, dry, fire-safe area.
	Unload and handle all roofing material with care. Dropping rolls, roof insulation, cans of adhesives, and roof accessories can damage these components enough to cause unsatisfactory application and performance.
	Notify manufacturers immediately if materials are not received in satisfactory condition. Note any damage on the trucking firm's bill of lading.
	Keep the temperature of all roll goods above 40°F for 24 hours prior to application. MBR products are considerably easier to install when maintained at temperatures above 50°F.
	Securely cover, with tarpaulins, all roofing materials remaining on the roof at the end of the work day.
✓	Observe all fire precautions involving the storage and handling of roofing materials.
	Do not heat any supplied BUR materials with an open flame.

FIGURE 5-2 Handling, protecting, and storing materials.

spilling from the containers, contamination, and/or moisture absorption. The latter causes foaming when heated in the kettle, which can result in injury to the roofing mechanic.

- Store solvents and solvent-containing material in a cool, dry, fire-safe area.

- Unload and handle all roofing material with care. Dropping rolls, roof insulation, cans of adhesives, and roof accessories can damage these components enough to cause unsatisfactory application and performance.

- Notify manufacturers immediately if materials are not received in satisfactory condition. Note any damage on the trucking firm's bill of lading.

- Keep the temperature of all roll goods above 40°F for 24 hours prior to application. MBR products are considerably easier to install when maintained at temperatures above 50°F.

- Securely cover, with tarpaulins, all roofing materials remaining on the roof at the end of the work day.

- Observe all fire precautions involving the storage and handling of roofing materials.

- Do not heat any supplied BUR materials with an open flame.

Applying BUR Materials

The proper application of roofing materials is as important as the materials themselves. The following are recommended guidelines for the application of all roofing materials. (See also Fig. 5-3.)

- Never use wet or damaged materials.

- Never apply any roofing materials during rain or snow, or to wet surfaces. Moisture trapped within the roofing system can cause severe damage to the roofing membrane and insulation.

- Review the application guidelines in this book.

- Never mop more than 4 feet ahead of the roll in temperatures below 50°F.

✓	Item
	Never use wet or damaged materials.
	Never apply any roofing materials during rain or snow, or to wet surfaces. Moisture trapped within the roofing system can cause severe damage to the roofing membrane and insulation.
	Review the application guidelines in this book.
	Never mop more than 4 feet ahead of the roll in temperatures below 50°F.
	Observe EVT guidelines for asphalt application.
	When using mechanical felt-laying equipment, be sure all orifices are open.
	Lightly squeegee all fibrous felts into the hot asphalt.
	Roll all felts into the hot asphalt. Do not "fly" felts into the asphalt.
	Take special care when applying any roofing felts in cold weather. Check the temperature of the asphalt at the mop or spreader to determine if it is at the proper temperature.
	Do not mix different grades of asphalt or dilute asphalt with any materials.
	Do not use coal-tar pitch or bitumen with any MBR products.
	Do not use cutback asphalt cements under any MBR products. The use of the mastics over the top of the product to strip in or cover nailheads is acceptable; however, MBR cement products are preferred.
	Heat the asphalt according to the manufacturer's recommendations. Check the temperature of the asphalt at the kettle and at the point of application.
	Adhere to the guidelines for heating asphalt in this book.

FIGURE 5-3 Applying BUR materials.

✓	Item
	Use the proper grade asphalt as specified by the membrane manufacturer. A good guideline to follow is to use the softest grade of asphalt commensurate with the slope and climate conditions.
	Install water cutoffs at the end of each day's work to prevent moisture from getting into and under the completed roof membrane. Be sure to remove the cutoffs prior to continuing the application process.
	Heed the specific cold-weather application procedures given in this chapter.
	Install the complete roofing system at one time. Phased construction can result in the felts slipping because of excessive amounts of asphalt between the phased felt plies. Blisters, due to entrapment of moisture, are also a common problem, as is poor adhesion due to dust or foreign materials collecting on the exposed felts of an incomplete roofing system.
	During the asphalt setting time, which can be as long as 45 minutes, it is essential that traffic be minimized so that the interply mopping is not displaced.
	Comply with published safety procedures for all products. Refer to appropriate instructions and product safety information.

FIGURE 5-3 (Continued)

■ Observe EVT guidelines for asphalt application.

■ When using mechanical felt-laying equipment, be sure all orifices are open.

■ Lightly squeegee all fibrous felts into the hot asphalt.

■ Roll all felts into the hot asphalt. Do not "fly" felts into the asphalt.

■ Take special care when applying any roofing felts in cold weather. Check the temperature of the asphalt at the mop or spreader to determine if it is at the proper temperature.

■ Do not mix different grades of asphalt or dilute asphalt with any materials.

■ Do not use coal-tar pitch or bitumen with any MBR products.

■ Do not use cutback asphalt cements under any MBR products. The use of the mastics over the top of the product to strip in or cover nailheads is acceptable; however, MBR cement products are preferred.

■ Heat the asphalt according to the manufacturer's recommendations. Check the temperature of the asphalt at the kettle and at the point of application.

■ Adhere to the guidelines for heating asphalt in this book.

■ Use the proper grade asphalt as specified by the membrane manufacturer. A good guideline to follow is to use the softest grade of asphalt commensurate with the slope and climate conditions.

■ Install water cutoffs at the end of each day's work to prevent moisture from getting into and under the completed roof membrane. Be sure to remove the cutoffs prior to continuing the application process.

■ Heed the specific cold-weather application procedures given in this chapter.

■ Install the complete roofing system at one time. Phased construction can result in the felts slipping because of excessive amounts of asphalt between the phased felt plies. Blisters, due to entrapment of moisture, are also a common problem, as is poor adhesion due to dust or foreign materials collecting on the exposed felts of an incomplete roofing system.

■ During the asphalt setting time, which can be as long as 45 minutes, it is essential that traffic be minimized so that the interply mopping is not displaced.

■ Comply with published safety procedures for all products. Refer to appropriate instructions and product safety information.

Using BUR Application Equipment

Bitumens provide waterproofing qualities and act as the glue that holds the BUR system together. Improperly heated bitumen can spell an early death to an otherwise good roof. Inefficient handling of hot stuff can rip a bottom line to pieces.

Methods for handling hot stuff have improved dramatically since the pioneering days of BUR. In the beginning, natural tars and asphalts were heated by wood fires in drums or homemade containers. Sometime near the turn of the century, oil-fired burners began to replace the homemade containers and store-bought roofing kettles.

Heating with Kettles

The variety of kettles offered today is vast, and the capacity range is wide. As always, make the manufacturer's attention to safety part of your purchasing decision. For example, before transporting trailered equipment, verify that your vehicle can safely handle the load. Check the kettle's suspension, connections, and safety lights just as you would those on your own car. A list of vendors is provided in App. A.

The kettle burner should not require full power to melt any dried material. Even the strongest kettles are not designed to have their limits tested with every use. It should take only a few minutes of low temperatures to create a vent around the heat riser. The attendant needs to be patient.

Much of the heat produced by the kettle is invisible. Remove all equipment and materials from the designated kettle area to keep them from absorbing the extreme heat. The attendant should keep a constant watch, but should never linger around the kettle. The combined heat from the kettle and the sun can cause heatstroke or heat cramps in a short time.

The best operator, of course, is one who has worked with the burner pump before and knows its behavior and peculiar habits. Still, no amount of experience can tell a roofer when hot bitumen is going to pop, splash, or get swept up in the wind. Make sure that everyone working around the material properly protects his or her hands, face, and other body areas at all times.

Always maintain and clean kettles properly. Some of the most common reasons for kettle fires are trash and debris in and around the kettle, a heavy bitumen layer coated on the outside, leaking or unignited fuel, and improper handling.

Most kettle fires can be contained easily in their early stages by simply closing the lid or shutting off the fuel supply. If the fire goes beyond that point, get help. A dry-chemical extinguisher is good for putting out kettle fires and is the best all-around fire extinguisher for the roofer.

A clean kettle saves time because carbon that builds up around the firing flues insulates them from the asphalt. Flues that are insulated with a buildup of carbon do not transfer heat to the melt as evenly or as rapidly as clean flues. A clean kettle is simply faster than a dirty one. It also saves energy for the same reason. When the flue system has to overcome carbon to get to the asphalt, more time and fuel are required.

A clean kettle produces a better quality of asphalt because of the absence of massive quantities of carbon and sludge. A dirty kettle may have to be overheated to melt through and draw off cocks or to make up for the poorer overall heating characteristics of the flue system. Manufacturers of roofing asphalt recommend heating asphalt slowly and emphasize the importance of a clean kettle.

Regular cleaning also increases the service life of the asphalt kettle. The carbon that clings to the flues in the kettle not only insulates the flues from the melt, but also causes the extra heat to stay in the flues.

A clean kettle is safer to operate. The stalactites that form on the underside of the lid and vat cover are porous and can trap oxygen and light volatile oils coming off the melt. This combination of light oils and oxygen can be a fire hazard. If ignited, such a fire can be exceedingly difficult to extinguish. Flashing occurs more often in a dirty kettle.

Storing in Tankers

When large amounts of asphalt or coal tar are transported between the processing plant and the roofer's storage yard or the jobsite, tankers are often used; they represent a sizable investment and, because of their size, present unique safety concerns.

Keep the tanker's ventilation system free and clear at all times. Failure to keep the ventilation system clean could cause a fire or explosion.

Your kettle and tanker came with a set of operating and cleaning instructions. If the instructions have been lost or misplaced, contact your dealer or the manufacturer for a duplicate set.

Pumping Hot Stuff

Getting roofing bitumens from the manufacturing plant to the tops of buildings and then sandwiched between plies of felt has never been an easy job. The one piece of equipment that has played a major role in high-production roofing has been the hot-stuff pump. With the ability to pump hot stuff at the jobsite, roofers are able to use other productive pieces of equipment. Tankers, felt-layers, gravel spreaders, and many other tools would be useless if the hot stuff was still lifted up 5 gallons at a time.

Virtually all pumping systems built into today's kettles and tankers have at least a few things in common: a small gasoline or light-propane gas-powered engine that drives a rotary gear pump that is submerged in the vat and an internal recirculating-type plumbing system. You could say that these are the three Ps of the hot-stuff pumping system: power, plumbing, and pump.

The drawings in the manufacturer's instruction manual explain how a particular kettle or tanker is plumbed.

Check the pipes that transport the hot stuff to the roof for leaks and built-up material constantly. The most important reason for these safety checks, other than the obvious one of protection of the workers, is that they maintain the performance of the equipment and keep the operation running smoothly. Remember to keep pumping at all times when the engine is running.

Once the hot stuff is on the roof, it is generally transported to where it is needed by a lugger or roof cart. The hot stuff is poured in buckets for mop application.

Kettle Safety

The following are safety precautions and practices that must be followed when working with bitumen and a kettle. (See also Fig. 5-4.)

- Place the kettle so that the lid opens away from the building and so that fumes are not blown into air ducts.

- Light burners by placing them on a noncombustible surface and pointing them away from equipment and material.

- Start the kettle at low heat and keep the flame straight down the tubes.

- Keep the kettle tubes covered with bitumen.

- Keep an ABC-rated fire extinguisher nearby.

- Keep the kettle clean by dipping it. This helps prevent kettle fires.

- Inspect hoses, fittings, and regulators daily before using the kettle to ensure proper functioning.

- Shut off burners before refueling the kettle engine.

- Secure propane cylinders in an upright position at a minimum of 15 feet from the kettle.

- Wear proper protective equipment while working with hot bitumen, including gloves, a long-sleeve shirt, long pants, and over-the-ankle boots.

- Avoid direct contact with asphalt and pitch fumes.

- Keep hot bitumen carrying or storage equipment free of any water or ice.

- Rope off the kettle area when working in populated areas.

- Avoid walking in hot stuff. It is slippery and can cause a fall and a burn.

At the end of the day, follow these cleaning steps.

- Turn over mop tubs and mini moppers and close lugger lids to prevent the accumulation of ice and water.

- Remove used mops from the roof, spin them out, and then place them on a noncombustible surface.

✓	Item
	Place the kettle so that the lid opens away from the building and so that fumes are not blown into air ducts.
	Light burners by placing them on a noncombustible surface and pointing them away from equipment and material.
	Start the kettle at low heat and keep the flame straight down the tubes.
	Keep the kettle tubes covered with bitumen.
	Keep an ABC-rated fire extinguisher nearby.
	Keep the kettle clean by dipping it. This helps prevent kettle fires.
	Inspect hoses, fittings, and regulators daily before using the kettle to ensure proper functioning.
	Shut off burners before refueling the kettle engine.
	Secure propane cylinders in an upright position at a minimum of 15 feet from the kettle.
	Wear proper protective equipment while working with hot bitumen, including gloves, a long-sleeve shirt, long pants, and over-the-ankle boots.
	Avoid direct contact with asphalt and pitch fumes.
	Keep hot bitumen carrying or storage equipment free of any water or ice.
	Rope off the kettle area when working in populated areas.
	Avoid walking in hot stuff. It is slippery and can cause a fall and a burn.
	At the End of the Day
	Turn over mop tubs and mini moppers and close lugger lids to prevent the accumulation of ice and water.
	Remove used mops from the roof, spin them out, and then place them on a noncombustible surface.
	Lock up kettle burners or remove them from the jobsite.
	Lock the kettle lid and outlet valve in populated areas.

FIGURE 5-4   Kettle safety.

- Lock up kettle burners or remove them from the jobsite.
- Lock the kettle lid and outlet valve in populated areas.

Operating in Cold Weather

During cold-weather operations, insulate the hot-stuff line that extends from the kettle to the rooftop with either a pipe covering or galvanized stove pipe. Do not overheat the asphalt. Make sure that the rooftop transport equipment, including the hot luggers and mop buckets, has adequate insulation.

Make sure that the hot stuff is hot enough at the point of application to permit good adhesion. This is especially important when it has been transported over a long distance.

During cold weather, it is absolutely essential that the entire roofing operation be completed as you progress on a day-to-day basis. Do not succumb to the temptation to phase your application unless it is a temporary roof.

Make sure that you are getting ply adhesion. Holidays or fishmouths do not have an opportunity to reseal themselves during cold weather. Broom in all felts.

It is absolutely mandatary that materials be delivered to the jobsite and installed in a dry condition. Cold weather places a special premium on this requirement. Do not place materials in small piles on the deck. The removal of snow, should it become necessary, from numerous small piles, is difficult. Piles of material make good snow fences. Store material in a heated warehouse or closed trailer just prior to installation.

A yard storage tank can be a real advantage in cold weather because a decision as to whether or not to bring hot stuff out on the job can be made at the last minute, depending on climatic conditions.

Keep in mind that base sheets and other roofing materials are more brittle and less flexible in cold weather. Give additional care and attention to their application. When there are different grades of material specified for summer and winter installations, be sure to use the correct grade such as winter-grade, plastic cement.

The completed roof is more brittle in cold weather, especially the coal-tar pitch. Try to set the job up in such a way that you are not continually transporting men and materials over the completed roof.

Have the kettle worker draw the kettle down at the end of the day. Hot stuff heats faster the next day if a full cold kettle does not have to be completely reheated.

Since it is desirable to complete the roof as one goes along, it is necessary to see that the general contractor coordinates the work of the other trades and installs the nailers, curbs, and other items needed to complete the work. If possible, do not let the stall deck get too far ahead. While snow is not too hard to remove, ice is difficult.

When job conditions permit, consider using larger crews so that you can take full advantage of good weather.

When concrete decks are poured in cold weather, it is difficult to wait until they cure. It helps to prime the deck with asphalt primer immediately after it is poured and give it a glaze coat of steep roofing asphalt. This makes ice much easier to remove because it does not stick to the asphalt. Take care, however, that the combined weight of this glaze coat and the asphalt used to apply the first layer of the subsequent roofing system does not exceed the recommended maximum for proper adhesion. Too much asphalt can cause later slippage in hot weather.

Remove any moisture from the deck surface. Moisture can cause poor adhesion or skips in the mopping asphalt that can entrap moisture within the roofing system. Squeegee all fiberglass ply felts to assure adhesion.

Use Table 5-2 as a guide to determine the true temperature of materials and personnel. Consider the use of temporary roofs if construction schedules require roof applications in cold or rainy weather.

Installing BUR Roofs

It is impossible in a book this size to give all the possible installation details for BTU applications. Fortunately, BUR manufacturers offer such information, and their technical representatives are available to help roofing contractors with specific problems. Let's look at a few of the more common application details.

Covering Roofs Temporarily

At times an owner or general contractor may require that the building be closed-in at a time when the weather is not conducive to good roof

TABLE 5.2 Wind Chill/Dry

	Bulb ambient temperature (°F)								
	50	41	32	23	14	5	−4	−13	−22
Wind velocity (mph)	Equivalent temperature (°F) in cooling power on exposed flesh								
Calm	50	41	32	23	14	5	−4	−13	−22
5	48	39	28	19	10	1	−9	−18	−27
10	41	30	18	7	−4	−15	−26	−36	−49
15	36	23	12	0	−13	−26	−36	−49	−62
20	32	19	7	−6	−18	−31	−44	−58	−71
25	30	16	3	−9	−24	−36	−51	−63	−78
30	28	14	1	−13	−27	−40	−54	−69	−81
35	27	12	0	−15	−29	−44	−58	−71	−85
40	27	12	−2	−17	−31	−45	−60	−74	−89
45	25	10	−4	−18	−33	−47	−60	−76	−89
50	25	10	−4	−18	−33	−47	−62	−76	−90

Little danger	Increasing danger	Great danger

Danger from freezing of exposed flesh (for properly clothed persons)

construction. Or, the roof area might have to be used as a work platform during construction. In these situations, it is recommended that you keep in mind the following requirements. (See also Fig. 5-5.)

- The components of this temporary roof system are all asphalt-related materials.

- The roof deck must be secured to the structure in such a way that it is able to support the designed live- and dead-load factors required for the building.

✓	Item
	The components of this temporary roof system are all asphalt-related materials.
	Secure the roof deck to the structure in such a way that it is able to support the designed live- and dead-load factors required for the building.
	A temporary roof is a roof membrane constructed of a minimum amount of material and placed over a building to permit limited construction to progress during inclement weather conditions. As such, it is not expected to exhibit the same watertight integrity as a new roof.
	Put other tradespeople on notice that care must be exercised when working on or over a temporary roof membrane or when placing penetrations through a temporary roof membrane.
	Raise heating, ventilation, and air conditioning (HVAC) units above the plane of the temporary roof to allow for flashing beneath the HVAC units.

FIGURE 5-5 Covering roofs temporarily.

■ A temporary roof is a roof membrane constructed of a minimum amount of material and placed over a building to permit limited construction to progress during inclement weather conditions. Thus, it is not expected to exhibit the same watertight integrity as a new roof.

■ Other tradespeople must be put on notice that care must be exercised when working on or over a temporary roof membrane or when placing penetrations through a temporary roof membrane.

■ Heating, ventilation, and air conditioning (HVAC) units must be raised above the plane of the temporary roof to allow for flashing beneath the HVAC units.

When putting a temporary roof over a nailable deck, install two plies of No. 15 asphalt-saturated, perforated organic felt. Start at the low point of the roof deck, such as a valley or drip edge. Overlap these plies 19 inches, leaving an exposure of 17 inches, and nail.

Embed the plies into a fluid, continuous application of asphalt. Apply the asphalt so that felt does not touch felt and no asphalt is applied to the substrate. Nail the back half of each sheet and cover it with a fluid, continuous application of asphalt. Embed the next ply of felt into the covered sheet. All plies of felt are to be *broomed* into place as they are applied to aid adhesion. Heat the asphalt in accordance with EVT standards and apply within the temperature range (EVT ± 25°F).

When applying a temporary roof over a steel deck, use one thin layer of perlitic-board roof insulation. The board should have a minimum thickness sufficient to span the flutes of the steel deck. Mechanically fasten the board to the deck with approved mechanical fasteners. Install two plies of No. 15 asphalt-saturated, perforated organic felt over the primed concrete deck. Follow these same guidelines for installing a temporary roof over a nailable deck. Prime the concrete with a coating of asphalt primer applied at the rate of ¾ gallon of primer per 100 square feet.

Attaching Insulation to Roof Decks

There are two basic ways to attach insulation to roof decks. You can use either the preferred two-layer system or the one-layer system.

TWO-LAYER SYSTEM

To install a two-layer system, first apply a base ply to the nailable roof deck. This serves as an attachment layer, upon which the insulation can be installed. Start the base ply at the low point of the roof deck, such as a valley or drip edge. Sidelap each course of felt 3 inches over the preceding course. Nail the laps on 9-inch centers. Extend two rows of nails, nailed on 18-inch centers, down the field of the felt.

Lay the first layer of insulation in hot asphalt or adhesive over the base ply. Generally, the long dimension of the insulation boards should run perpendicular to the roof slope. The long joints of the insulation boards should be laid in a continuous straight line, with end joints staggered. If hot asphalt is used, heat the asphalt in accordance with EVT standards and apply within the temperature range.

Lay the second layer of insulation in a mopping of hot asphalt or adhesive with the joints staggered from (not aligned with) the joints of the first layer. Heat the asphalt in accordance with EVT standards and apply within the temperature range.

ONE-LAYER SYSTEM

Apply a base ply to the nailable roof deck following the instructions for the two-layer system. Lay the first layer of insulation in hot asphalt or adhesive over the base ply. Generally, the long dimension of the insulation boards should run perpendicular to the roof slope. The long joints of the insulation boards should be laid in a continuous straight line, with end joints staggered. If hot asphalt is used, heat the asphalt in accordance with EVT standards and apply within the temperature range.

Figure 5-6 shows the attachment of insulation to a steel deck, while Fig. 5-7 illustrates two- and one-layer installation over a nonnailable concrete roof deck.

Figure 5-8 shows loose-laid insulation, which can be installed on any surface. Generally, the long dimension of the insulation boards runs perpendicular to the roof slope. The long joints of the insulation boards are laid in a continuous straight line, with end joints staggered. The National Roofing Contractors Association (NRCA) recommends that consideration be given to mechanically fastening or spot adhering the first layer of insulation to the roof deck under loose-laid systems.

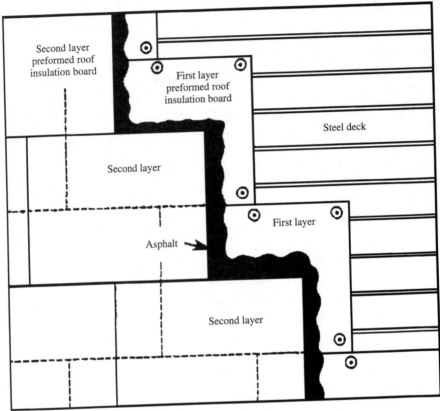

FIGURE 5-6 Attachment of insulation to steel deck.

Use only enough mechanical fasteners or adhesive to hold the insulation boards in place during the installation of the roofing membrane. Attach the second layer of insulation, if used, to the first layer with a compatible adhesive or hot asphalt. As an alternative procedure for nailable roof decks, both layers can be mechanically fastened to the deck at the same time.

Nailing Requirements

Roof insulation can be applied on inclines up to 6 inches per 12 inches, but it must be acknowledged that nailer usage violates the vapor retarder as a result of fastener penetrations, as does the perimeter nailing of insulation for uplift resistance.

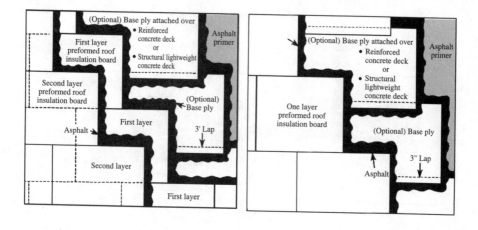

A   B

FIGURE 5-7 Two- and one- layer installation over nonnailable concrete roof deck.

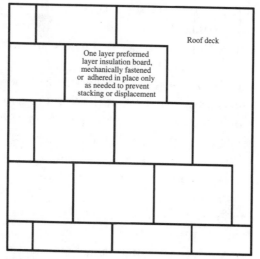

FIGURE 5-8 Loosely laid insulation on any roof surface.

When the deck incline is such that roofing felts must be nailed (2 inches and over for smooth-surfaced roofs and 1 inch and over for gravel- and mineral-cap sheet-surfaced roofs), use wooden nailing strips (4-inch minimum nominal, 3½-inch minimum actual width) at the ridge and at the approximate intermediate points detailed in Table 5-3.

TABLE 5-3 Nailer Spacing

Incline (in.)	Smooth	Gravel	Cap sheet
0–1	Not required	Not required	Not required
1–2	Not required	20' face to face	20' face to face
2–3	20' face to face	10' face to face	10' face to face
3–4	10' face to face	Not recommended	4' face to face
4–6	4' face to face	Not recommended	4' face to face

Attach nailing strips to the deck that are the same thickness as the insulation and at least 3½ inches wide. Run them at right angles to the incline to receive the insulation and retain the nails securing the felts. Run the felts parallel to the incline, at right angles to the nailers. Nails must have at least a 1-inch-diameter cap. When capped nails are not used, place fasteners through caps that have a minimum diameter of 1 inch.

For three-, four-, and five-ply roofs, locate a nail at each nailer and space ¾ inch from the leading edge of the felt. For cap-sheet application, locate one nail at each nailer and space approximately ¾ inch from the leading edge of each ply felt. Figure 5-9 shows the pattern for the nailers and spacing for the four-ply system.

The termination of a continuous cap sheet must occur at a nailer. At points of termination, locate five nails at each nailer (Fig. 5-10). Space the first nail ¾ inch from the leading edge of the cap sheet and the remaining four nails approximately 8½ inches on center. Stagger the nails across the width of the nailer to reduce the chance of the cap sheet tearing along the nail line.

It is recommended that each succeeding row of fasteners be staggered from the preceding row. For example, nail one row of fasteners on the lower half of the nailer board, then nail the succeeding felt on the upper half to avoid tearing the felts. Table 5-4 shows nailer spacing for asphalt/cap sheet, asphalt/smooth-surfaced, and asphalt/gravel-surfaced roofs.

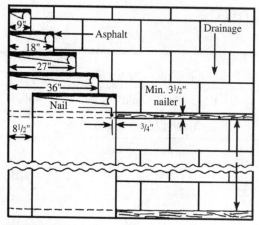

FIGURE 5-9 Pattern for nailer and spacing with four-ply system.

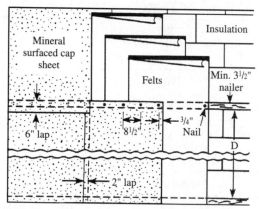

FIGURE 5-10 At points of termination, locate five nails at each nailer.

Installing MBR Membranes

Various numbers of plies can be used in the construction of a BUR system. Figure 5-11 illustrates the use of three-, four-, or five-ply membranes over a structural deck (without insulation) that can receive and adequately retain nails or other types of mechanical fasteners, as recommended by the deck manufacturer. Examples of such decks are wooden plank and plywood. This procedure is not for use directly over gypsum, lightweight insulating concrete decks either poured or precast, or over fill made of lightweight insulating concrete.

TABLE 5-4 Nailer Spacing for Various Roof Types

Incline (in.)	Nailer spacing (D)	Asphalt type
Asphalt/cap sheet roofs		
0–½	Not required	II
½–1	Not required	III
1–2	20' face to face	III
2–3	10' face to face	III
3–6	4' face to face	IV
Asphalt/smooth-surfaced roofs		
0–½	Not required	II
½–1	Not required	II
1–2	Not required	III
2–3	20' face to face	III
3–4	10' face to face	IV
4–6	4' face to face	IV
Asphalt/gravel-surfaced roof		
0–½	Not required	II
½–1	Not required	III
1–2	20' face to face	III
2–3	10' face to face	III

Figure 5-12 shows a four-ply gravel-surfaced fiberglass mat BUR application that can be used over any type of structural deck that is not nailable and that offers a suitable surface to receive the roof. Prime poured and precast concrete decks.

Flashing

Flashings prevent water intrusion at any roof area where the membrane is interrupted, is terminated, or joins an area or projection having

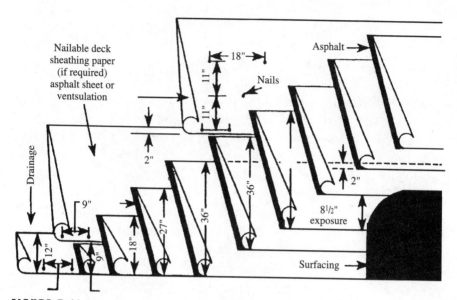

FIGURE 5-11 Use of three-, four-, or five-ply membranes.

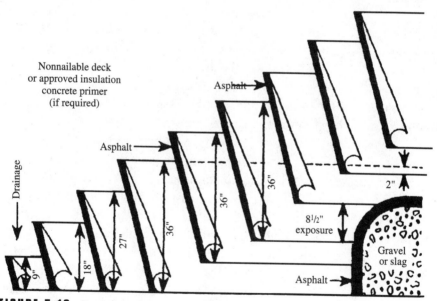

FIGURE 5-12 Four-ply gravel-surfaced fiberglass mat.

a marked change in slope or direction. This condition can occur at gravel stops, curbs, vents, parapets, walls, expansion joints, skylights, drains, and built-in gutters.

Flashing is generally divided into two categories, base flashings and counterflashings. Base flashings are, in a sense, a continuation of the membrane that is turned up on a surface and installed as a separate operation. Base flashings are usually of a nonmetallic material, such as a plastic, or an asphalt-impregnated product. Counterflashings or cap flashings can be made of metal, plastic, or impregnated felt. They shield or seal the exposed edges of the base flashing.

Combining nonmetallic materials and metal brings out the best in each material. Since the bituminous-based flashings have the same movement capabilities as the roof membrane, they work together as a unit and are recommended as the only type of material for base flashings. Because of the rigidity of metal and its extreme movement with temperature changes, its use for base flashing is not recommended. Metal cap or counterflashings, installed where there is no possibility of standing water, are acceptable only after composition base flashing is properly sealed.

Flashing can be applied by hot or cold application methods. Asphalt-coated fiberglass can be applied in a hot mopping of bitumen. The advantages of hot-applied flashing are speed, economy, and convenience. The proper grade of asphalt must be used, however. Do not use coal-tar pitch or low-softening-point asphalt because flashings embedded in these bitumens sag. Secure hot-applied flashings at the top edge with mechanical fasteners spaced 4 to 8 inches on center or by some other positive means to prevent sliding or sagging.

Use reinforced base flashing in conjunction with a ply of fiberglass-ply felt as a backer felt and set in a mopping of hot asphalt to the primed masonry surface. The desired guarantee period to be offered for the roof's life determines the need for a backer felt.

The cold-application system consists of alternate layers of ply felt set in and covered with trowelings of bestile flashing cement. The thinner ply felt conforms readily to the angle between deck and parapet. Bestile flashing cement provides adhesion to the primed wall and to each ply, enhances interply waterproofing, and provides more time for careful application.

Flashing systems of this type set up quite hard, so they do not have the tendency to sag or run under hot temperatures. They are especially practical when no nailing facilities are available. A major benefit to the bestile flashing system is its compatibility with any bituminous roofing system.

There are two main types of bituminous flashing cements for cold application. Industrial roof cement is a nonasbestos, fiber-reinforced, solvent-cutback, all-purpose plastic cement. This cement remains in a somewhat plastic state and is used where some movement is expected, such as under a metal roof edge, where the differential movement between metal and felt is absorbed in the plastic cement. It also is used to repair blisters and to accomplish general roof maintenance by itself and in conjunction with fiberglass-ply felt on horizontal surfaces.

Bestile flashing cement is basically the same as industrial roofing cement, but it contains certain nonasbestos reinforcing fibers. This material sets up quite hard instead of remaining plastic. Bestile is used almost exclusively for flashing work on vertical surfaces where differential movement is not expected. It should not be used with metal flashing.

The solid fastening of all flashing accessories to wooden nailers, or the substrate, is mandatory. Unequal movement due to poorly secured flashing results in splits in the felt stripping and subsequent leaks.

Finishing Flashing Details

There are many flashing designs that the specifier or architect might desire. Figures 5-13 through 5-15 show the more common designs. When bidding on a project, pay close attention to all complex roof joints and flashing details. These areas are very common triggers of roof leaks and resultant damage.

If the details are not drawn in a precise manner, address the issue with the contractor, design anchor, and principal. Have acceptable drawings completed or include a clause in the addendum to the contract that requires a signed inspection slip from the architect during execution. Remember that this pause for the architect will add to the job costs. In Fig. 5-13, for example, the roofing felts must extend to the top of the cant. Starting just below the point where the base flashing terminates, mop the wall and the roofing felts on the surface of

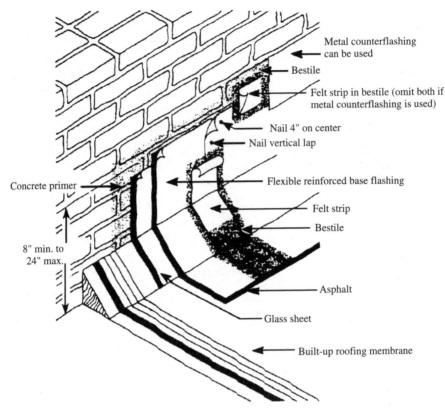

Metal counterflashing can be used

Bestile

Felt strip in bestile (omit both if metal counterflashing is used)

Nail 4" on center

Nail vertical lap

Concrete primer

Flexible reinforced base flashing

Felt strip

Bestile

8" min. to 24" max.

Asphalt

Glass sheet

Built-up roofing membrane

FIGURE 5-13 Flashing design.

the cant with hot Type III or IV asphalt. Immediately set the backer felt, if required, into the hot asphalt. Smooth the felt to set it firmly into the asphalt. Terminate the bottom edge of the backer felt at the bottom edge (base) of the cant.

Starting just above the top edge of the backer felt, mop the wall and the backer felt with hot Type III or IV asphalt. Back mop a 6- to 8-foot fiberglass section of flexible reinforced base flashing. Holding the upper corners of the base sheet, position its horizontal edge on the roof membrane and roll it into place over the cant strip and up the wall.

Mechanically fasten the base flashing on 4- to 8-inch centers along the top edge and at the laps. Drive fasteners through caps that are at least 1 inch in diameter, unless they have an integral flat cap at least 1 inch across. Cover the vertical laps with a 4-inch-wide strip of ply felt

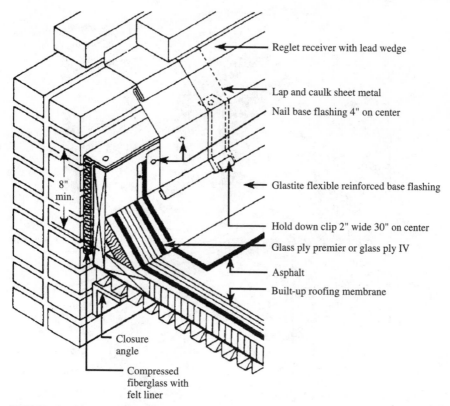

Reglet receiver with lead wedge

Lap and caulk sheet metal

Nail base flashing 4" on center

8" min.

Glastite flexible reinforced base flashing

Hold down clip 2" wide 30" on center

Glass ply premier or glass ply IV

Asphalt

Built-up roofing membrane

Closure angle

Compressed fiberglass with felt liner

FIGURE 5-14 Flashing design.

embedded in and troweled over with a ⅛-inch-thick layer of bestile or roofing cement.

Applying BUR Penetration Flashing

The penetration expansion joint covers, stacks, supports, and vents can be handled on BUR applications as follows.

Expansion joint cover. This type of expansion joint cover, when curb-mounted, allows for building movement in all directions (Fig. 5-16A).

Stack flashing. Provide curbs for any accessory passing through the roof deck such as a vent pipe, stack (Fig. 5-16B), or conduit. Also provide curbs for roof fixtures such as skylights or ventilating fans.

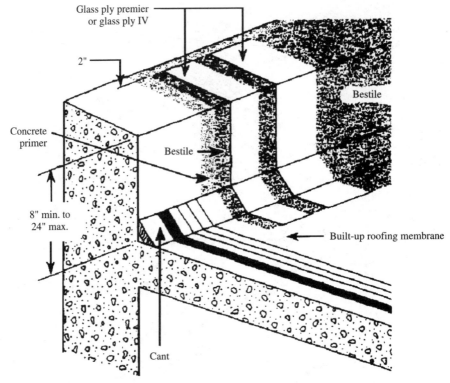

Glass ply premier or glass ply IV

2"

Bestile

Concrete primer

Bestile

8" min. to 24" max.

Built-up roofing membrane

Cant

FIGURE 5-15 Flashing design.

Equipment or sign support. Raising supports above the roof level simplifies maintenance (Fig. 5-16C). For heavy loads, locate continuous supports over structural members such as girders or columns.

One-way roof vents. Use roof vents (Fig. 5-16D) to reduce the moisture vapor content of insulation or certain types of wet-fill poured decks. Vents also can be used in conjunction with a ventsulation felt when parapet or roof-edge venting is not practical. The placement of vents depends on job conditions, but there should be a minimum of one vent for each 10 squares of roof area.

Maintaining BUR Membranes

Both BUR and MBR membranes are relatively fragile components. The membrane is subjected to wind, hail, snow, rain, and ultraviolet

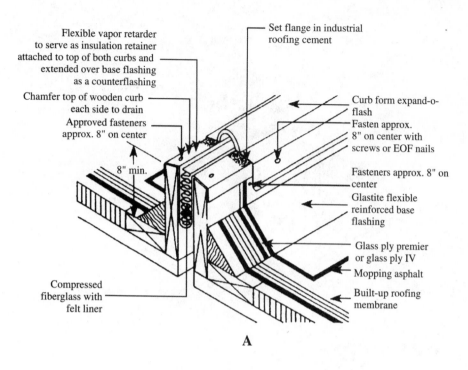

Flexible vapor retarder to serve as insulation retainer attached to top of both curbs and extended over base flashing as a counterflashing

Chamfer top of wooden curb each side to drain

Approved fasteners approx. 8" on center

8" min.

Compressed fiberglass with felt liner

Set flange in industrial roofing cement

Curb form expand-o-flash

Fasten approx. 8" on center with screws or EOF nails

Fasteners approx. 8" on center

Glastite flexible reinforced base flashing

Glass ply premier or glass ply IV

Mopping asphalt

Built-up roofing membrane

A

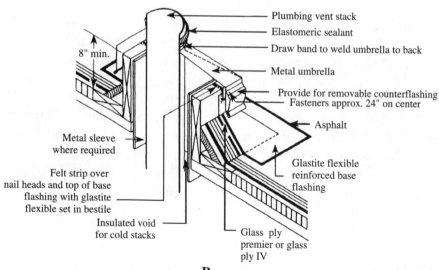

8" min.

Metal sleeve where required

Felt strip over nail heads and top of base flashing with glastite flexible set in bestile

Insulated void for cold stacks

Plumbing vent stack

Elastomeric sealant

Draw band to weld umbrella to back

Metal umbrella

Provide for removable counterflashing Fasteners approx. 24" on center

Asphalt

Glastite flexible reinforced base flashing

Glass ply premier or glass ply IV

B

FIGURE 5-16 (*A*) Expansion joint cover. (*B*) Stack flashing. (*C*) Equipment or sign support. (*D*) One-way roof vents.

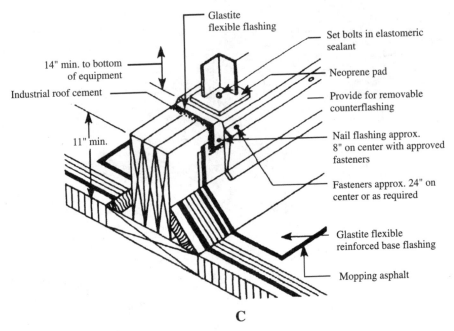

Glastite flexible flashing

Set bolts in elastomeric sealant

14" min. to bottom of equipment

Neoprene pad

Industrial roof cement

Provide for removable counterflashing

11" min.

Nail flashing approx. 8" on center with approved fasteners

Fasteners approx. 24" on center or as required

Glastite flexible reinforced base flashing

Mopping asphalt

C

FIGURE 5-16 *(Continued)*

energy, as well as foot traffic and various forms of abuse. While other parts of the building are more visible, and thus tend to be maintained on a regular basis, the roofing system often is overlooked until it leaks. By that time, extensive and costly damage may already have occurred.

Inspections, maintenance, and repair are an excellent opportunity for securing an ancillary cash flow. This should be treated as a separate business. If it is well managed, it can be a simple cash cow that also markets your company's other services.

The most important reason to establish a program of periodic maintenance inspection is to protect the owner's investment. A properly executed maintenance program can add years to the life of a roof. It can also detect minor problems before damage is widespread, which in turn avoids interruption of the internal functions of the building. Because of this, more and more building owners and management firms are hiring qualified roofers to inspect their roofs twice a year. During such an inspection, carefully check every component

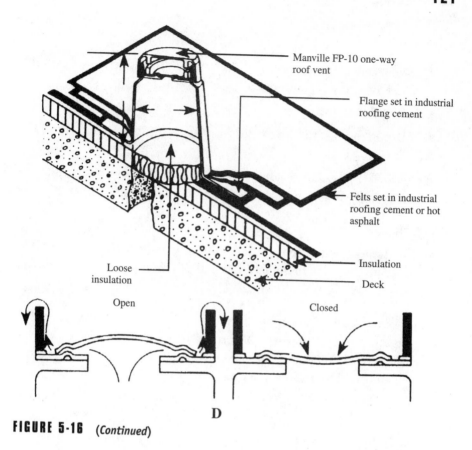

Manville FP-10 one-way roof vent

Flange set in industrial roofing cement

Felts set in industrial roofing cement or hot asphalt

Insulation

Deck

Loose insulation

Open

Closed

D

FIGURE 5-16 (*Continued*)

of the roofing system for signs of deterioration. The following are some of the most common problems encountered on roofs.

Blisters

Blisters are usually associated with skips or voids between two impermeable surfaces. The entrapped air contains moisture, which expands as the temperature rises and creates sufficient pressure to push the felts apart. Small, unbroken blisters that are less than 12 inches in diameter are best left alone. Blisters holding pressure do not leak water. If the felts are organic, recoat the bare spots over unbroken blisters with industrial roofing cement to protect against deterioration. Repair large or numerous blisters with hot asphalt.

Repair broken blisters using the following steps.

■ Remove the entire blister to the point at the edge where remaining felts are well adhered.

■ Let the remaining surfaces dry thoroughly.

■ Fill the depression level with the surface of the BUR using roofing cement.

■ Trowel the cement smoothly over the depression patch and feather the edges so that water flows easily over the patch.

Ridging

This problem is sometimes called buckling or wrinkling and is the result of dimensional changes in the BUR membrane. Some of the common causes of ridges follow.

■ Interior moisture vapor migrates to the cool underside of the membrane, condenses, and then is absorbed by the organic fibers of the felt, causing a swell or ridge over each joint of insulation in a picture-frame pattern.

■ An unattached or poorly attached membrane or insulation shifts during seasonal changes due to expansion and contraction.

■ Insecurely attached deck units shift out of position.

■ Slippage of the membrane.

Ridges flex during seasonal changes or with structural movement. This can cause them to fatigue and crack on or near the crown. Recoat unbroken ridges in organic felt membranes wherever the coating has slipped and left bare felt.

Splitting

This can be due to damage during or after construction or to conditions that cause a concentrated stress in the BUR membrane. Possible causes include stress from differential movement or cracking of the supporting structural deck or substrate, which can occur where the deck changes direction, where two types of deck are installed adjacent to each other, or where a BUR is solidly cemented to a deck that develops shrinkage cracks.

Stress from the contraction of the membrane in an area where the membrane or a component of the roofing system is poorly attached can also cause splitting. This stress can occur between the deck and insulation, the layers of insulation, or the BUR membrane and insulation.

Another common cause of splitting is breakage of the attachment between components by rolling loads across the roof, heavy loads that cause deflection, or the drumming of the roofing system by high winds. Membrane splits will recur as long as the basic substrate stress is not accommodated. For a long-term solution, proceed as follows.

1. Clean the surface of the membrane a minimum of 16 inches beyond the split edges. Completely remove the aggregate surface if present.

2. Prime the cleaned surface if it is dirty or contaminated.

3. Position dry, unattached 4-inch-wide felt centered over the split.

4. Cover and seal the split. Do not re-embed the aggregate so that the repair can be monitored.

To provide some flexibility to a split repair with considerable movement, follow the first three steps above, and then:

1. Over a 4-inch strip, cement a 16-inch-wide strip of elastomeric film embedded in a ⅛-inch troweling of cement.

2. Strip the edges of the elastomeric film with a 6- and 10-inch strip of inorganic felt, each embedded and coated in a ⅛-inch troweling of roof cement or hot asphalt, as illustrated in Fig. 5-17.

Loose or Unbonded Edges

When a roof membrane is mopped together, the felts are arranged so that the uppermost ply overlaps the lowest ply by 2 inches. If the looseness is less than 2 inches, a smooth-surfaced roof can easily be repaired by slitting away the loose material. For slight looseness that extends more than 2 inches, insert roof cement with the point of a trowel, and step in the felts.

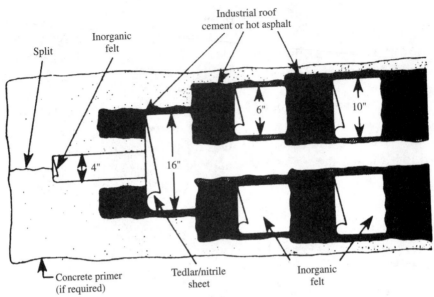

FIGURE 5-17 Stripping edges of elastomeric film.

If edges are more widely curled or torn, more extensive patching may be necessary in order to restore the roof to the original number of plies. Clean and prime the surfaces before attempting either hot or cold patches.

Fishmouths

Fishmouths are raised tunnel-type openings at the ply-felt laps. These can allow water to enter deep within the plies of the membrane and should be sealed without delay. Clean or spud the surface along the ridge of the fishmouth as necessary. Split open the ridge crown, trim away all excess material, and allow the surfaces to dry. Then coat the cavity surfaces with roof cement.

Punctures

Membrane punctures can cause potentially serious deterioration and premature failure. They can be caused by dropped tools, hail, thrown or wind-blown objects, or popped fasteners. If a small amount of moisture has entered the system through the puncture, it is preferable to vent it before attempting to make a repair.

Parapet Wall Flashings and Copings

Excessive movement, water absorption, or age-hardening can cause the wall-joint grout and caulking to fail. A nonhardening caulking compound with an acrylic or other nonhardening synthetic base designed especially for roofs is most satisfactory for metal joints. Follow the priming instructions carefully, as failure through adhesion loss is common.

For general masonry repair and to reset loose masonry, use a hard-setting cement, such as flashing cement. Remove loose dirt carefully using a wire brush, and then prime with a thin application of asphalt primer. Apply the cement generously and reset the copings. Joints can be reinforced with courses of flashing cement and woven glass cloth or glass felt.

Efflorescence on the masonry walls often indicates that moisture is entering the wall elsewhere and carrying salts to the surface. Sealing with asphalt can cause spalling. Find the entry point, such as the outside face of the wall. Point this out to the principals and require that needed testing be done before fixing. Then dampproof with special coatings designed for this purpose.

Porous surfaces can be primed with asphalt primer and damp-proofed with an asphalt emulsion. Clay emulsions have a unique ability to release vapor while resisting liquid water. Reset loose fasteners, reglet plugs, etc., as necessary to hold flashings or counterflashings in place.

Reroofing BUR Systems

Each retrofit or recover project has its own specific problems that require one of three basic approaches: total replacement of the roofing system down to the deck, which is generally the best recommendation; replacement of the damaged sections; or a complete recover of the existing roofing system. The best method of repair depends on an evaluation of the evidence presented.

Since architectural drawings, specifications, and engineering calculations are often unavailable for reroofing projects, evaluate roof-system damage and required repairs with care. The first evaluation method is to make a nondestructive moisture survey (see Chap. 14) to determine the existence and location of moisture in the membrane or roof insulation.

Before you begin, it is helpful if you can obtain a set of original plans and any records of later changes. These modifications might have a bearing on the current roof problems and could indicate whether the structural system is capable of supporting the additional loads that might be required for reroofing purposes. Also review the building's interior to evaluate changes in structure or occupancy.

Inspect the entire roof system from above and below in order to develop a history of past and present problems. Also consider the owner's plans for the future of the building.

Inspecting the Underside

Building plans and specifications can help establish structural details during the inspection process. Use a camera to record existing problems. Some of the basic questions that need to be answered include: What is the age of the roof, and are leaks recent or part of a long history of problems? Is there evidence of building settlement? Are water stains in evidence? Are there any new through-deck penetrations that do not appear on the plans? Can the structural system support the additional weight of new roofing components or systems?

Inspecting the Topside

In order to perform satisfactorily, roof systems must have proper drainage, be secured, and not trap moisture. When you inspect the top of the roof, consider the condition of the following building elements and roof-system components.

Parapet or fire walls. Inspect for efflorescence, cracks, open joints in mortar, loose or broken copings, or any other points where water can enter.

Flashing. Wrinkling indicates differential movement between the deck and the wall. Inspect flashings for damage from foot traffic, tears, splits, or deterioration. Examine metal or membrane counterflashings to assure tight joints and positive attachment to the wall.

The top edge of the counterflashing must be well secured by face fastening or, if inserted in a reglet or mortar joint, completely caulked.

Roof edges. Wooden nailers should not show signs of deterioration and should provide satisfactory nail or fastener retention. Insufficiently fastened metal roof edges can move, which in turn can cause defects. Raise all roof edges above the roof level to impede the flow of water and eliminate ponding.

Drains. The critical considerations for drains are number, placement, and condition. Is there a backup drain in every drainage area? Are the drains located at the low points in the deck when the deck is under maximum load? If delayed drainage is required by local codes, can the deck tolerate the extra loading? Are drain flanges properly flashed? Is there a buildup of flashing thickness around drains that prevents water from flowing freely into the drains? If drains are not in sumps, is the insulation around the drains thin enough to allow heat from within the building to prevent freezing?

Expansion and relief joints. Consider these accessories at points of stress when roof rehabilitation is undertaken and when inspection indicates their necessity. If joints are in place on the existing roof, are they doing the job for which they were intended?

Taking Test Cuts

Once you establish the condition of the roof, flashing, accessories, and the deck by visual examination, make several test cuts. The cuts should be no less than 8 × 42 inches and should be made at right angles to the length of the roofing felts. Cut all the way down to the deck. These test cuts reveal the following information.

- The construction of the existing roof system
- Whether more than one roof has been applied
- The thickness, type, and condition of the existing insulation
- The condition of the bond between the insulation and the vapor retarder and/or deck

■ The condition of the bond between the membrane and the insulation or deck

■ The existence, type, and condition of the vapor retarder

■ The bond between the vapor retarder and the deck

■ The condition of the deck from the topside

■ The type of bitumen, asphalt or coal-tar pitch, and the amount of interply mopping used

Use the information obtained from the visual inspection and from an analysis of the test cuts to decide which course of action is best: total replacement, partial replacement, or a complete reroofing. If a decision cannot be made based on this information, review the following criteria.

TOTAL REPLACEMENT

If the deck structure is unsound or inadequately secured to the supporting system, remove and replace the entire roof and deck. This extreme step must be planned and executed carefully to protect the building's contents. Be certain to protect yourself— walk away from any "job" on which the owner requests a shoddy work product in order to save money. Lawsuits are rampant in this day and age.

If the structural deck is sound, but the insulation has deteriorated, replace both the membrane and the insulation. This operation also requires the replacement of all base flashings. Metal counterflashings could conceivably be reused if they are in good condition.

Consider adding insulation to improve the roof system's overall thermal efficiency. Although roof insulation does not add greatly to the weight of the system (about 1 pound per inch of thickness per square foot), this additional loading must be considered. The owner must be convinced that the cost of increasing the roof's thermal efficiency can be justified by reduced heating and cooling costs projected for the life expectancy of the new roof.

If the building use indicates the need, consider using a one- or two-ply fiberglass felt and asphalt vapor retarder. Install the insulation in two layers and stagger the joints between the layers. If the deck permits, install the first layer with mechanically secured fasteners and install the second layer in a mopping of hot bitumen.

PARTIAL REPLACEMENT

If the survey indicates that many areas of the roof are salvageable, make repairs to keep the system watertight. The cause of problems in the substrate is often moisture infiltration, either through a damaged membrane or from within. If the membrane is not damaged and the insulation is wet, the problem is most likely a damaged or missing vapor retarder.

If the deck is insulated, investigate moisture or structural problems and make corrections to eliminate any further deterioration. Repair any blisters or splits. If the splits are due to structural movement, consider constructing expansion joints.

Before you apply any new materials, prepare the existing BUR substrate. Remove dirt, debris, and wet, damaged, and deteriorated materials from the existing roof system. Repair major defects. Ensure that the existing roof system is thoroughly secured to the roof deck. Remove high spots and fill in low spots so that insulation boards lie flat. Remove loose aggregate from existing aggregate-surfaced BURs.

The application techniques for complete or partial BUR reroofing are basically the same as those described earlier for BUR installation.

Figure 5-18*A* and *B* illustrates techniques for applying new roofing material over an existing membrane and on a nonnailable deck. Figure 5-18*C* details an application over an existing membrane on a nailable deck. Figure 5-18*D* shows the installation over an existing membrane on a steel deck.

There are two ways to install insulation over an existing BUR application. In the adhered insulation procedure, prime the existing BUR membrane surface and install the first layer in hot asphalt or adhesive. Generally, the long dimensions of the insulation boards should run perpendicular to the roof slope, and the long joints of the insulation boards should be laid in a continuous straight line with end joints staggered (Fig. 5-19). Lay the second layer of insulation in hot asphalt or adhesive with the joints staggered from (not aligned with) the joints of the first layer.

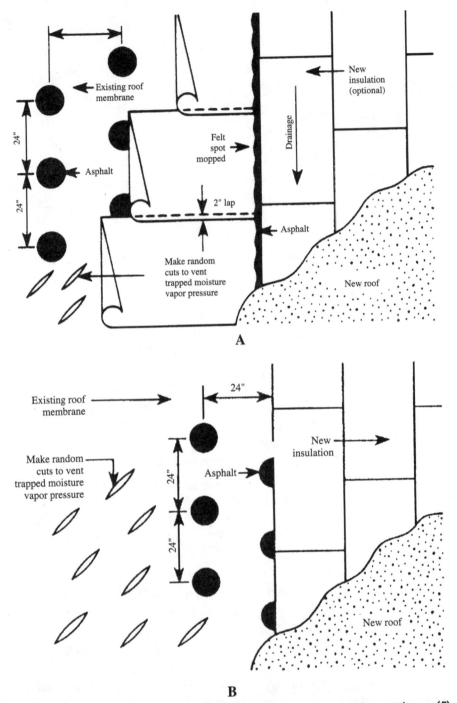

FIGURE 5-18 (*A*) Techniques for applying new roofing over an existing membrane. (*B*) Techniques for applying new roofing over a nonnailable deck. (*C*) Application over an existing membrane on a nailable deck. (*D*) Installation over an existing membrane on a steel deck.

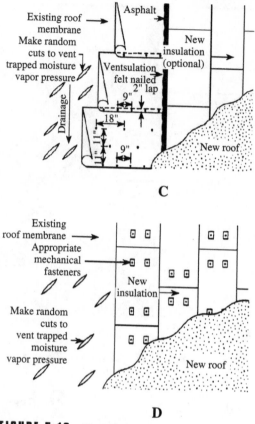

FIGURE 5-18 *(Continued)*

To mechanically fasten insulation (Fig. 5-20*A*), select the size, type, and number of fasteners based on the type of deck and the thickness of the original BUR system. Then mechanically fasten the first layer of insulation through the existing BUR membrane and into the nailable roof deck.

Generally, the long dimensions of the insulation boards should run perpendicular to the roof slope, and the long joints of the insulation boards should be laid in a continuous straight line with end joints staggered. Lay the second layer of insulation in hot asphalt or

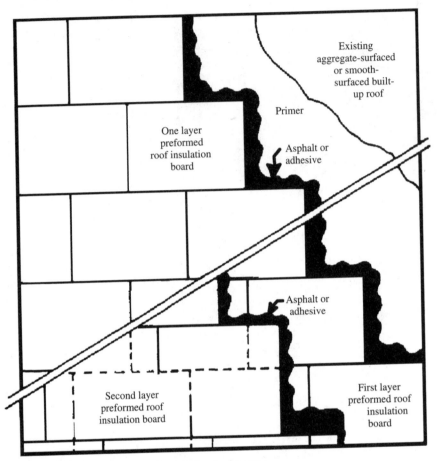

FIGURE 5-19 Insulation boards laid in straight line.

adhesive, with the joints staggered from (not aligned with) the joints of the first layer (Fig. 5-20*B*).

As an alternative procedure, both layers can be mechanically fastened to the roof deck. Select the size, type, and number of fasteners based on the type of deck, the thickness of the new insulation and original BUR system, and the desired wind-uplift resistance.

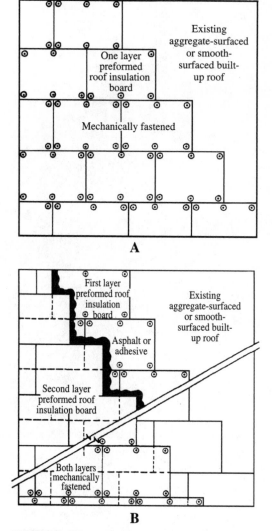

FIGURE 5-20 (*A*) Mechanical fasteners are installed.
(*B*) Second layer of insulation with joints staggered.

6

Single-Ply Roofing

T he single-ply membrane roofing system was developed in Europe in the late 1950s, partially in response to difficulties in obtaining asphaltic materials for conventional roofing systems. Single-ply membranes were first introduced in the United States in the mid-1960s by European manufacturers. Since their first appearance in our country, single-ply materials have become increasingly popular. Whether imported from Europe or produced domestically, these products have proven themselves in a wide variety of climates over more than four decades of use.

Built-up roofs (BUR) are literally constructed on the roof by the contractor, using component materials such as felts and asphalt (Fig. 6-1). Thus, they are subject to problems caused by weather, worker error, and material inconsistencies. Single-ply membranes, however, are flexible sheets of compounded synthetic materials that are manufactured in a factory to strict quality control requirements. This process minimizes the risks inherent in BUR systems.

Primary among the many physical and performance properties that these materials provide are strength, flexibility, and long-lasting durability. The prefabricated sheets are inherently flexible, can be used with a variety of attachment methods, and are compounded to provide watertight integrity for years of life. These various factors are

FIGURE 6-1 A single-ply installation.

the reason for the popularity of single-ply systems, which account for up to 40 percent of the commercial roofing market.

There are many different single-ply roofing products. Although their chemistry and composition are complex, the Single-Ply Roofing Institute (SPRI) classifies them into three main groups: thermosets, thermoplastics, and modified bitumens. Each of these types of single-ply membranes includes of a number of individual products.

Casting Thermosets

Single-ply thermoset materials are chemical crosslinkages of polymer that cannot be changed once the sheet material is cast. Two types of thermosets are used for roofing: vulcanized, or cured, elastomer and nonvulcanized, or noncured, elastomer.

The advantages associated with elastomeric roofing include performance, cost benefits, conservation, the substitution of materials, and adaptability to a wide range of roof configurations. The membrane is able to elongate and accommodate movement in the substrate. The extension or elongation of certain systems at room temperature can be as high as 700 to 800 percent. Elastomeric membranes can bridge non-working joints and cracks in the substrate without cracking and splitting, provided they are not bonded or are reinforced at these locations.

The ability of the membrane to remain flexible at low temperatures is another feature. Some elastomeric membranes remain flexible at temperatures as low as −50°F, whereas conventional bituminous membranes become brittle within a range of about 0° to 45°F. Some elastomeric membranes retain their ability to elongate at low temperatures, although the elongation is reduced from that at room temperature, or approximately 68°F.

Some elastomeric roofing systems weigh less than 10 pounds per 100 square feet of roof area. This is a minimal weight when compared to that of smooth-surfaced bituminous systems, which weigh approximately 150 pounds per square foot.

Roof designs for many modern buildings must be architecturally attractive and functional. These roofs include a variety of configurations, such as domes, barrels, and hyperbolic paraboloids. Conventional materials usually are not suitable for these shapes. Elastomeric membranes are able to conform to a variety of shapes and contours. This makes them suitable roofing materials for these modern architectural designs.

In addition, some elastomeric roof membranes are available in a variety of colors that can enhance the attractiveness of roofs. Colors can also be reflective, which reduces the absorption of solar radiation and results in lower roof temperatures.

Vulcanized Elastomers

Ethylene propylene diene terpolymer (EPDM) and chlorinated polyethylene (CPE) are the two most popular vulcanized elastomers.

ETHYLENE PROPYLENE DIENE TERPOLYMER (EPDM)

This thermoset membrane is compounded from rubber polymer and is often referred to as *rubber roofing*. The American Society for Testing and Materials (ASTM) classifies this material as an M class polymer. The M in EPDM is sometimes taken to mean *monomer*, which is technically incorrect.

Ethylene and propylene, derived from oil and natural gas, are the organic building blocks of EPDM. When these are combined with diene to form the basic rubber matrix, the result is a long-chain hydrocarbon with a backbone of saturated molecules and pendant double bonds. The practical translation is that ethylene, propylene, and diene combine to form a large molecule that is very stable when exposed to sunlight, heat, ozone, and moisture.

These molecules can be cured, or *vulcanized*, into a rubber sheet that permits elongation of more than 400 percent without structural damage. In other words, EPDM is the binder material, or the basic rubber matrix that gives the final membrane rubber properties. Carbon black, oils, processing aids, and curatives are added to increase tensile strength, flexibility, mixing, and dimensional stability.

Features that contribute to the popularity of EPDM single-ply roofing systems include:

- Long-term weatherability, including excellent resistance to temperature extremes, sunlight, ozone, and moisture

- Ease, speed, and cleanliness of installation

- Flexural stability; an elongation factor enables EPDM membranes to accommodate roof deck movement and displacement

- Compatibility with a wider variety of polystyrene insulation products than asphaltic materials

- Ease of maintenance

- Proven long-term performance

Another feature is adaptability. Various application techniques, such as ballasted, fully adhered, and mechanically fastened, allow EPDM roofing systems to be applied to virtually any roof surface: flat, spherical, curved, or slanted.

EPDM also offers reroofing flexibility. When the roof deck is moisture-free, EPDM can be installed over existing roof membranes in reroof situations. Finally, code and standardization progress by the Rubber Manufacturers Association (RMA), Underwriters' Laboratories, Inc. (UL), and Factory Mutual (FM) contribute to the demand.

Ballasted: Ballasted systems are the most common for rubber-based, single-ply roofs. The thermal insulation and rubber membrane are loosely laid over the roof deck and then covered with ballast, which is usually round, washed river rock as specified by ASTM. The major advantages of this method include low installation costs, ease of installation, a UL Class A fire rating, and separation of the membrane from the deck, which allows for maximum independent movement.

Fully Adhered: This is the second most popular application method (Fig. 6-2). Ideal for contoured roofs, sloped surfaces that cannot withstand the weight of a ballasted system, or reroofing applications over existing material, fully adhered membranes are completely bonded to the substrate using contact adhesives. Major advantages include their light weight, durability, ease of maintenance, and aesthetically clean, smooth appearance.

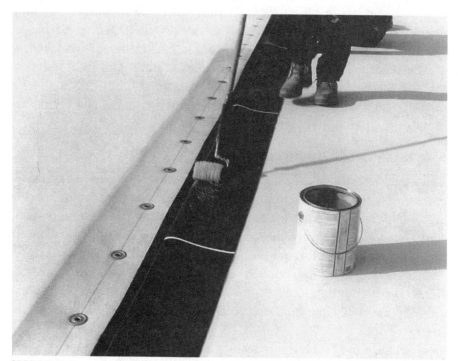

FIGURE 6-2 Apply adhesive on a fully adhered EPDM roof.

Mechanically Fastened: With this method, the membrane is loosely laid over the substrate and then anchored to the deck using fasteners. Many types of mechanically fastened systems are available. Most common are those attached directly to the deck. Nonpenetrating systems are also available. Mechanically fastened systems are lightweight, easy to install, and relatively inexpensive. As discussed later in this chapter, a recent development in EPDM is a sheet that can be *heat welded.*

CHLORINATED POLYETHYLENE (CPE)

CPE exhibits properties of both vulcanized and nonvulcanized elastomers. CPE is manufactured as a thermoplastic, but over time it cures as a thermoset. This means that most CPE systems are typically hot-air- or solvent-welded. The features and benefits of CPE are discussed in greater detail in the nonvulcanized, uncured elastomers section.

Polychloroprene or Neoprene

Neoprene is a generic name for polymers of chloroprene. It was the first commercially produced synthetic rubber and exhibits resistance to petroleum oils, solvents, heat, and weathering. It is available in sheet and liquid-applied forms, and in weathering and nonweathering grades. Weathering grade is black; nonweathering is light-colored. Nonweathering grade must be protected from sunlight, normally by applying a coating of usually chlorinated polyethylene. Neoprene formulations are no longer widely used for roofing.

Nonvulcanized, Uncured Elastomers

The manufacturing process differentiates these uncured elastomeric membranes from vulcanized polymers. During the production of vulcanized polymers, combinations of chemicals, primarily polymers, fillers, and additives, are processed together and cured by heating so that chains of molecules are permanently crosslinked.

This curing method results in a *thermoset* membrane with low tensile strength and high elongation values. In the field, cured elastomers must be applied and repaired with adhesives. New chemical bonds cannot be formed.

Nonvulcanized, uncured elastomers, in contrast, are manufactured without any crosslinks between chains of polymer molecules. Although exposure to the elements can naturally cure some of these polymers during their lifespan, all nonvulcanized elastomers can be heat-welded during the initial installation.

The generic classification of polymers known as nonvulcanized elastomers includes chlorosulfoned polyethylene (CSPE), chlorinated polyethylene (CPE), polyisobutylene (PIB), and nitrile alloy with butadiene-acrylonitrile copolymers (NBP).

CHLOROSULFONED POLYETHYLENE (CSPE)

CSPE, sold under the DuPont trademark Hypalon, is a polymer that has enjoyed increased popularity over the years because of its attractive white appearance and energy-efficient, heat-reflective properties. Other factors contributing to the popularity of CSPE and other rubber-based, single-ply roofing systems include

- Long-term weatherability; excellent resistance to temperature extremes, sunlight, ozone, and moisture
- Ease, speed, and cleanliness of installation
- Reduced labor costs in many cases
- Flexural stability; accommodation of deck movement and displacement
- Routine reroof installation over existing moisture-free membranes
- Compatibility with many insulation materials
- Code and standardization progress by the RMA and others
- Ease of maintenance
- Proven long-term performance

Another benefit is adaptability. Application techniques allow rubber-based roofing systems to be applied to virtually any roof surface: flat, spherical, curved, or slanted.

CSPE is a saturated polymer that contains chlorine and sulfonyl chloride groups attached to a polyethylene backbone. It is an elastomeric material. Typical elongations range from 200 to 400 percent, depending on the type and amount of reinforcing fillers present in the compound. This means that CSPE can stretch far beyond its original length and return to its original configuration without loss of structural integrity.

Actually, a unique feature of CSPE is that it is manufactured as a thermoplastic, but over time it cures as a thermoset. This means that most CSPE systems are typically hot-air- or solvent-welded. The advantages of welded seams include relatively quick procedures, good strength, and the lack of additional seaming material. There are two primary methods of application for CSPE rubber-based, single-ply systems, each of which offers certain advantages in specific applications.

Mechanically fastened. The membrane is loosely laid over the substrate and then anchored to the deck using fasteners. Many types of mechanically fastened systems are available. The most common are those that attach directly to the deck. Nonpenetrating systems also are available. Advantages include the fact that the systems are lightweight, easy to install, and relatively inexpensive.

Fully adhered. Ideal for contoured roofs or sloped surfaces that cannot withstand the weight of a ballasted system, fully adhered membranes are completely bonded to the substrate with contact adhesives. The major advantages include the fact that these systems are lightweight, durable, and easy to maintain. They also have an aesthetically clean and smooth appearance.

Only a handful of CSPE systems are installed using the ballasted technique popular with other rubber-based roofing systems because most building owners prefer not to cover the attractive white membrane. In fact, CSPE systems are often specified because of their attractive white exterior.

CHLORINATED POLYETHYLENE (CPE)

This material is formulated around its prime polymers and blended with pigment and processing aids, which serve as release agents and antioxidants. In fact, one manufacturer produces 15 grades of cured and uncured raw CPE polymer.

CPE made its roofing debut in 1967. Prior to that, it emerged in various military and civilian applications, most notably as a pond liner.

The majority of today's CPE roof membranes are offered in an uncured composition and are reinforced with a polyester scrim by individual roofing manufacturers. Standard thicknesses are 40 to 48 mils. Both CPE and CSPE normally are formulated without plasticizers because of their inherent flexibility as an elastomer.

As closely related elastomers, CSPE and CPE share similar behavioral characteristics. An aged CSPE, however, cannot be heat-welded. Adhesives must be applied if field repairs are necessary. The methods of application are the same for CPE and CSPE.

POLYISOBUTYLENE (PIB)

PIB is usually a 60-mil membrane made from synthetic rubber polymer, or polyisobutylene, pigments, fillers, and processing aids within several quality-control parameters that relate to thickness, density, elongation, hardness, and so on. The underside of the membrane is generally laminated with a 40-mil, needle-punched, nonwoven, rot-proof polyester fabric. The membrane is finished with a 2-inch-wide, self-sealing edge material that is protected by a strip of release paper.

PIB is compatible with hot asphalt and shows excellent resistance to weathering, radiant heat, and ultraviolet (UV) light. Used primarily as a final waterproofing membrane over existing flat or low-sloped roof assemblies, PIB can also be used as a waterproofing membrane for new-construction roof assemblies.

PIB is a lightweight system that requires no ballast. It can be installed quickly and economically. PIB passes UL's Class A fire rating and FM's I-90 wind-resistance rating.

These systems offer both the building owner and the roofing contractor numerous advantages. The self-sealing edge offers assured seam strength, long-term waterproofing, lower installation costs, faster application per worker hour, and no need for special capital equipment. The fact that the system is unballasted can eliminate the need to structurally reinforce the building before it is reroofed. PIB also minimizes the deck load and eliminates the logistics of adding 1000 to 1200 pounds of gravel for every 10-foot-square area.

PIB has a short but successful history in the United States. As more people become familiar with the system, realize that it is a solution for many roofing problems, and view the evidence of product support and longevity, the PIB system continues to show steady growth in the single-ply market.

NITRILE ALLOYS (NBP)

These membranes, as thermoset elastometers, are compounded from butadiene-acrylonitrile copolymers, nonvolatile polymeric plasticizers, and other patented ingredients.

These membranes typically are made by coating the compound on a heavy-duty polyester fabric. They range in thickness from 30 to 40 mils. Seams are hot-air welded. Used for almost 30 years for such diverse products as exterior door gaskets and footwear, NPB is recognized for its weather-protectant and waterproofing capabilities. The material exhibits good chemical resistance and low-temperature flexibility, but is sensitive to aromatic hydrocarbons.

Heating Thermoplastics

When heat is applied to a thermoplastic, its polymer chains slide freely over one another. This makes the plastic more pliable and heat weldable. When returned to ambient temperature, the polymer chains again intertwine and regain their original properties. This process can be repeated again and again with the same results, which explains why seaming is so excellent and easily done with a thermoplastic.

Thermoplastic roofing systems tend to be lighter in color, which can add value in terms of aesthetics. They are especially popular in multitiered roofing that can be seen from above by building occupants or neighbors.

The two most common chlorinated hydrocarbon thermoplastics are polyvinyl chloride (PVC) and CPE.

Polyvinyl Chloride (PVC)

PVCs, originally produced in Germany more than 30 years ago, are among the most versatile thermoplastics for industrial and commercial applications. PVC is synthesized from vinyl chloride and is a member of a larger group of polymers designated as *vinyls*. These polymers are composed of intertwined molecular chains. This, in part, is what gives PVC its unusually good physical properties.

PVC is one of the easiest materials to use. In its uncompounded state, PVC is a rigid material. Historically, PVC has been used in residential and commercial piping, plumbing, window frames, and exterior siding. When blended with plasticizers, PVC becomes soft and pliable. The use of the proper plasticizers enables membranes based on PVC to be used over a wide temperature range without substantial change. Because PVC is a plastic and not a rubber, it is unaffected by ozone.

Ingredients are added to PVC to protect the polymer during processing and manufacture, or to achieve specific requirements such as increased flame retardance, resistance to microbiological attack, and resistance to UV light. PVC is an excellent choice when high performance and economy of cost are the primary factors in roof selection.

PVC is forgiving. Installation mistakes can be easily corrected, and alterations, such as the addition of air-conditioning systems at a later date, are easily accomplished. For this reason, PVC conforms to nonstandard details with exceptional ease. With regard to seam integrity, thermoplastic roofing membranes can be welded together with heat or solvents. Once welded, they develop bond strengths that equal or surpass the strength of the base material.

PVC membranes can be installed by three different methods: loose-laid, partial bonding, and fully adhered. The simplest installation procedure is to loosely lay the material on the substrate. Attach the membrane only around the perimeter of the roof and at any penetrations. The insulation does not need to be fastened in place, as the roofing system is ballasted to resist wind uplift. Since the membrane is not attached to the substrate, stresses in the substrate are not transferred to the membrane. Gravel ballast must be of sufficient size and free of sharp edges, so that it does not puncture the membrane.

The partial bonding method uses elements of both the loose-laid and fully adhered systems. Firmly fasten insulation material beneath the membrane to the substrate. Then mechanically fasten round plates firmly to the substrate in a predetermined pattern and spacing. Finally, bond the membrane to the pattern. Partial bonding allows the membrane to float free over a crack or joint in the substrate. This distributes stress in the membrane between adhered areas. It is not intended to serve as or replace an expansion joint. On partially adhered systems, there are voids between membranes and substrates. If moisture enters the void and vapor pressure builds, blistering can occur.

The fully adhered method completely fastens the roofing system to the roof deck. For sheet systems, total bonding is usually achieved by applying an adhesive. The substrate for adhesive-bonded systems must be smooth, clean, dry, and free from dust, dirt, grease, oil, wax, and loose particles. Many manufacturers recommend that the substrate be primed before adhesive is applied to achieve greater bond.

Contact adhesives are typically used in total bonding. They are applied to the top surface of the substrate and the bottom of the sheet. Some sheets are self-adhering, however, so that the bottom of the sheet adheres to the substrate without an adhesive.

The fully adhered method is suited to reroofing projects when the old roofing is not removed. It can also be used on steep roofs that cannot contain ballast and where partial bonding allows the membrane to sag between fastening points. Mechanically fasten a layer of hardboard or insulation to the deck, since PVC is incompatible with asphalt or coal tar.

The system designer must consider all roofing materials when specifying PVC. Common roof materials like asphalt and coal-tar pitch cannot come in contact with PVC single-ply roofing. If they do, the PVC membrane may fail. All PVC roofing manufacturers stress that the membrane must be completely isolated from any contact with bituminous materials. Even fumes from coal tar must be avoided. Bituminous materials have the effect of leaching the plasticizers out of the PVC, leaving it weak and brittle.

Polystyrene insulation must not come in direct contact with PVC, since it can rapidly extract the plasticizers from the membrane. The treatment used for wooden blocking and nailers must be carefully considered because only waterborne wood preservatives can be used.

Chlorinated Polyethylene (CPE)

Thermoplastic elastomers, such as CPE, are unique in that they have elastomeric properties, have rubber-like elasticity, and are easy to install, like PVC. In fact, CPE can be applied in the same manner as PVC. Welds can be made with heat or solvent and detailing is accomplished easily because of the thermoplastic-like properties of CPE.

With noncuring elastomers, such as CPE, the chemical properties are such that the polymer chains are designed not to crosslink during the material's useful lifetime. Such a material has most of the desirable performance properties of a true rubber and easily conforms to any subsequent alteration to the roof. Such membranes seam almost as easily after being in place several years as when they were initially installed. Noncuring elastomers exhibit a mix of the qualities of both rubber and thermoplastic.

CPE resin in a pure form offers a soft membrane with poor physical properties. In order to be used as a roofing material, CPE polymer must be enhanced with other reinforcing chemicals, such as processing aids that ease polymer processing and stabilizer packages that protect and enhance the polymer when it is installed on the roof.

CPE offers the best chemical resistance of any conventionally available, single-ply roofing material. Depending on the given application, CPE is resistant to acetic acid, asphalt, bleach, chlorine, coal tar, fuel oils, animal fats and oils, fertilizer, sulfuric acid (acid rain), and a broad range of other corrosives. An excellent all-around choice as a roofing membrane, CPE has superior resistance to weathering, UV radiation, ozone, and microbiological attack. In addition, CPE has good fire resistance qualities.

Featuring Modified Bitumen Roofing

Modified bituminous roofing (MBR) systems have their technical roots in Italy, France, and Germany. They first appeared on the American market in 1975. Consisting of asphalt, a conventional base material, and any one of a number of polymeric modifiers, MBR membranes bridge the gap between conventional hot-applied BUR and nonconventional single-ply membranes.

MBR systems cover the full range of inplace system specifications. These systems currently are being applied using loose-laid, partially attached, and totally adhered methods, and as protected membrane systems. In addition, MBRs can be applied over a wide range of roof decks and roof insulations.

The requirements for deck selection, roof insulation, vapor retarder application, drainage, and flashing details that apply to conventional hot-BUR applications are generally applicable to most MBR systems as well. This makes the selection of an MBR system particularly attractive to the roofing contractor who might be skeptical of other single-ply systems' application requirements and techniques.

The MBR system is also versatile in its application method. These systems can be applied by hot-asphalt application, heat-welding or torching, or cold adhesive. Some systems have self-adhering properties. Again, this versatility provides the roofing contractor with a wide range of application choices that can be tailored to specific job requirements.

The MBR membrane is similar to hot-BUR systems, but represents a refinement in conventional asphalt technology. Though both systems have distinct softening points and proper installation temperatures, MBRs are composite sheets modified by either plastic or elastomeric material, such as atactic polypropylene (APP), styrene block copolymer (SBC), styrene-butadiene-styrene (SBS), or styrene-butadiene-rubber (SBR).

Both the SBR-modified and APP-modified bitumen systems produce good lap-joint strength. Lap-joint construction for the APP-modified membranes normally is accomplished by torching. SBR- and SBS-modified membranes normally are adhered using hot asphalt as the adhesive. They also can be torched, although the APP-modified systems are more conducive to torching. The APP modification allows the bitumen material to flow easily when heat is applied directly, thus the term *torching felt*.

Like BUR membranes, the MBR membrane consists of layers of modified asphalt that waterproof a reinforcing material. The reinforcing in MBR might be glass-fiber mats, polyester scrim, or a combination of the two. Each type of reinforcing material imparts different properties to the membrane. The manufacturing process usually dictates the location of the reinforcing material within the membrane.

Membranes with reinforcing material sandwiched in the center are the easiest to fabricate. Properties of the reinforcement material, however, also affect its location. Because polyester is not UV resistant, it must be buried in the mat, but not too close to the bottom because heat causes it to shrink and melt. Glass-fiber reinforcement material generally is placed close to the top of the mat to serve as a wearing surface. It resists foot traffic and UV degradation and provides a fire rating. One argument against this location is the potential for delamination during application.

Manufacturers using both types of reinforcement put the glass fiber at the top and the polyester in the middle. Roofing contractors like reinforcement in the middle because the top melts slightly when the next layer is applied, which fuses it to the bottom of the layer above. MBR membranes range in thickness from 40 to 60 mils.

Some MBR membranes are best covered. For example, SBS must be covered at all times because its UV and ozone resistance is low. Most surfacing consists of granules provided by the manufacturer. This granule surfacing does not add much weight to a system. Other coverings include manufacturer-installed metals or applied coatings of acrylics, asphalt emulsions, or fibrated aluminum. Of these, fibrated aluminum is the most popular because of its reflective properties. The differential movement of metal applied over asphalt is solved by fabricating tiny expansion joints into the metal.

UV-resistant APP can be left uncoated. It usually is coated, however, to promote longer life. Coatings include fibrated aluminum, acrylics, and asphalt emulsions. Mineral surfacings are available, too, but generally are used only for their aesthetic effect. Surfacings also are used to achieve fire ratings because most systems, with the notable exception of glass-fiber-reinforced membranes, cannot attain a fire rating alone.

MBR systems generally are applied in either a single-ply or multi-ply application (Fig. 6-3), frequently using conventional organic or fiberglass felts as the base ply. Currently, the heat welding and hot-asphalt mopping application methods seem to enjoy the most prominence.

MBR systems are suitable for application in new construction as well as reroofing jobs. Because of their superior load-elongation behavior, MBRs are frequently applied directly over existing conventional roofing materials. This is possible only if the existing substrate is not badly deteriorated or heavily moisture laden.

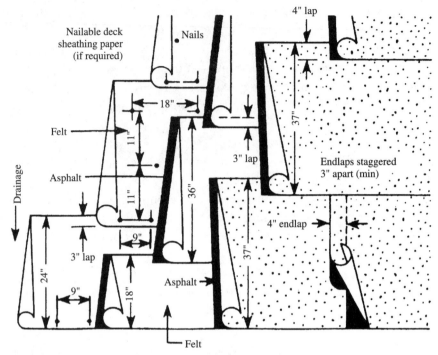

FIGURE 6-3 MBR can be applied with multiple sheets.

Preparing the Roof

The following are some general guidelines for the care and use of single-ply membrane materials. (See also Fig. 6-4.)

- Deliver materials in the manufacturer's original unopened packaging with labels intact.

- Store materials onsite under protective coverings and off the ground.

- Do not store material on the deck in concentrations that impose excessive strain on the deck or structural members.

- Proceed with roofing work only when existing and forecasted weather conditions permit installation in accordance with the manufacturer's recommendations and warranty requirements.

- Take special precautions, as recommended by the manufacturer, when applying roofing at temperatures below 40°F.

✓	Item
	Deliver materials in the manufacturer's original unopened packaging with labels intact.
	Store materials onsite under protective coverings and off the ground.
	Do not store material on the deck in concentrations that impose excessive strain on the deck or structural members.
	Proceed with roofing work only when existing and forecasted weather conditions permit installation in accordance with the manufacturer's recommendations and warranty requirements.
	Take special precautions, as recommended by the manufacturer, when applying roofing at temperatures below 40°F.
	Do not torch-fuse membranes directly to flammable substrates.
	Check with the manufacturer about application requirements.
	Check with the manufacturer about slope requirements.
	The National Roofing Contractors Association (NRCA) recommends that all roofing materials be installed on roofs with positive slope to drainage.
	If UL-classified roofing membranes are required, carefully evaluate the manufacturer's test data to determine compliance with code or other fire-related performance characteristics.
	Ensure that edge nailers, curbs, and penetrations, including drain bases, are in place and properly secured before starting the job so that the roof system can be installed as continuously as possible.
	Ensure that the insulation or base ply is positively attached to the roof deck.

FIGURE 6-4 Preparing the roof.

✓	Item
	Ensure that the bonding agent used to assemble laps and to adhere the single-ply membrane to the roof substrate is permitted by and acceptable to the material supplier.
	Ensure that positive attachment is achieved with partially adhered and mechanically fastened MBR systems that preclude a continuous film.
	Ensure that the roof membrane is applied in such a way that water can run over, or that the membrane has sidelaps and/or endlaps.
	Check that sheets are aligned so that minimum required end- and sidelap widths are maintained.
	Check that membrane laps are watertight. Repair voids and fishmouths within the lap as soon as possible.
	Install temporary water cutoffs at the end of each day's work. Remove them before installing additional insulation or membrane.
	Apply surfacings, when required, in accordance with the membrane supplier's requirements and so that the entire membrane surface is covered.

FIGURE 6-4 (Continued)

- Do not torch-fuse membranes directly to flammable substrates.

- Check with the manufacturer about application requirements.

- Check with the manufacturer about slope requirements.

- The National Roofing Contractors Association (NRCA) recommends that all roofing materials be installed on roofs with positive slope to drainage.

- If UL-classified roofing membranes are required, carefully evaluate the manufacturer's test data to determine compliance with code or other fire-related performance characteristics.

- Ensure that edge nailers, curbs, and penetrations, including drain bases, are in place and properly secured before starting the job so that the roof system can be installed as continuously as possible.

- Ensure that the insulation or base ply is positively attached to the roof deck.

- Ensure that the bonding agent used to assemble laps and to adhere the single-ply membrane to the roof substrate is permitted by and acceptable to the material supplier.

- Ensure that positive attachment is achieved with partially adhered and mechanically fastened MBR systems that preclude a continuous film.

- Ensure that the roof membrane is applied in such a way that water can run over, or that the membrane has sidelaps and/or endlaps.

- Check that sheets are aligned so that minimum required end- and sidelap widths are maintained.

- Check that membrane laps are watertight. Repair voids and fish-mouths within the lap as soon as possible.

- Install temporary water cutoffs at the end of each day's work. Remove them before installing additional insulation or membrane.

- Apply surfacings, when required, in accordance with the membrane supplier's requirements and so that the entire membrane surface is covered.

Applying Substrate Materials

Suggested substrate materials for MBR systems include light steel, which does not have the weight of gravel or ballast; primed concrete, which requires torch welding, particularly when not insulated; mechanically fastened base sheets; and uninsulated assemblies when the deck is wood, gypsum, lightweight insulating concrete, or cementitious wood fiber.

MBR is not recommended for roofs where there are concentrations of acids, hydrocarbons, or oils. It also is not cost effective for wide-open spaces because large rolls of EPDM can be installed less expensively.

The condition of the substrate is crucial to the performance of the membrane. The surface of the deck must be clean, firm, smooth, and visibly and sufficiently dry for proper application. There are different preparation requirements for new construction, replacement, or recovering over each type of substrate. To avoid defects and litigation, the material supplier and the roofing contractor can be responsible only for acceptance of the surface of the substrate that is to receive the roofing system. Design considerations and questions about structural soundness are the responsibility of the owner or the owner's representative. Put this in the contract.

If there is a question about the suitability of the deck and/or its surface, reach an agreement with the material supplier, the roofing contractor, and the owner before proceeding with the work.

When a primer is required, coat the surface with an amount sufficient to cover the entire surface. The amount required varies, depending on the nature of the surface. Allow the primer to dry before the membrane is applied. If visual examination indicates deficiencies in primer coverage, apply additional primer for complete coverage and then allow it to dry before applying the membrane.

The existing roof surface must be suitable to receive a single-ply membrane. Both the NRCA and the RMA recommend putting a recover board over existing gravel-surfaced roofs before the membrane system is applied. Remove loose and protruding aggregate by spudding, vacuuming, or power brooming to achieve a firm, even surface to receive the recover board.

When the existing roof is smooth or mineral-surfaced, refer to the material supplier's specifications. Some specifications permit fully adhering by heat welding, hot mopping, or adhesives. Others require base sheets over the existing roof, while still others permit spot mopping, strip attachment, and/or mechanical fastening.

Laying Insulation

Manufacturing tolerances, dimensional stability, application variables, and the nature of insulation boards make it difficult to obtain tightly butted joints. Some variance is expected, but the spacing between insulation boards should range from no measurable space to no more than ¼ inch or the space specified by the material supplier.

Fill insulation gaps between adhered or mechanically fastened insulation boards in excess of ¼ inch with roof insulation. Reduce gaps between loosely laid insulation boards by adjusting the boards or adding insulation. If the insulation boards appear to be out of square, make a diagonal measurement to confirm the squareness. Do not use defective material.

When composite-board, polyisocyanurate foam-board, polyurethane foam-board, perlitic board, or wood fiber-board insulation is used as the insulation substrate under a torch-applied MBR membrane, install a base sheet as the first layer below the roof membrane. This protects the insulation substrate from the flame and heat of the torch. Lap each base sheet a minimum of 2 inches over the preceding sheet; end-laps should be a minimum of 4 inches. Adhere the base sheets to the insulation in accordance with the manufacturer's specifications.

Fastening Insulation and Base Plies

Various mechanical fasteners are used to apply insulation and base plies. They also can be used to attach the single-ply membrane to the roof deck and to attach flashing materials. The fasteners often are specified by type, number, and spacing distance to fulfill the attachment functions. Practical considerations often prevent exact spacing, such as 6 inches on center. Reasonable variances from specified spacing distances are expected, but the minimum number of fasteners specified should be used.

The fastener type and spacing should be as specified by the material supplier with the understanding that the spacings are average values and the spacing between any two fasteners can vary. Should fastener deficiencies be discovered, install additional fasteners as needed and space them appropriately.

On all nailable roof decks, install a base sheet as the first layer below the roof membrane. This base sheet serves as a separating layer between the nailable deck and the roof membrane. Lap each base sheet a minimum of 2 inches over the preceding sheet. Endlaps should be a minimum of 4 inches. Fasten the base sheets according to the manufacturer's recommendations.

Applying Single-Ply Materials

As stated earlier, there are several attachment methods, depending on the single-ply material. System configurations change from specification to specification and are predicated not only on the manufacturer's procedure, but on site conditions. This can and does leave some questions about the proper method to use.

Nonetheless, fundamental procedures for the correct installation of all single-ply membranes remain the same, regardless of the system or specification. The strength and waterproofing characteristics of single-ply membrane systems depend on the construction of the laps and seams between membrane sheets and between the membrane and flashing materials.

Lap dimensions typically are specified as an amount of overlap between adjacent sheets. Because of the many construction variables already mentioned, some variance can occur. An overlap exceeding the specification is not considered detrimental. The material supplier's minimum lap values must be maintained, however, so as not to affect the strength and waterproof integrity of the membrane. Fabricate membrane laps so that they are watertight. Repair voids and fishmouths within the lap.

Products are frequently supplied with laying lines. Use these lines to position the subsequent course of material for compliance with nominal lap requirements. Because of the variation in substrates, products, and systems, refer to the material supplier's specifications for minimum requirements.

If the product does not have laying lines, use the material supplier's stated lap dimension as the minimum lap requirement. If examination reveals insufficient lap width, install a strip no narrower than twice the required lap over the deficient lap. Use the appropriate attachment method. Check all laps of heat-welded systems for adequate bonding. Seal any unbonded area.

Several factors determine the point at which work begins. For water runoff, begin work at a high point on the roof and move toward the low point. This prevents any drainage water from overnight rains from working its way under the new roof. Always shingle membrane laps with the flow of water or parallel to the flow of water.

Consider the deck type. Metal decks have corrugations that fall in 6-inch modular spacings. As with rigid insulation, fasten the membrane to the top of the flutes to meet Factory Mutual Engineering (FME) specifications. Because the exposed width of most full rolls of exposed membrane is 67½ inches, the edge of the sheet eventually falls between the flutes when the membrane is run parallel to the flutes. When this happens, trim the excess back to the nearest flute top (Fig. 6-5A).

When the membrane is run perpendicular to the metal deck corrugations, the amount of trimming and membrane waste is less than it would be if the membrane ran parallel to the corrugations (Fig. 6-5B).

Applying Thermoset Materials

Thermoset single-ply material can be applied loose-laid and ballasted, partially adhered, or fully adhered. Always roll out the sheet and allow it to relax. Inspect the sheet for defects as it is being rolled out.

To seal the laps, fold the sheet back onto itself and clean the underside of the top sheet and the upperside of the bottom sheet with the recommended solvent to remove dirt and talc from the lap area. Apply adhesive evenly to both surfaces and allow it to set according to the manufacturer's application instructions. Then roll the top sheet over the bottom sheet in a manner that minimizes voids and wrinkles. Apply pressure to the lap to ensure contact.

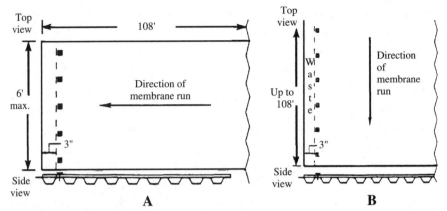

FIGURE 6-5 (*A*) Membrane running parallel to metal decking might require trimming along entire length of roll. (*B*) Membrane running perpendicular to the metal deck corrugations. The maximum trim is across the width of the roll.

LOOSE-LAID AND BALLASTED METHOD

With this method, the membrane is placed on a substrate without bonding and held in place by ballast. NRCA recommends that consideration be given to mechanically fastening or spot adhering the first layer of insulation to the roof deck under loosely laid roof systems. Only enough mechanical fasteners or adhesive need be used to hold the insulation boards in place during installation of the roofing membrane and to prevent stacking or displacement.

Attach the second layer of insulation, if used, to the first layer with a compatible adhesive. As an alternative procedure for nailable roof decks, both layers can be mechanically fastened to the deck at the same time.

Adhesive application quantities, mechanical fastener spacing, lap dimensions, and other minimum installation requirements vary from manufacturer to manufacturer. Over cellular glass insulation and any other insulation substrate where rough surfaces, protrusions, or sharp edges might affect the service of the membrane, install a separation sheet between the insulation and the roof membrane. Fasten the separation sheet only often enough to hold the sheets in place until the roof membrane is applied.

Roll out and align the sheet so that it overlaps the previous sheet by the required lap width. Seal all membrane laps together. Fabricate laps to shed water whenever possible.

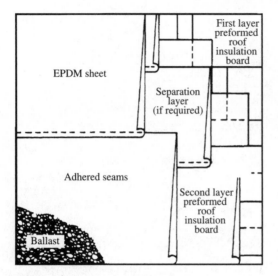

FIGURE 6-6 Loose-laid membrane system.

Ballast the loose-laid membrane with rounded stone, such as washed river gravel (Fig. 6-6). The amount of needed ballast varies with the location and height of the building. Substitute concrete pavers for rounded stone ballast if a protective underlayment is installed to protect the roof membrane from the abrasive surface of the paver.

PARTIALLY ATTACHED SYSTEMS

There are three generic ways to partially attach a thermoset (EPDM) membrane (Fig. 6-7). For the first method, roll out the sheet and align it so that it overlaps the previous sheet by the required lap width. Install mechanical fasteners with large washers or install bars with mechanical fasteners through the membrane and into the deck. Cover these fasteners or bars with membrane pieces that are bonded to the membrane in a manner similar to the methods used for sealing laps. Then seal all membrane laps together.

For the second method, roll out the sheet and install mechanical fasteners with large washers or bars with mechanical fasteners through the membrane within the area of the lap and into the deck. Seal all membrane laps together so that the fasteners are covered.

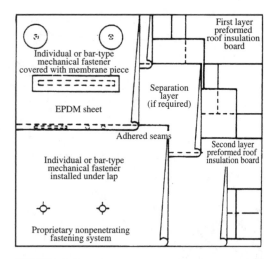

FIGURE 6-7 Partially attached roof system.

For the third method, mechanically fasten a proprietary fastening system in a predetermined pattern over the separation layer if used. Install the roof membrane over the fasteners. Roll out and align the sheet so that it overlaps the previous sheet by the required lap width. Sandwich the membrane between the parts of the fastening system by installing the second part of the mechanical fastener. Seal all membrane laps together.

FULLY ATTACHED SYSTEMS

For sheet systems, total bonding is usually achieved by applying an adhesive. The substrate for adhesive-bonded systems must be smooth, clean, dry, and free from dust, dirt, grease, oil, wax, and loose particles. Prime the substrate before applying the adhesive to achieve a greater bond.

Contact adhesives usually are used in total bonding. They are applied to the top surface of the substrate and the bottom of the sheet. Some sheets are self-adhering, however, so that the bottom of the sheet adheres to the substrate without an adhesive.

To apply a fully attached or total-bonded application, roll out the sheet and align it so that it overlaps the previous sheet by the required lap width. Fold the sheet back onto itself and coat the bottom side of

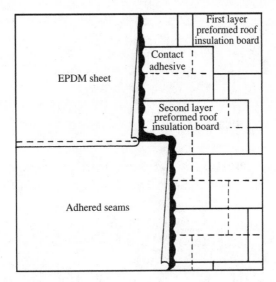

FIGURE 6-8 The mechanics of a fully adhered roof system.

the membrane and the top side of the deck with adhesive. Avoid getting adhesive on the lap joint area.

After the adhesive has set according to the adhesive manufacturer's application instructions, roll the membrane into the adhesive in a manner that minimizes the occurrence of voids and wrinkles. Repeat this for the other half of the sheet. Take care to adhere the very middle of the sheet (Fig. 6-8). Then seal all membrane laps together. Fabricate caps to shed water wherever possible.

Seaming Single-Ply Materials

A single-ply system is only as strong as its seams. There is more than a little truth to this cliche, especially with elastomerics. Unlike hot BUR, elastomeric and plastomeric sheets are preformed in the factory. Most of the problems with these systems can be traced back to field workmanship, of which lap seaming is a critical part. Most contractors feel that poor workmanship is the most common reason for single-ply roof failure. Industry experts are inclined to agree.

Most rubber-based, single-ply membranes require adhesive for lap seaming. A new generation of adhesives and sealants that speed application and improve the performance of rubber-based systems is providing simplified design, ease of inspection, proven performance, and low maintenance.

Bonding Adhesives

All-purpose bonding adhesives, geared toward roofing applications, adhere EPDM, neoprene, and butyl roof membranes to wood, concrete, metal, and certain insulation surfaces. These are contact-type adhesives.

Their use involves the application of a uniform coat to the back of the rubber membrane, as well as to the surface to which it is to be bonded. These successful products have earned wind-test ratings from FM and are accepted by most code authorities. Manufacturers advise that the mating surfaces be dry, clean, and free from oil, grease, and other contaminants. The best working temperatures are 40°F or above.

Many EPDM suppliers currently use one-part butyl adhesives to seal their 45- and 60-mil membranes. These adhesives normally give consistent peel strengths in the range of 3 to 4½ pounds per inch. The application of one-part butyls is somewhat less forgiving than that of other adhesive systems, however. The open time or window, defined as the time from when the seam can first be closed to when it no longer grabs, is shorter and varies with weather conditions. Keep in mind that one-part butyl requires a heavier adhesive coating. It can skin over and give a false indication that it is dry. In addition, it has been found that improperly stirring the adhesive results in less-than-adequate peel strengths.

Butyl cement begins to cure when exposed to heat and moisture. Once a can of adhesive is opened, do not reseal it or use it again because a substantial thickening of the cement occurs in about 48 hours. On the other hand, if butyl goes bad, the applicator knows it because the material does not stick. The pass/fail criteria for other adhesives are not as obvious.

Splicing Adhesives

There are lap splicing adhesives for EPDM, neoprene, and butyl membrane roofing. These adhesives provide long-lasting, weather-resistant bonding between the single-ply membranes used in various systems. Technology in the last five years has further enhanced the long-term performance of these adhesives. The size of the splice determines the application. In most cases, for flashing and expansion joint applications, the minimum splice width is 3 inches.

These also are contact adhesives that must be applied to both of the surfaces being bonded. The splice area must be dry and completely free of dust, talc, and other contaminants. The best working temperatures are between 40° and 120°F.

Sealants

Adhesive-seamed, cured, and uncured elastomer membranes all require the use of a lap sealant at the seam edge. Most adhesives are vulnerable to moisture until fully cured. Some materials take as long as a week to become fully moisture resistant. In addition to sealing out moisture and other contaminants, the sealants absorb thermal expansion and resist fatigue, vibration, and biological attack. They are ideal for sealing roofing protrusions, cracks, duct work, and exterior seams.

In order to allow the solvents to escape from inside the seam, do not apply the lap sealant for at least 2 hours. If rain is imminent or there is a need to leave the jobsite, apply the lap sealant earlier rather than later. Repairing minor blistering in the caulked edge is always easier than dealing with ruined seams.

It is also easy to wander off the membrane edge when you install the lap sealant. Because of this, it is important to tool the sealant. Ideally, one worker should install the material, while a second worker follows behind and tools the edge. If one waits too long to tool, the material can harden and start building up on the tool. Sometimes a primer is applied to the sheet before the adhesive to enhance seam performance.

Unlike splice or primer washes, the solvent is applied at full strength. The combination of primer and neoprene adhesive usually offers performance similar to that of one-part butyl.

To apply the adhesive system, first use a splice cleaner on the seam. After the area dries, roll the primary butyl adhesive over the entire width of the splice. Next, apply a bead of inseam sealant tube on top of the butyl and roll the seam closed. Finally, secure the splice with a lap sealant. The inseam sealant is also designed to serve as a redundant adhesive. If the installer misses a small section of the seam with the butyl, the inseam sealant should still adhere to the EPDM and seal the seam. Because the sealant has a greater mass than the butyl adhesive, it is capable of filling any voids or gaps in the seam created by an uneven substrate.

Be aware that the inseam sealant does not have the *green strength* to hold the seam together during setup time. Therefore, the butyl adhesive must be installed in conjunction with the sealant to avoid fishmouths at the edge of the seam.

Tape

Butyl-based tape offers the extra mass of adhesive needed when three layers of membrane come together, called a T-joint. According to many roofers, the tape is somewhat more sensitive to dust, dirt, and contaminants than the liquid adhesive. Applying the liquid adhesive with a bristle brush can lift contaminants into the adhesive mass. On the other hand, tapes have speed and labor-saving advantages, if the applicator is skilled.

While many tapes used in the past were made of uncured material, today's tape adhesives are cured. Uncured tapes move independently of the roof during expansion and contraction. Cured tape is able to move with the sheet.

Heat Welding

Until recently, one of the easiest ways to tell if an applied single-ply membrane was thermoset or thermoplastic was to look at the seams. Thermoset seams were installed with adhesive tape or sealer, while thermoplastic seams were usually heat-welded.

Several manufacturers are now producing a hot-air weldable EPDM membrane, however, which eliminates the use of adhesives or tapes. One product consists of a patented blend of polypropylene and vulcanized rubber. The polypropylene gives the membrane heat-welded characteristics and greater oil resistance, while EPDM and carbon black protect the polypropylene from UV degradation. During manufacture, a codispersion between the EPDM and polypropylene takes place,with the molecular crosslinking of the polypropylene already in the EPDM formulation.

Another heat-weldable EPDM is made of polypropylene and a thermoplastic rubber. A coextrusion process causes the crosslinking of the two materials, with the top layer containing more vulcanized rubber and the bottom layer more polypropylene.

The 45-mil membrane is mechanically fastened in seam using 2-inch-diameter washers set 18 inches on center. A UL Class 90 wind-uplift rating and Class B fire rating (smooth surfaced) have already been achieved. Tear strength is a consideration in a nonreinforced, mechanically attached sheet. The crosslinking process gives the membrane a 360-pounds-per-square-inch tear strength across the sheet, which is where wind uplift forces are greatest, according to the manufacturer. CSPE, as mentioned earlier in the chapter, is heat-welded.

Applying Thermoplastic Materials

PVC, CPE, and PVC blends can be installed by loose-laid, partially attached, or fully adhered methods.

Using the Loose-Laid Method

On all nailable roof decks, install a separator sheet as the first layer below the roof membrane to serve as a separating layer between the nailable deck and the roof membrane (Fig. 6-9*A* and *B*). Lap each separator sheet a minimum of 2 inches over the preceding sheet. Endlaps should lap a minimum of 4 inches. Fasten the separator sheets only as necessary to hold the sheets in place until the roof membrane is applied.

Roll out the sheet over the separator sheet and inspect the membrane for defects. Align the sheet so that it overlaps the previous sheet by the required lap width. Seal all membrane laps together using either heat welding or chemical fusion. Fabricate laps to shed water wherever possible. Apply pressure to the lap to improve the bond.

Carefully inspect lap edges. Repair unsealed areas, voids, and fishmouths. Generally, manufacturers require that exposed membrane edges be finished with a sealant. Fasten all membrane perimeters, terminations, and penetrations as required by the manufacturer.

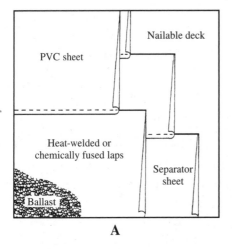

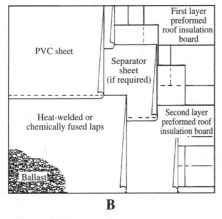

FIGURE 6-9 (*A*) Loosely laid PVC sheet over a nailable deck. (*B*) Loosely laid PVC over an insulated deck.

Ballast the loosely laid membrane with rounded stone, such as washed river gravel. The amount of ballast needed varies with the location and height of the building. Concrete pavers can be substituted for rounded stone ballast, provided that a protective underlayment is installed to protect the roof membrane from the abrasive surface of the paver.

Using the Partially Attached Method

There are three generic methods for partially attaching thermoplastic membranes (Fig. 6-10). For the first method, mechanically fasten PVC-coated metal discs in a predetermined pattern over the separator sheet. Roll out the membrane and inspect it for defects. Heat weld or chemically weld the membrane to the PVC-coated metal discs as the membrane is rolled out onto the deck. Align the sheet so that it overlaps the previous sheet by the required lap width. Then seal all membrane laps using heat welding or chemical fusion. Fabricate laps to shed water wherever possible. Apply pressure to the lap to improve the bond.

The second method calls for the installation of mechanical fasteners with large washers through the membrane and into the deck. Cover these fasteners with membrane patches and heat weld or chemically fuse the patches to the membrane. Seal the membrane laps using heat welding or chemical fusion. Fabricate the laps to shed water and apply pressure to the lap to improve the bond.

The third method follows the same initial steps. Roll out and inspect the membrane and seal the laps using heat welding or chemical fusion. Fabricate the laps to shed water, and apply pressure to the lap to improve the bond. Then place continuous metal bars in a pre-engineered pattern on top of the membrane and mechanically fasten them to the nailable roof deck. Cover the bars with membrane patches that are heat welded or chemically fused to the membrane.

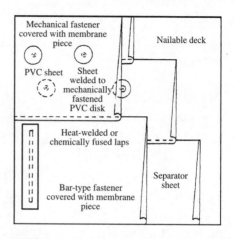

FIGURE 6-10 Partially attached PVC roof system.

Using the Fully Adhering Method

There are two generic methods for fully adhering thermoplastic membranes (Fig. 6-11). For solvent-based adhesives, roll out the sheet and inspect it for defects. Align the sheet so that it overlaps the previous sheet by the required lap width. Fold the sheet back onto itself and coat the bottom side of the membrane and the top side of the deck with adhesive. Avoid getting adhesive on the lap joint area.

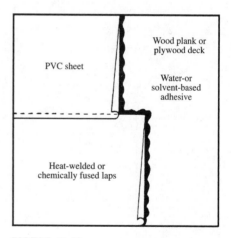

FIGURE 6-11 A fully adhered PVC roof system.

After the adhesive has set according to the manufacturer's application instructions, roll the membrane into the adhesive in a manner that minimizes the occurrence of voids and wrinkles. Repeat this for the other half of the sheet. Carefully adhere the middle of the sheet.

For water-based adhesives, follow the steps for solvent-based adhesives, but avoid getting the adhesive on the lap joint area when you coat the deck. After the adhesive sets according to the manufacturer's application instructions, roll the membrane into the adhesive in a manner that minimizes the occurrence of voids and wrinkles.

Seaming Thermoplastics

PVC, CPE, and the PVC blends are inherently heat weldable with typical lap seam strengths of more than 20 pounds per inch. While hot-air and solvent welding procedures are quicker than adhesive seaming, a considerable amount of applicator skill is still required. For example, CPE exhibits a higher melting temperature than most PVCs and a narrower welding window. It is especially important to conduct test welds when ambient roof conditions change substantially.

There are two basic types of welders available: automatic and hand. The automatic welder rides on the top sheet, which is being welded to the bottom sheet. For this reason, it is possible to start welding in one of two roof corners: the upper right-hand corner or the lower left-hand corner (Fig. 6-12). To determine these two corners, face the same direction as the run of the membrane. The roof corners to your upper right and lower left are the possible starting points.

Use the hand welder to weld seams when automatic welding is not possible, such as near a parapet wall or curb detail. Hand welding is also used for many flashing details, applying membrane patches, and repairing poor or unfinished machine-welded seams (Fig. 6-13).

When the gun heats the membrane to a semimolten state, use a hand roller to bond the lap seam. A number of interchangeable nozzle tips are designed to match any hand-welding situation. For heavily reinforced PVC blends, applying pressure at the right temperature is crucial to a good

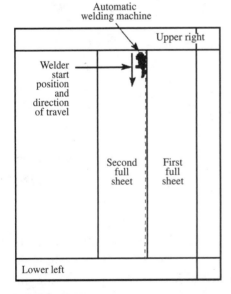

FIGURE 6-12 Welding can begin at one of two points on the roof, the upper right or the lower left.

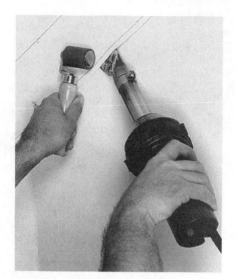

FIGURE 6-13 A hand welder in use.

lap seal. Because there are several models and designs, always follow the adjustment procedures (Fig. 6-14) outlined in the service manual (see App. A for vendors).

Checking Welded Seams

Working conditions vary from job to job, from day to day, and even from hour to hour. Varying conditions require different heat and speed settings on the welding machine. Because conditions vary, it is extremely important that you inspect all welded seams with care. Make adjustments to welder speed, temperature, and setup as needed.

Wait several minutes before testing seam strength, since even good welds can be separated while they are still hot. A good weld is one where the top sheet does not separate from the bottom sheet without damaging the membrane.

Proper seam width is approximately 1½ inches. The welded seam should be smooth and continuously bonded without voids or air pockets. The lap edge should be bonded completely together. Check for gaps in this seam by running a cotter-pin puller along the seam.

The automatic welder usually misses the first few inches of a seam, so pull back the lap after it has cooled and mark off the area that requires hand welding. If there are voids in the weld seam, or if the

FIGURE 6-14 Making an adjustment on a hand welder.

finished seam is weak, it probably means that the hot-air nozzle did not have enough time to heat the membrane to the necessary molten state. To correct this problem, slightly lower the machine's welding speed and make another test run.

Other problems, such as dirt on the membrane, moisture on the membrane, or variances in the power supply, also can produce voids and poor bonding. Inspect the welded seam and the area around the weld. A light browning or burning of the membrane indicates that the nozzle is overheating the weld area. To correct this, increase the welding speed slightly and test again.

If the membrane still burns at the highest speed setting, lower the nozzle temperature by adjusting the heat control dial. Brown streaks at the edge of the overlap usually are caused by the outer edge of the nozzle dragging heated dirt and membrane particles across the membrane. To prevent this, clean the nozzle with a wire brush.

Pinch wrinkles can be pulled into the membrane sheets when the automatic welder is not set up properly. If pinch wrinkles occur, review the setup and operating procedures given earlier in this section and those provided with the welder. Repair pinch wrinkles with membrane patches.

Applying Modified Bitumen Materials

MBR can be applied using torch-applied, hot-mopped asphalt or a self-adhered method.

Torch-Applied or Heat-Weld Method

To apply the membrane, roll out the sheet and inspect it for defects. Align the sheet so that it overlays the previous sheet by the required lap width and then reroll it. Heat the underside of the roll to soften the bitumen coating. Take care not to overheat the top surface of the sheet.

Most torch-applied membranes come with a factory-installed *burn-off* film. This film is usually polyethylene, though it can be polypropylene. Burn-off films are applied as a release agent and as a compatible material that blends with the MBR coating when heated. This creates an excellent cohesive bond at the lap or seam. Films also facilitate bonding to other surfaces. Some torch-down products incorporate talc or sand as a release instead of burn-off films.

To obtain *flow* of the MBR coatings, apply heat to the burn-off film surface in sufficient amounts to soften and melt the coating. When the roll is unwound you should see a flow ahead of the roll and at the sidelap that minimizes voids and wrinkles. At the sidelap and all seams, flow should be from a minimum of ¼ inch to a maximum of ½ inch. It should be continuous and uninterrupted. No flow at all indicates that insufficient heat was applied. A flow more than ½ inch indicates that too much heat was applied.

Take care to ensure proper lap alignment. Membrane laps are sealed together as the sheet is adhered to the deck. Fabricate laps to shed water wherever possible. Check all seams for proper bonding. Check any lap or seam that has no flow for integrity. Some manufacturers require that all seams be *buttered* or troweled. In most cases, however, this is not necessary.

Any installation areas that have an observable flow in excess of ½ inch are not likely to have bonding surfaces. This yields the maximum effect because of excess coating flow. In some cases, the coating can be almost completely forced away from the bonding surfaces, which reduces maximum bonding potential.

Excess heat also can affect polyester-reinforced membranes. These membranes are susceptible to stretch and thermal breakdown if too much heat is applied. Polyester burns and melts, which dramatically reduces performance characteristics. Fiberglass-reinforced membranes do not stretch under heat. They can, however, experience thermal breakdown of binders, which can result in a loss of performance characteristics.

Hot-Mopped Method

Roll out and inspect the sheet. Then align the sheet so that it overlaps the previous sheet by the required lap width. Reroll the membrane. Apply a mopping of hot asphalt immediately in front of the roll. Follow the guidelines for heating asphalt as predicated by the equiviscous temperature (EVT) printed on the carton or wrapper, or review Table 5-1. Never heat the asphalt to its flash-point temperature.

Align the membrane plies carefully. Stagger or offset endlaps a minimum of 3 feet. Apply the first membrane course at the lowest point of the deck. This prevents water from flowing against the lap seams. Mop consistently at about 25 pounds per square. The mopping action should be continuous and uninterrupted when fully adhered systems are installed. Do not feather or taper moppings.

When applying the hot-asphalt-applied MBR membrane, advance the roll into hot asphalt that has been applied no more than 4 feet in front of the roll. Moppings more than 4 feet from the advancing roll can cool and cause false bonding. Check all laps and seams for proper bonding. Fabricate the laps to shed water whenever possible.

Flow from the mopping asphalt at laps and seams should be from the torch-applied membrane. Use a thermometer to check asphalt temperatures at each point before application.

For added protection and aesthetics, mineral granules can be sprinkled on the asphalt flow at all laps and seams while the asphalt is still hot. Other materials can also be used to surface MBR membranes. In addition to their primary function to protect the membrane from the elements, these surfacings often serve to increase the fire and impact resistance of the roofing system.

Liquid-applied surfacing materials vary in physical properties and in formulation. Surfacing materials and aggregate are applied by various techniques, such as hand spreading and mechanical application. They also are applied on a variety of roofs in many climatic conditions. These factors and others preclude a high degree of uniformity in applying liquid-applied surfacing materials and aggregate over a roof area.

Hot-asphalt-applied membranes come with a sand release that is used as an antiblocking agent and as a surface that facilitates adhesion to the asphalt. Hot-asphalt-applied MBR membranes are always SBS-modified bitumens with a sanded bonding surface.

Self-Adhered Method

Cold adhesives are used with some MBR membranes to bond the sheet to the substrate. Use the adhesive specified by the material supplier. During the application of a fully adhered system, you should observe a continuous, firmly bonding film of adhesive. The adhesive should flow out from the membrane to form a seal. If aesthetic appearance is a factor or is dictated by specification, matching granules can be used to cover the flow-out area.

When using the fully self-adhered application, prime the surface of the roof deck with asphalt primer and allow it to dry. Primers are not required if a base sheet is used. Roll out the sheet, inspect it for defects, and align it so that the membrane overlaps the previous sheet by the required lap width. Then reroll it.

Remove the release paper from the underside of the membrane and roll the membrane over the primed surface in a manner that minimizes voids and wrinkles. Take care to ensure proper lap alignment. Membrane laps are sealed together as the sheet is adhered to the deck. Fabricate laps to shed water whenever possible. Apply pressure to both the membrane and lap areas to ensure contact.

In a partially adhered membrane application, MBR compounds, asphalt, and adhesives are used to secure the roof assembly to the structure. These materials are installed in a configuration prescribed by the material supplier, such as spot or strip bonding. They often are specified by amount and spacing distances. Practical considerations often prevent the application of the exact amount and spacing, such as 12 inches in diameter and 24 inches on center of the bonding agent. Reasonable variances are expected.

The application should result in a firmly bonded membrane. The spacing of the bonding agent should be specified by the material supplier. The design team and owner should understand that the spacings are average values and that the distance between any two adhesive locations can vary. Correct deficiencies by installing an additional bonding agent as needed. Space appropriately. Determine the scope of the discrepancy and take appropriate remedial action.

Flashing at Terminations

Single-ply membrane terminations at roof edges, parapets, and flashing are treated in a variety of ways depending on the material used.

Cured Elastomeric Membrane Flashing

Cured elastomeric membrane that is identical to the roof membrane material can be used as flashing material. Cut the membrane to size and coat the bottom side of the flashing membrane and the area to be flashed with adhesive. After the adhesive has set according to the manufacturer's application instructions, roll the membrane into the adhesive in a manner that minimizes the occurrence of voids and wrinkles.

Exercise care that the flashing does not bridge where there is a change of direction, i.e., where the parapet wall intersects the roof deck. Seal joints where the cured elastomeric membrane flashing material meets with roof membrane in a manner similar to that used for membrane laps. Apply pressure to the laps to ensure contact.

Uncured Elastomeric Membrane Flashing

Uncured elastomeric membrane can be used in lieu of cured elastomeric material. It is especially suited for penetrations, corners, and other flashings where the membrane must be formed or must change direction. The uncured membrane is used in a manner similar to cured membrane material. Upon curing, however, it permanently shapes to conform to the substrate being flashed.

PVC-Coated or PVC-Clad Metal Flashing

PVC-coated or PVC-clad metal can be used as flashing material. The flat metal stock is cut and fabricated to the proper shape to meet field requirements. The sheet metal pieces should not exceed 10 feet in length. Fasten the metal in place with fasteners typically spaced 4 to 6 inches on center. Round the corners and the edges of the sheet metal flashing to protect the membrane from punctures or chafing. Cover the gap with tape.

Splice the two pieces of metal together with a minimum 5-inch-wide piece of PVC membrane. Center the piece on the gap so that it extends from the top of the metal flashing and out onto the field of the roof. Either heat weld or chemically fuse the PVC membrane to the PVC-clad or PVC-coated metal flashing.

Reinforced PVC Membrane Flashing

Reinforced PVC membrane flashing can be used in lieu of PVC-coated or PVC-clad metal flashing provided the membrane is first mechanically fastened to the nailable roof deck at the flashing. Fasten the reinforced PVC membrane flashing at its upper edge, and shape it to conform to the parapet wall, curb, or edging. Then extend the reinforced flashing onto the field of the roof a minimum of 4 inches. Heat-weld or chemically fuse it to the roof membrane. Endlaps are to be a minimum of 4 inches. Heat-weld or chemically fuse them to the adjoining piece of flashing. Various PVC flashing details are shown in Fig. 6-15.

MBR Flashing Systems

MBR flashing materials are applied using a variety of methods, including heat welding, hot asphalt, and adhesive applications. The degree of heat required for appropriate heat welding, the variables affecting the application rates and temperatures of any asphalt used, and the amount of adhesive used are subject to many variables, including weather conditions, job conditions, material type, and application method.

In addition, the material supplier's requirements must be considered. Membrane termination follows the same basic procedure at roof edges, parapets, and other flashing applications as the BUR materials described earlier. It can be handled in one of three ways.

Torch-applied MBR membranes can be used as flashing material for masonry, concrete walls, curbs, or other noncombustible substrates. Cut the membrane material to size and prime the wall or curb and the area of the roof membrane to be flashed. Place the MBR flashing membrane

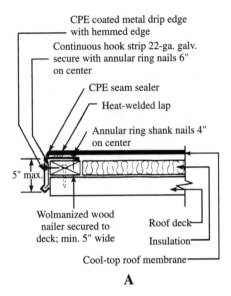

CPE coated metal drip edge
with hemmed edge

Continuous hook strip 22-ga. galv.
secure with annular ring nails 6"
on center

CPE seam sealer

Heat-welded lap

Annular ring shank nails 4"
on center

5" max.

Wolmanized wood
nailer secured to
deck; min. 5" wide

Roof deck

Insulation

Cool-top roof membrane

A

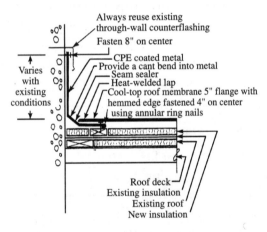

Always reuse existing
through-wall counterflashing

Fasten 8" on center

CPE coated metal

Provide a cant bend into metal

Seam sealer

Heat-welded lap

Cool-top roof membrane 5" flange with
hemmed edge fastened 4" on center
using annular ring nails

Varies
with
existing
conditions

Roof deck

Existing insulation

Existing roof

New insulation

B

FIGURE 6-15 (*A*) CPE-coated metal drip edge.
(*B*) Installing coated metal flashing under existing
wall counter. (*C*) Heated vent-pipe flashing.

bottom-side up onto a plywood or other suitable platform. Heat the underside of the membrane with a torch to soften the coating bitumen.

Set the heated membrane material to the wall or curb in a manner that minimizes the occurrence of voids and wrinkles. Take care that the flashing does not bridge where there is a change of direction, such

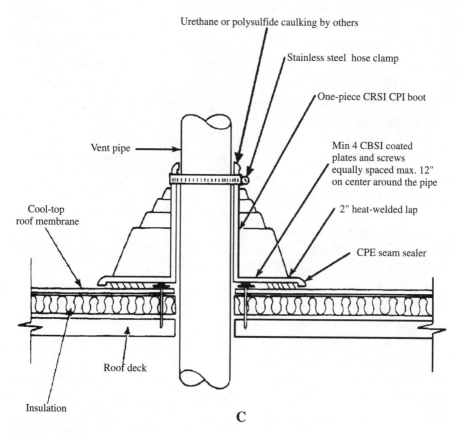

Urethane or polysulfide caulking by others

Stainless steel hose clamp

One-piece CRSI CPI boot

Min 4 CBSI coated plates and screws equally spaced max. 12" on center around the pipe

Vent pipe

2" heat-welded lap

Cool-top roof membrane

CPE seam sealer

Roof deck

Insulation

C

FIGURE 6-15 *(Continued)*

as where the parapet wall intersects the roof deck. Seal the joints where the MBR flashing material meets with the roof membrane in a manner similar to that used for membrane laps. Apply pressure to the laps to ensure contact.

Hot-mopped MBR membranes can also be used as flashing material. Follow the directions for torch-applied membranes. Instead of torching, apply hot asphalt to the area to be flashed and then to the underside of the membrane. Seal the joints where the MBR flashing material meets with the roof membrane in a manner similar to that used for membrane laps. Apply pressure to the laps to ensure contact.

The third method is to use self-adhered MBR membranes as flashing material. Cut the membrane material to size and prime the wall or curb and the area of the roof membrane to be flashed. Remove the release paper and set the membrane to the wall or curb in a manner that minimizes the occurrence of voids and wrinkles. Take care that the flashing does not bridge where there is a change of direction. Seal the joints where the MBR flashing material meets with the roof membrane in a manner similar to that used for membrane laps. Apply pressure to the laps to ensure contact.

Locate the flashing terminations a minimum of 8 inches above the level on the roof. Counterflashing should follow the material supplier's specifications. Some material suppliers have a maximum height limitation for flashings. Refer to the material supplier's recommendations.

A cant strip can be installed to modify the angle between the roof deck and the vertical surface. For heat-welded applications, make cant strips from a flame-resistant material and cover them with a base felt. Consult the manufacturer's instructions for recommendations regarding the use of cant strips.

Install the membrane before the flashing is applied and above the plane of the roof, above the cant, and up the vertical surface as specified by the material supplier. Do not run the roof membrane up the vertical surface to act solely as a base flashing. Remove and reinstall areas of loose, inadequately or improperly bonded flashings using the same materials used in the original application. Cut out, rebond, and reinstall laps that are inadequately bonded.

Carefully inspect flashing seams; repair unsealed areas, voids, and fishmouths. Smoothly finish exposed edges or torch-applied MBR membrane with a hot, rounded trowel.

Surfacing Single-Ply Roofs

Some MBR membranes have factory-applied surfacings that provide weather resistance or improve the appearance of the roof membrane. Some membranes require that the surfacings or coatings that provide these properties be applied in the field. The following is a description of the more common surfacing options for MBR systems.

Ballast. Loosely laid MBR systems require the use of ballast to build resistance to wind-uplift forces. The ballast is usually rounded stone, such as washed river gravel. The amount of ballast needed varies with the location and height of the building. Concrete pavers can be substituted for rounded stone ballast, provided that the roof membrane is protected from the abrasive surface of the paver.

Aggregate. For fire resistance, the application of a flood coat and aggregate surfacing can be required. The aggregate should be essentially opaque and a nominal ⅜ inch in diameter. Embed the aggregate in a flood coat of hot asphalt.

Mineral granules. For fire resistance, some membranes require the field application of mineral granule surfacing. For each roof square, embed approximately 50 pounds of No. 11 roofing granules in an emulsion coating that has been applied to the membrane at the approximate rate of 2 gallons per roof square (see manufacturer's instructions).

Emulsion coatings. For aesthetics, some membranes require the application of an asphalt emulsion top coating. Application rates and techniques vary from manufacturer to manufacturer.

Reflective coatings. Fibered and nonfibered aluminum/asphalt coatings can be applied to most smooth-surfaced MBR membranes. Application rates and techniques vary depending on the manufacturer.

Factory-applied/self-surfaced membranes. Some MBR membranes might not require the field application of additional surfacing materials. These include membranes with factory-applied mineral granules or metal foils and some smooth-surfaced systems.

Asphalt Roofing

Asphalt products are the most commonly used low- and steep-pitched roofing material for both residential and commercial buildings. Asphalt roofing can be classified into three groups: roll roofing, shingles, and underlayment.

Roll roofing and shingles are outer roof coverings. They are exposed to the weather and designed to withstand the elements. Saturated felts are inner roof coverings. They provide the necessary underlayment protection for the exposed roofing materials.

Asphalt flux is used to manufacture asphalt roll roofing products and asphalt shingles. It is obtained from the fractional distillation of petroleum that occurs toward the end of the refining process. Asphalt flux is sometimes processed at the oil refinery and delivered to the asphalt material manufacturer in a state that satisfies the manufacturer's specifications. Many manufacturers, however, purchase the flux and do their own refining.

Organic felts and glass-fiber base mats are the most common reinforcement materials in asphalt roofing materials. Organic felts are produced from various combinations of rag, wood, and other cellulose fibers. Glass-fiber base mats are composed of inorganic, continuous or random, thin glass fibers that are bonded firmly together with plastic binders. Generally, the glass fibers in these mats are reinforced with additional chopped glass-fiber strands or with continuous, random, or parallel glass-fiber filaments.

The surfacing found on asphalt products is of two types. Fine mineral granules are dusted on the surfaces of smooth asphalt roll roofing materials and on the back side of mineral-surfaced asphalt roll roofing materials to prevent the convolutions of the roll from sticking together after the material is wound into rolls. Finely ground minerals are also dusted on the back side of asphalt shingles to prevent the shingles from sticking together in the package.

Talc and mica are the most frequently used mineral surfacings. They are not intended to be a permanent part of the finished product and gradually depart from exposed surfaces after the roofing materials are applied.

Coarse mineral granules are used on some asphalt roll roofing products and on asphalt shingles to protect the underlying asphalt coating from the impact of light rays. The granules are opaque, dense, and graded for maximum coverage. Coarse mineral granules also increase the fire-resistance rating of the asphalt roofing product.

This surfacing is available in a wide range of colors and color blends, which improves the adaptability of asphalt roofing materials to different types of buildings. The range of colors also provides a large variety of design possibilities.

The materials most frequently used for coarse mineral surfacing are either naturally colored slate, naturally colored rock granules, or rock granules colored by a ceramic process. The ceramic-coated mineral granules are applied to the top surface of the shingles during the manufacturing process. This provides the widest range of colors available in any type of roofing material. They range from white and black to hues of red, brown, and green.

Using Asphalt Shingles

Asphalt shingles are the most common roofing material used today. They are manufactured as strip shingles, interlocking shingles, and large individual shingles in a variety of colors and weights.

Strip shingles are rectangular. They measure approximately 12 inches wide and 36 inches long, and can have as many as five cutouts along the long dimension. Cutouts separate the shingle's tabs, which are exposed to the weather and give the roof the appearance of being made up of a larger number of individual units. Strip shingles also are manufactured without cutouts to produce a much different effect. The three-tab shingle is the most common strip shingle.

Most shingles are available with strips or dabs of a factory-applied, self-sealing adhesive, which is a thermoplastic material that is activated by the sun's heat after the shingle is installed. Exposure to the sun's heat bonds each shingle securely to the one below for greater wind resistance. This self-sealing action takes place within a few days during the spring, summer, and fall. In winter, the self-sealing action varies depending on geographical location, roof slope, and the orientation of the house on the site.

Strip-shingle tabs can be trimmed or offset to obtain straight or staggered buttlines, respectively. They also can be embossed or built up from a number of laminations of base material to give a three-dimensional effect. Each of these shingle characteristics—staggered buttlines, embossing, and laminations—can be combined in various ways to create textures on the finished roof surface that resemble wood, slate, or tile.

In recent years, the laminated two-ply asphalt shingle has become very popular. Because two layers of shingles are laminated together, there is more weight and better protection than with a one-ply, three-tab shingle. Most three-tab shingles have a 15- to 20-year limited warranty compared to 25 to 30 years for laminated shingles. That is 5 to 15 years more service.

In addition, fiberglass laminated shingles carry a Class A fire rating from Underwriters' Laboratories, Inc. (UL). Single-ply asphalt shingles usually have Class C fire ratings. Fiberglass also offers greater resistance to warping, rotting, blistering, and curling. It contains more coating asphalt to better protect the structure from weather elements.

Interlocking shingles provide resistance to strong winds. The shingles come in various shapes and with different locking devices that provide a mechanical interlock. Large individual shingles are generally rectangular or hexagonal in shape.

Most manufacturers certify asphalt shingles to wind-performance test standards on a continual basis through independent third-party testing laboratories. In this standard test, which has been used for many years, shingles are applied to a roof deck according to the manufacturer's specifications, sealed under controlled conditions, and then subjected to ongoing wind conditions for 2 hours. The asphalt shingles are monitored continuously throughout the test.

Using Roll Asphalt Roofing Materials

Roll roofing is sometimes used for low-cost housing and utility buildings, such as sheds, garages, and barns, in place of asphalt shingles. Roll roofing is an asphalted material that is manufactured and surfaced like shingles. It comes in rolls that are 36 to 38 feet long and approximately 36 inches wide.

Some mineral-surfaced roll roofings are manufactured with a granule-free selvage edge that indicates the amount by which each succeeding course should overlap the preceding course. Follow the manufacturer's recommendations with respect to toplap, sidelap, and endlap because the amount of overlap determines how much of the material is exposed to the weather and the extent of roof surface coverage, i.e., whether most of the surface is covered by a single or a double layer of roll roofing. In addition to its use as a roof covering, roll roofing is also a common flashing material.

Using Saturated Felt Underlayments

This type of asphalt roofing product consists of a dry felt impregnated with an asphalt saturant. It is used primarily as an underlayment for asphalt shingles, roll roofing, and other types of roofing materials, and as sheathing paper. Saturated felts are manufactured in different weights for use as regular or heavy-duty underlayment.

Preparing for Asphalt Roofing Materials

Apply asphalt roofing materials over a solid-sheathed deck surface. If plywood is used as the deck material, use exterior plywood either $^{15}/_{32}$ inch thick for 16-inch rafter separation or $^{5}/_{8}$ inch thick for 24-inch rafter separation. Separate the plywood panels by at least $^{1}/_{16}$ inch to allow for expansion.

The underlying structure should provide a rigid deck surface that does not sag, shift, or deflect under the weight of the roofing, the workers installing the materials, or the snow loads the roof might have to support. An unstable structure could lead to movements that affect the integrity of the roofing materials and flashings.

Warping, as well as other problems, can result even with well seasoned lumber if the attic space under the roof deck or the spaces between the rafters and beams are improperly ventilated in a cathedral ceiling or mansard roof. Poor ventilation can result in an accumulation of moisture that eventually condenses on the underside of the roof deck.

Prior to the application of asphalt roofing materials on plywood or wooden planks, inspect the deck for delaminated areas on the plywood, warped boards, loose knots, improper nailing, and excessively resinous areas. Wood resin can chemically affect asphalt roofing materials.

The thickness, type, grade, and installation of other deck materials should conform to local building code regulations. Asphalt roofing materials can be applied to cement roof decks, but the manufacturers' recommendations for attachment should be researched and followed.

Selecting Nails

Use large-headed, sharp-pointed, hot-dipped, galvanized, or the equivalent, steel or aluminum nails with barbed or otherwise deformed shanks to apply asphalt shingles. The best roofing nails are made from 11- or 12-gauge wire, have heads that are $^{3}/_{8}$ or $^{7}/_{16}$ inch in diameter, and are long enough to penetrate the roofing materials. The nails should extend through plywood decks and at least $^{3}/_{4}$ inch into wooden plank decks.

TABLE 7-1. Recommended Nail Sizes

Purpose	Nail length (in.)
Roll roofing materials on new decks	1
Strip or individual shingles on new decks	$1\frac{1}{4}$
Reroofing over old asphalt roofing materials	$1\frac{1}{4}$–$1\frac{1}{2}$
Reroofing over old wooden shingles	$1\frac{3}{4}$
Reroofing over 300-pound or heavier asphalt shingles	$1\frac{1}{2}$–$1\frac{3}{4}$

If a fastener does not penetrate the deck properly, remove the fastener and repair the hole in the shingle with asphalt plastic cement or replace the entire shingle. Then place another fastener nearby. To satisfy most wooden decking requirements for asphalt shingles, use nails of appropriate lengths as indicated in Table 7-1.

Do not nail into or above factory-applied adhesives. Carefully align each shingle. Whenever possible, make sure that no cutout or end joint is less than 2 inches from a nail in an underlaying course. To prevent buckling, start nailing from the end nearest the shingle just laid and proceed across. To prevent distortion, do not attempt to realign a shingle by shifting the free end after two nails are in place.

Place the fasteners according to the shingle manufacturer's specifications. Align the shingles properly to avoid exposing fasteners in the course below. Drive the nails straight and stop them flush with the shingle surface. Do not sink the nail into or break the shingle's surface.

Using Staples Instead

Staples can be used in place of roofing nails on a one-for-one basis. If there is any doubt about the acceptability of staples in a particular application, consult the shingle manufacturer.

Roofing staples have a 1-inch crown, legs long enough to penetrate ¾ inch into sheathing, and chisel points. Pneumatic guns and special hammers can be adjusted for staples with leg lengths up to 1½ inches.

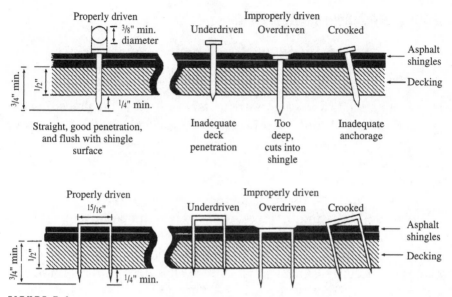

FIGURE 7-1 Good and bad nailing and stapling positions.

Some types of guns have a guide that helps the roofer apply shingles with uniform exposure.

Staples can be used if the roof slope is 4 in 12 or greater. At least six staples per strip are required. At the eaves, rakes, hips, and ridges, and where shingles touch flashing, use nails instead of staples because of their greater holding power. It also is important that the staples be driven parallel to the length of the shingle with properly adjusted pneumatic equipment in order to obtain the required penetration into the deck. Figure 7-1 illustrates examples of good and bad nailing and stapling.

Following Manufacturer's Instructions

Gypsum concrete plank and tile, fiber board, or similar nonwood deck materials require special fasteners and/or details to adequately anchor asphalt roofing materials. In such cases, it is recommended that the deck manufacturer's fastening specifications be followed in order to assure that the responsibility for the deck material's performance is placed upon the deck material manufacturer.

Specifying Underlayments

When a single-layer underlayment is required, apply one layer of No. 15 asphalt-saturated, nonperforated felt horizontally. Heavier underlayment is not necessary, but might be required by codes in some areas. Lap all felt sheets a minimum of 2 inches over the preceding felt sheet. Endlaps should be a minimum of 4 inches. Nail the felts under the lap only as necessary to hold the felts in place until the asphalt roofing material is applied. Laps can be sealed with plastic asphalt cement as required.

If specifications call for a double-layer underlayment, apply two layers of No. 15 (minimum) asphalt-saturated, nonperforated felt horizontally. Apply a 19-inch-wide starter sheet to the eaves. Cover the starter sheet with a full-width sheet. Lap succeeding sheets 19 inches over the preceding sheets, with a 17-inch exposure. Endlaps should be a minimum of 6 inches. Backnail the felts under the laps only as necessary to hold the felts in place until the asphalt roofing material is applied. Laps can be sealed with plastic asphalt cement as required.

In locations where the January mean temperature is 30°F or lower, apply two plies of No. 15 felt or one ply of No. 50 felt. Set this in hot asphalt or mastic, or an adhered bitumen membrane underlayment to roof decks with slopes less than 4 inches per foot, regardless of the slope. Work from the eaves to a point 24 inches inside the building's inside wall line to serve as an ice shield.

The limitations imposed by the slope, or pitch, of the roof deck are factors that must be considered when the roof is designed and underlayments specified.

For instance, self-sealing strip shingles, with tabs, can be applied on roof decks that have a slope of 4 inches per 12 inches or more if at least one layer of No. 15 asphalt-saturated, nonperforated felt is applied horizontally to serve as the underlayment.

Self-sealing strip shingles, with tabs, also can be applied to roof decks with a slope of 3 inches per 12 inches or more if at least two layers of No. 15 asphalt-saturated, nonperforated felt are applied horizontally to serve as the underlayment.

Roof decks with a slope of 2½ inches per 12 inches or more must have at least two layers of No. 15 asphalt-saturated felt set in hot asphalt or mastic to serve as the underlayment before self-sealing strip shingles, with tabs, can be installed.

Laminated asphalt shingles, individual lock-down shingles, and self-sealing strip shingles without tabs can be applied on roof decks with a slope of 4 inches per foot or more if at least one layer of No. 15 asphalt-saturated, nonperforated felt is applied horizontally to serve as the underlayment.

Fastening Asphalt Roofing Materials

When using roofing nails to fasten shingles, drive the nail shaft into the center of all four fastener locations. When using staples, drive the staple crown into the center of both inside locations. Apply the outermost staple leg on center at both end locations.

Attaching Three-Tab Strip Shingles

In areas with normal weather conditions, fasten each of these shingles, both the conventional 36-inch-long shingles and the metric shingles, with four fasteners. When the shingles are applied with a 5-inch exposure (5⅝ inches for metric), apply the fasteners on a line ⅝ inch above the top of the cutouts, 1 inch in from each end, and centered over each cutout.

For high-wind areas, use six fasteners instead of four. The fastener locations should be on a line ⅝ inch above the top of the cutouts, 1 inch in from each end, and 1 inch to the left and right of center of each cutout.

Attaching No-Cutout Strip Shingles

Fasten each of these shingles, both the conventional 36-inch-long shingles and the metric shingles, with four fasteners. When the shingles are applied with a 5-inch exposure (5⅝ inches for metric), the fastener locations should be on a line 5⅝ inches (6⅛ inches for metric) above the butt edge and located 1 inch and 12 inches (1 inch and 13 inches for metric) from each end.

For high-wind areas, fasten no-cutout strip shingles with six fasteners. When the shingles are applied with an exposure of 5 inches

(5⅝ inches for metric), the fastener locations should be on a line 5⅝ inches (6⅛ inches for metric) above the butt edge and located 1 inch, 11 inches, and 13 inches (1 inch, 12 inches, and 14 inches for metric) from each end.

Installing Asphalt Shingles

If a roof surface is broken by a dormer or valley, start applying the shingles from a rake edge of the roof and work toward the dormer or valley. If the roof surface is unbroken, start at the rake that is most visible. If both rakes are equally visible, start at the center and work both ways. On hip roofs, start at the center and work out toward the edges in both directions. No matter where the application begins, apply the shingles across and diagonally up the roof. This ensures that each shingle is fastened properly.

Straight-up application or *racking* can result in less than the recommended number of nails being used because of the manner in which the shingles are applied. Racking requires that part of the shingles in some courses be placed under those already applied in the course above. Because part of the shingle is hidden, it can be overlooked when the shingle is fastened. When tiles are installed diagonally, each shingle is completely visible until it is covered by the course above. Racking can also accentuate shading tendencies.

Applying the Starter Course

The starter course can be a 7-inch or wider strip of mineral-surfaced asphalt roll roofing material or a row of shingles trimmed to the manufacturer's specifications. The starter course or strip protects the roof by filling in the spaces under the cutouts and joints of the first course of shingles. It should overhang the eaves and rake edges by ¼ to ⅜ inch.

If self-sealing shingles are used for the starter strip, remove the tab portion of each shingle and position the remaining strip with the factory-applied adhesive face up along the eaves. Trim at least 3 inches from the end of the first shingle in the starter strip. This ensures that the cutouts of the first course of shingles are not placed over the starter strip joints. Fasten starter strips parallel to the eaves along a line 3 to 4

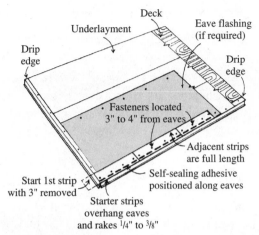

FIGURE 7-2 Starter course.

inches above the eaves. Position the fasteners so that they are not exposed under the cutouts in the first course (Fig. 7-2).

If shingles without a self-sealing adhesive are used for the starter strip, remove the tab portion of each shingle and position the remaining strip along the eaves. Complete the procedure by following the instructions above.

If roll roofing is used for the starter strip, nail along a line 3 to 4 inches above the eaves. Space the nails 12 inches apart. If more than one piece of roll roofing must be used, lap the end joint 2 inches and cement it.

Applying First and Successive Courses

The first course is the most crucial. Be sure it is laid perfectly straight. Check it regularly during application against a horizontal chalkline. A few vertical chalklines aligned with the ends of the shingles in the first course ensure the proper alignment of the cutouts.

If applying three-tab shingles or roll roofing for the starter strip, bond the tabs of each shingle in the first course to the starter strip by placing a spot of asphalt plastic cement about the size of a quarter on the starter strip beneath each tab. Then press the tabs firmly into the cement. Avoid excessive use of cement, as this can cause blistering.

Creating Three-Tab Strip Shingle Architectural Patterns

There are several different methods of applying three-tab strip shingles. The designs usually correspond to the amount removed from the first shingle in each successive course. Removing different amounts from the first shingle staggers the cutouts so that one course of shingles does not line up directly with those of the course below. Start first courses of shingles with full or cut shingles, depending on the pattern desired.

SIX-INCH PATTERN

This design is often called "the cutouts that break on joint halves." The six-inch pattern starts each succeeding course with shingles that are 6 inches smaller than the starter shingle in the preceding course. The first course starts with a full-length shingle. The second course starts with a shingle that has had 6 inches removed. The third course starts with a shingle that has had 12 inches removed, and so on through the sixth course, which starts with a shingle that has had 30 inches removed.

Adjacent shingles in each course are all full length. The seventh course again starts with a full-length shingle, and the pattern is repeated every six courses. Figure 7-3 illustrates the pattern.

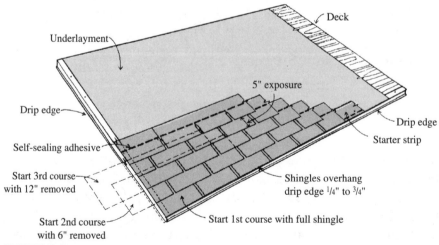

FIGURE 7-3 Six-inch pattern.

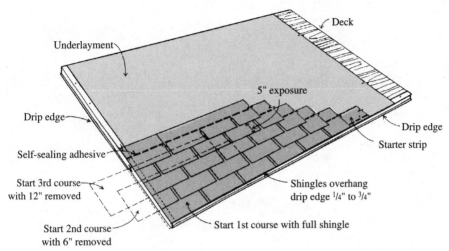

Deck

Underlayment

5" exposure

Drip edge

Drip edge

Starter strip

Self-sealing adhesive

Start 3rd course
with 12" removed

Shingles overhang
drip edge ¼" to ¾"

Start 2nd course
with 6" removed

Start 1st course with full shingle

FIGURE 7-4 Five-inch pattern.

FIVE-INCH PATTERN

This application is similar to the six-inch pattern, except that the second through seventh courses are each 5 inches smaller than the preceding one. The first course begins with a full shingle. The second course starts with 5 inches removed from the first shingle, and so on through the seventh course, which has 30 inches removed from the first shingle. Adjacent shingles in each course are full length. The pattern is repeated every seven courses. One alternative method calls for starting the eighth course with 11 inches removed from the first shingle (Fig. 7-4).

FOUR-INCH PATTERN

This one is often called "the cutouts that break joints on thirds." The four-inch pattern design is illustrated in Fig. 7-5. Start the first course with a full shingle. Start the second course with 4 inches removed from the first shingle, the third course with 8 inches removed, and so on through the ninth course, which has 32 inches removed from the first shingle. Adjacent shingles in each course are full length. The tenth course again begins with a full-length shingle and the pattern is repeated every nine courses.

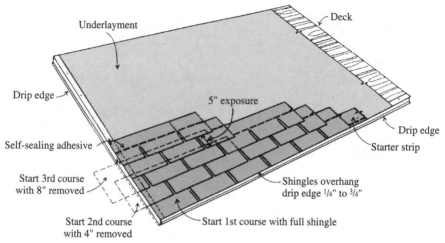

Underlayment

Deck

Drip edge

5" exposure

Self-sealing adhesive

Drip edge

Starter strip

Start 3rd course
with 8" removed

Shingles overhang
drip edge ¼" to ¾"

Start 2nd course
with 4" removed

Start 1st course with full shingle

FIGURE 7-5 Four-inch pattern.

Note: Never use an alignment system where shingle joints are closer than 4 inches to one another.

RANDOM SPACING

This pattern can be achieved by removing different amounts of material from the rake tab of succeeding courses in accordance with the following principles.

- The width of any rake tab should be at least 3 inches.

- The cutout centerlines of any course should be located at least 3 inches laterally from the cutout centerlines in both the courses above and the courses below.

- The rake tab widths should not repeat closely enough to cause the eye to follow a cutout alignment.

Starting the first course with a full-length strip indicates the length of the starting tab for each succeeding course. This procedure is necessary to produce satisfactory random spacing.

RIBBON COURSES

The use of a ribbon course every fifth course strengthens the horizontal roof lines and adds a distinctive, massive appearance that many homeowners find desirable. A preferred method of ribbon course application involves a special starting procedure that is repeated every fifth course.

- From the tip of a 12-inch-wide strip shingle, cut away 4 inches. This provides an unbroken strip 4 × 36 inches in size and a strip 8 × 36 inches in size that contains the cutouts.
- Lay the 4-×-36-inch strip along the eaves.
- Cover this strip with the 8-×-36-inch strip, with the bottom of the cutouts laid down to the eaves.
- Lay this first course of full shingles (12 × 36 inches) over the other layers (see Fig. 7-6) with the bottom of the cutouts laid down to the eaves.

Note: Offset cutouts between the first two layers as explained for six-inch, four-inch, or random spacing patterns.

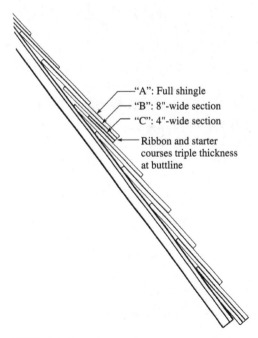

"A": Full shingle
"B": 8"-wide section
"C": 4"-wide section
Ribbon and starter courses triple thickness at buttline

FIGURE 7-6 Ribbon courses.

Repeat this procedure every fifth course, or after a predetermined number of courses, by aligning the ribbon course with the tops of the cutouts of the fifth course. This method produces a triple-thickness buttline every fifth course.

Note: Very steep slopes tend to reduce the effectiveness of factory-applied, self-sealing strip adhesives, especially on the colder or shaded areas of the roof.

Applying Flashing

Flashing controls moisture entry at ridges; hips; and valleys; around pipes that protrude through the roof's surface; at chimneys and dormers; and wherever a roof butts against a wall. Careful attention to flashing is essential to good roof performance, regardless of the type of roof construction or its cost.

Several materials can be used for flashing. Roll roofing is the most widely used, and is generally acceptable at ridges, hips, and valleys. The best flashing for any purpose, however, is corrosion-resistant metal. The metal can be 0.019-inch-thick aluminum, 26-gauge galvanized steel, or 16-ounce copper. Vinyl flashing and plastic sleeves around pipes also are used. Make upkeep a part of your maintenance business.

When using metal flashing, remember that metals react to one another in the presence of water. This reaction, called galvanic action, causes one of the metals to corrode. Metals are ranked in the order in which they react to other metals when they are in direct contact. This ranking is called the electromotive series. Here is the ranking of metals most often used in construction.

1. Aluminum
2. Zinc
3. Steel
4. Tin
5. Lead
6. Brass
7. Copper
8. Bronze

If aluminum flashing is applied with steel nails, the aluminum corrodes because it is higher than steel in the electromotive series. If copper flashing is applied with steel nails, the steel corrodes. The farther apart the metals are in the series, the sooner and faster corrosion occurs. The simplest way to avoid galvanic action is to attach metals with fasteners of the same metal. One alternative is to use a nonmetallic washer, such as neoprene or asphalt, which keeps the metals from touching each other. However, the washers can degrade and must be maintained.

Building Valleys

Construct valley flashings only after the deck has been properly prepared by applying a saturated felt underlayment, as shown in Fig. 7-7. Regardless of the type used, the valley must be smooth, unobstructed, of sufficient capacity to carry water away rapidly, and capable of withstanding occasional water backup.

The construction procedures for building different types of valleys with various materials follow. As might be expected, several steps are

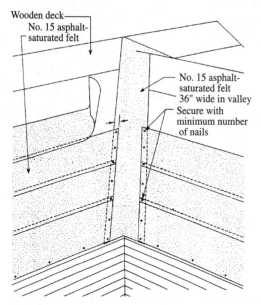

FIGURE 7-7 Valley flashing; felt underlayment.

the same for each type of valley. Unless indicated otherwise, follow these general rules for each application method.

- Use only enough nails to hold strips of felt or material smoothly in place.
- Nail the material along one side of the valley first.
- Place nails in a row 1 inch from the edge of the strip.
- As the nailing proceeds along the second side, press the roofing material firmly in place in the valley.

Before the roofing material is applied, snap two chalklines along the full length of the valley. Place one chalkline on each side, so that the chalklines are 6 inches apart at the ridge, i.e., 3 inches from the center of the valley along each of the intersecting roofs.

The chalklines should diverge at the rate of ⅛ inch per foot as they approach the eaves. For example, a valley 8 feet long is 7 inches wide at the eaves, and a valley 16 feet long is 8 inches wide at the eaves.

The chalklines serve as a guide by which to trim the last unit in order to ensure a clean, sharp edge.

OPEN VALLEY USING ROLL ROOFING MATERIAL

The application of mineral-surfaced asphalt roll roofing material, colored to match or contrast with the roof covering, is satisfactory for open-valley flashing. This type of application is illustrated in Fig. 7-8.

Center an 18-inch-wide layer of mineral-surfaced asphalt roll roofing material in the valley with the mineral-surfaced side down. Cut the lower edge to conform to and be flush with the eave flashing strip. When the flashing material must be lapped, lap the ends of the upper piece over the lower piece by at least 12 inches and secure it with asphalt plastic cement.

On top of the first strip of material, center another strip of 36-inch-wide material with the mineral-surfaced side up. Secure this strip with nails in the same manner as the underlying 18-inch strip of material. Follow the general instructions for holding the material in place.

Align the material along the chalklines and clip the upper corner of each end single to prevent water from penetrating under the courses.

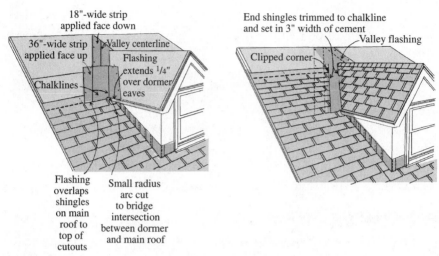

FIGURE 7-8 Open valley using mineral-surfaced asphalt roll roofing.

Cement the roofing material to the valley lining with a spot of plastic asphalt cement.

OPEN VALLEY USING METAL FLASHING

Open valleys can be formed by laying strips of at least 26-gauge galvanized metal, or an equivalent noncorrosive, nonstaining material, in the valley angle and lapping the roofing material over the metal on either side. This leaves a space between the edges of the roofing material to channel water down the valley.

Extend valley flashing metal at least 9 inches from the centerline of the valley in each direction at the ridge and make it wide enough at the eaves to extend at least 4 inches under the roofing material. Form a splash diverter rib not less than 1 inch high at the flowline from the metal. Flashing metal pieces should not be longer than 10 feet. Install metal flashing at open-valley locations as follows.

Center a 36-inch-wide layer of No. 15 asphalt-saturated, nonperforated felt in the valley over the previously applied underlayment. Cut the lower edge to conform to and be flush with the eaves flashing strip. When this sheet must be lapped, lap the ends of the upper piece over the lower piece by at least 12 inches and secure it with plastic asphalt cement. Follow the general directions for holding the materials in place.

Next, install the valley metal. Cut the lower edge of the first piece to conform to and be flush with the eaves flashing strip. Lap succeeding pieces over underlying pieces by at least 6 inches. Secure the metal flashing to the roof deck with metal cleats spaced from 8 to 12 inches apart. Use the chalklines to trim the last unit in order to ensure a clean sharp edge.

Do not puncture the flashing metal with the roofing nails. Instead, clip the upper corner of each end shingle to prevent water from penetrating under the courses and cement the roofing material to the valley metal with a spot of plastic asphalt cement or mastic.

OPEN VALLEY FOR DORMER ROOF

A special flashing procedure is recommended when an open valley occurs at a joint between a dormer roof and the main roof from which it projects.

After the underlayment is installed, apply the main roof shingles to a point just above the lower end of the valley. Fit the course last applied close to, and flashed against, the wall of the dormer under the projecting edge of the dormer eaves.

Apply the first strip of valley lining in the manner described for open valley flashing. Cut the bottom end of this first strip so that it extends ¼ inch below the edge of the dormer deck, with the lower edge of the section that lies on the main deck projecting at least 2 inches below the point where the two roofs join.

Then cut the second, or upper, strip on the dormer side to match the lower end of the underlying strip. Cut the side that lies on the main deck to overlap the nearest course of shingles in an identical manner to the other courses. Shape the lower end of the lining to form a small canopy over the joint between the two decks.

Apply the following courses of shingles over the valley lining, with the end shingles in each course cut to conform to the guidelines. Embed these courses of shingles into a 3-inch strip of plastic cement. Complete the valley in the usual manner.

After the shingles are applied to both sides of the dormer, apply the dormer ridge shingles. Start at the front of the dormer and work toward the main roof. Apply the last ridge shingle so that it extends at least 4 inches onto the main roof.

Slit the center of the portion attached to the main roof and nail it into place. Then apply the main roof courses to cover the portion of the last ridge shingle on the main roof. Snap a chalkline so that the shingles on the main roof continue the same alignment pattern on both sides of the dormer.

WOVEN VALLEY

The woven valley treatment is limited to strip-type shingles. Individual shingles cannot be used because nails might be required at or near the center of the valley lining.

In order to avoid placing a nail in an overlapped shingle too close to the center of the valley, it might be necessary to cut short a strip that would otherwise end near the center of the valley and then continue from this cut end over the valley with a full-length strip. This method has the advantage of increasing the coverage of the shingles throughout the length of the valley. This adds to the weather resistance of the roof in an area that might otherwise be vulnerable.

To install flashing at woven valley locations, center a 36-inch-wide strip of No. 15 asphalt-saturated, nonperforated felt in the valley over the existing No. 15 asphalt-saturated felt underlayment. Use only enough nails to hold the sheet smoothly in place.

Lay the first course of material along the eaves of one roof area up to and over the valley. Extend it along the adjoining roof area for at least 12 inches. Then lay the first course along the eaves of the intersecting roof area and extend it over the valley on top of the previously applied shingle.

Lay successive courses alternately, first along one roof area and then along the other, weaving the valley shingles over each other, as shown in Fig. 7-9. Tightly press the shingles into the valley and nail in place. Locate no nails closer than 6 inches to the valley's centerline. Locate two nails at the end of each terminal strip.

Woven valley shingles also can be laid over the lining by first covering each roof area to a point approximately 3 feet from the center of the valley and weaving the valley shingles in place later.

CLOSED-CUT VALLEY

This design can be used only with strip-type shingles and roll roofing materials. To install flashings at closed-cut valley locations,

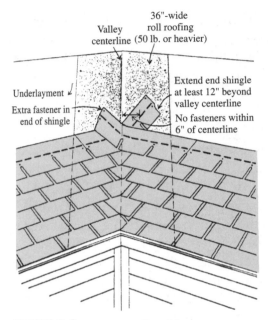

FIGURE 7-9 Weaving valley shingles.

center a 36-inch-wide strip of No. 15 asphalt-saturated, nonperforated felt in the valley over the existing No. 15 asphalt-saturated felt underlayment. Use only enough nails to hold the sheet smoothly in place.

Lay the first course of shingles along the eaves of one roof area up to and over the valley. Extend it along the adjoining roof area for at least 12 inches. Continue to apply the second and successive courses over the valley in the same manner as the first course. Press the shingles into the valley and nail in place. Locate no nail closer than 6 inches to the valley's centerline. Locate two nails at the end of each terminal sheet (Fig. 7-10).

Apply the first course of shingles along the eaves of the intersecting roof area and extend it over the previously applied shingles. Trim a minimum of 2 inches back from the centerline of the valley. Clip the upper corner of each end shingle to prevent water from penetrating under the courses, and then embed it in a 3-inch-wide strip of plastic asphalt cement.

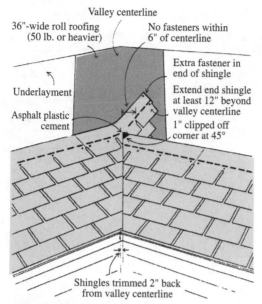

Valley centerline

36"-wide roll roofing
(50 lb. or heavier)

No fasteners within
6" of centerline

Underlayment

Asphalt plastic
cement

Extra fastener in
end of shingle

Extend end shingle
at least 12" beyond
valley centerline

1" clipped off
corner at 45°

Shingles trimmed 2" back
from valley centerline

FIGURE 7-10 Closed-cut valley; two nails.

Flashing Other Roof Structures

Place roof flashing around chimneys, soil stacks, dormers, or other structures that project above the roof surface. Use individual flashings, one for each course of shingles, along the sides that are perpendicular to the shingle courses. Bend each at a right angle so that part of the flashing lies flat on the roof on top of the shingle, while the other half rests against the side of the chimney or dormer. Embed the upper edge in a mortar joint or cover with the dormer finish. Flash and counter-flash the upper and lower sides, with one flashing running under the shingles and the other lying on top of them.

FLASHING AGAINST SIDE VERTICAL WALLS

Roof planes that butt against vertical walls at the end of shingle courses are best protected by metal flashing shingles placed over the end of each course. The method is called step flashing.

The metal flashing shingles are rectangular in shape, from 5 to 6 inches long, and wider than the exposed surface of the roofing

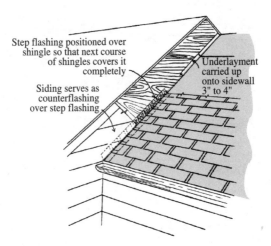

Step flashing positioned over
shingle so that next course
of shingles covers it
completely

Underlayment
carried up
onto sidewall
3" to 4"

Siding serves as
counterflashing
over step flashing

FIGURE 7-11 Flashing against side vertical walls.

shingles. With strip shingles laid 5 inches to the weather, use flashing shingles that are 6 × 7 inches in size.

Bend the flashing shingle so that it extends 2 inches out over the roof deck. Extend the remainder of the shingle up the wall surface. Place each flashing shingle just above the exposed edge of the shingle that overlaps it, and secure it to the wall sheathing with one nail in the top corner, as shown in Fig. 7-11. Because the metal is 7 inches wide and the roof shingles are laid 5 inches to the weather, each flashing shingle overlaps the next by 2 inches.

Bring the finish siding down over the flashing to serve as cap flashing, but hold it far enough away from the shingles so that the ends of the boards can be readily painted to keep the roofing materials from becoming damp and deteriorating.

FLASHING AGAINST VERTICAL FRONT WALLS

Apply the shingles up the roof until a course must be trimmed to fit at the base of the vertical wall. Plan ahead and adjust the exposure slightly in the preceding two courses so that the last course is at least 8 inches wide. Apply a continuous piece of metal flashing over the last course of shingles by embedding it in asphalt plastic cement and nailing it to the roof.

Bend the metal flashing strip so that it extends at least 5 inches up the vertical wall and at least 4 inches onto the last shingle course. Do not nail the strip to the wall. Apply an additional row of shingles over the metal flashing strip, and trim it to the width of the strip. Then bring the siding down over the vertical flashing to serve as cap flashing. Place wooden siding far enough away from the roof shingles to allow for painting. Do not nail siding into the vertical flashing.

If the vertical front wall meets a sidewall, as in dormer construction, cut the flashing so that it extends at least 7 inches around the corner. Then continue up the sidewall with the step flashing.

FLASHING AGAINST CHIMNEYS

To avoid stresses and distortions due to the uneven settling of roofing materials, the chimney usually sits on a separate foundation from the building and is normally subject to some differential settling. Therefore, flashing at the point where the chimney projects through the roof requires a type of construction that allows for movement without damaging the water seal. To satisfy this requirement, use base flashings that are secured to the roof deck and counterflashings that are secured to the masonry (Fig. 7-12).

Before any flashings are installed, apply shingles over the roofing felt up to the front face of the chimney and construct a cricket or saddle between the back face of the chimney and the roof deck. Design the cricket to prevent the accumulation of snow and ice and to deflect water around the chimney (Fig. 7-13).

Begin the flashing construction by installing at least 26-gauge galvanized metal, or an equivalent noncorrosive, nonstaining material, between the chimney and the roof deck on all sides. Apply the base flashing to the front first. Bend the base flashing so that the lower section extends at least 4 inches over the shingles and the upper section extends at least 12 inches up the face of the chimney. Set both sections of the base flashing in plastic asphalt cement.

Use metal step flashing for the sides of the chimney. Position the pieces in the same manner as flashing for a vertical sidewall. Secure each piece to the masonry with plastic asphalt cement and use nails to secure it to the roof deck. Embed the overlapping shingles in plastic asphalt cement.

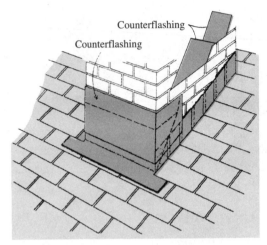

FIGURE 7-12 Flashing against chimney.

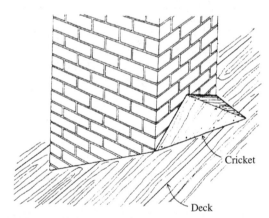

FIGURE 7-13 Location and configuration of chimney cricket.

Place the rear base flashing over the cricket and the back of the chimney. The metal base flashing should cover the cricket and extend onto the roof deck at least 6 inches. It should also extend 6 inches up the brickwork. Bring the asphalt shingles up to or over the cricket and cement in place.

Place counterflashings over all apron, cricket, and step flashings to keep water out of the joint. Begin by setting the metal counterflashing

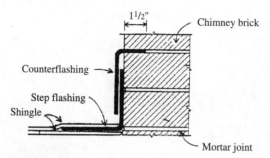

FIGURE 7-14 Metal counterflashing into brick-work.

into the brickwork as shown in Fig. 7-14. This is done by raking out the mortar joint to a depth of 1½ inches and inserting the bent edge of the flashing into the cleared joint. Once it is in place and has a slight amount of spring tension, the flashing cannot be dislodged easily. Refill the joint with portland cement mortar. Finally, bend the counter flashing down to cover the step flashing and to lie snugly against the masonry.

FLASHING SOIL STACKS AND VENTS

Practically all dwellings have vent pipes or ventilators that project though the roof. These circular protrusions require special flashing methods.

Apply shingles up to the vent pipe. Then cut a hole in a shingle to go over the pipe and set the shingle in asphalt plastic roofing cement (Fig. 7-15A). Place a preformed flashing flange that fits snugly over the pipe over the shingle and vent pipe, and set it in asphalt plastic cement. Place the flange over the pipe to lie flat on the roof (Fig. 7-15B).

After the flashing is in place, resume shingle application. Cut shingles in successive courses to fit around the pipe and embed them in asphalt plastic roofing cement where they overlay the flange. Avoid excessive use of cement as it can cause blistering. Do not drive fasteners close to the pipe. The completed installation should appear as shown in Fig. 7-15C, with the lower part of the flange overlapping the lower shingles and the side and upper shingles overlapping the flange.

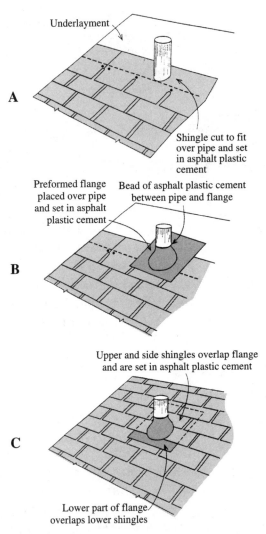

Underlayment

A

Shingle cut to fit
over pipe and set
in asphalt plastic
cement

Preformed flange
placed over pipe
and set in asphalt
plastic cement

Bead of asphalt plastic cement
between pipe and flange

B

Upper and side shingles overlap flange
and are set in asphalt plastic cement

C

Lower part of flange
overlaps lower shingles

FIGURE 7-15 (*A*) Flashing soil stacks and vents;
(*B*) flange over pipe; (*C*) complete installation.

Follow the same procedure when a ventilator or exhaust stack is located at the ridge, with one difference. Bring the shingles up to the pipe from both sides. Bend the flange over the ridge so that it lies in both roof planes and overlaps the roof shingles at all points. Position the ridge shingles to cover the flange. Embed the ridge shingles in asphalt plastic cement where they overlap the flange.

EAVE FLASHING OR ICE DAMS

Whenever there is the possibility of icing along the eaves, cement all laps in the underlayment courses from the eaves to a point at least 24 inches beyond the interior wall line of the building. The cemented double-ply underlayment serves as the eave flashing (Fig. 7-16).

To construct the eave flashing, cover the entire surface of the starter strip with a continuous layer of asphalt plastic cement applied at the rate of 2 gallons per 100 square feet. As an alternative, lap cement can be used at the rate of 1 gallon per 100 square feet. Place the first course over the starter course, and press it firmly into the cement. After the first course is in place, coat the upper 19 inches with cement.

Position the second course and press it into the cement. Repeat the procedure for each course that lies within the eave flashing distance. It is important to apply the cement uniformly so that the overlapping felt floats completely on the cement without touching the felt in the underlying course. Avoid excessive use of cement; it can cause blistering.

After completing the eave flashing, secure each successive course by using only enough fasteners to hold it in place until the shingles are applied.

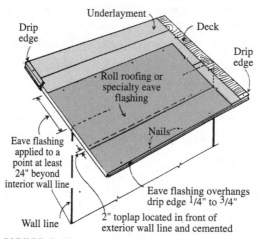

FIGURE 7-16 Eave flashing.

DRIP EDGES

Drip edges provide efficient watershedding at the eaves and rakes and keep the underlying wood or plywood deck from rotting. Use 28-gauge galvanized metal (at a minimum) or an equivalent noncorrosive, non-staining material to make drip edges along eaves and rakes and at the edges of the deck. Provide an underlayment between the metal edge and the roof deck along the rake and over the metal edge along the eave.

Extend the drip edge back from the edge of the deck not more than 3 inches and secure with appropriate nails spaced 8 to 10 inches apart along the inner edge. In high-wind areas, space nails 4 inches on center.

Remember that galvanized and most other flashings do deteriorate with time. Valleys and gutters, even on copper, can fill up with debris. Maintenance is very important on all roofs. Sell this service to clients rather than battling lawsuits.

Finishing Hips and Ridges

For hips and ridges, use individual shingles cut down to 12 × 12 inches from 12-×-36-inch three-tab shingles or to a minimum of 9 × 12 inches on two-tab or no-cutout asphalt shingles. Taper the lap portion of each cap shingle slightly so that it is narrower than the exposed portion. Some shingle manufacturers supply ready-cut hip and ridge shingles and specify how they should be applied.

To apply the cut ridge shingles, bend each shingle along its center-line so that it extends an equal distance on each side of the hip or ridge. Chalklines can assist in proper alignment. In cold weather, warm the shingle until it is pliable before bending. Apply the shingles with a 5-inch exposure, beginning at the bottom of the hip or from the end of the ridge opposite the direction of the prevailing winds.

Secure each shingle as illustrated in Fig. 7-17, with one fastener on each side, placed 5½ inches back from the exposed end and 1 inch up from the edge. The fastener length for hip and ridge shingles should be ¼ inch longer than that recommended for shingles.

Applying Strip Shingles on Low Slopes

Asphalt strip shingles can be used on slopes between 2 and 4 inches per foot if special procedures are followed. Never use shingles on

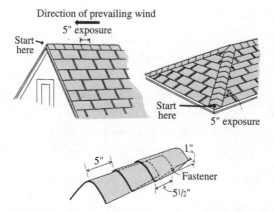

FIGURE 7-17 Secured shingle (special drawing of hip and ridge).

slopes lower than 2 inches per 12 inches. Water drains slowly from these slopes and creates a greater possibility of water backup and damage from ice dams. The special application method described here ensures that the low-slope roof remains weathertight.

UNDERLAYMENT

On low-slope applications, cover the deck with two layers of asphalt-saturated felt. Fasten a 19-inch-wide strip of underlayment and place it along the eaves so that it overhangs the drip edge by ¼ to ⅜ inch. Place a full 36-inch-wide sheet over the starter strip with the long edge placed along the eaves and completely overlapping the starter strip.

Position succeeding courses of 36-inch-wide sheets so that they overlap the preceding course by 19 inches. Secure each course using only enough fasteners to hold it in place until the shingles are applied. Endlaps should be 12 inches wide and located at least 6 feet from the endlaps in the preceding course.

EAVE FLASHING

Follow the instructions for applying eave flashing in the Applying Flashing section in this chapter. Special eaves flashing membranes also are available. If one of these specialty products is used, follow the manufacturer's application instructions with care.

SHINGLE APPLICATION

For increased wind resistance, either use self-sealing shingles that incorporate a factory-applied adhesive or cement the tabs of free-tab shingles to the underlying course. To cement a free-tab shingle to the underlying course, you place two spots of asphalt plastic cement about the size of a quarter under each tab. Then press the tab into the adhesive. Be sure to cement all tabs throughout the roof. Any of the shingle application methods discussed earlier in this chapter can be used on low slopes.

Reroofing with Strip Shingles

Many of the procedures used to apply shingles on new roofs also are followed in reroofing. Other procedures, specifically designed for reroofing applications, have been developed to meet the particular requirements of this type of work. The differences depend primarily on the type of material that is on the existing roof, its condition, and whether or not the new roofing can be placed over it.

Before reroofing, inspect the condition of the old roof and its supports. If the roof deck or its supports are found to be warped, rotted, or otherwise unsound, remove the old roofing and repair or replace the roof structure. If necessary, repoint the chimney and realign, clean, rebuild, or replace damaged gutters. Install flashing in valleys wherever needed and check that ventilation under the roof is adequate. Apply metal drip edges along the eaves and rakes and a flashing strip along the eaves.

In general, roofs covered with wooden shingles, asphalt shingles, asphalt roll roofing, slate, or built-up asphalt (BUR) roofing can be left in place. Depending on local codes, a maximum of three roofs, an original and two reroofs, can be installed before a tear-off becomes mandatory. Tile or cedar shake roofs must be removed because of their irregular surface and the difficulty in fastening through them.

The main consideration, however, is the condition of the framing beneath the existing roof, because new material installed on top of the old adds weight to the structure. The roof framing must be strong enough to support these additional loads, plus the weight of the roofers and their equipment.

Another important consideration is the deck material to which the new material is attached. If the deck has deteriorated to the point where it does not provide adequate anchorage for new roofing fasteners, replace it. Also, the roofing nails used to apply asphalt shingles over old roofs must be long enough to penetrate through the old layers of roofing and an additional ¾ inch into the wood of the roof decking below. They are longer than nails used in new construction or when the old roofing has been removed.

Removing Old Roofing

If existing roofing must be removed, strip the materials down to the roof deck. If the deck under wooden shingles or shakes consists of spaced sheathing, start removing the old shingles at the ridge so that broken material does not fall through the open sheathing into the attic where it can create a fire hazard and cleaning problems.

If metal cap flashings at the chimney and other vertical masonry wall intersections have not deteriorated, bend them up out of the way so that they can be used again. Carefully remove shingles in these areas to avoid damaging reusable flashing. For safety, keep the deck clear of waste material as the work progresses. Sweep the deck clean after all old roofing has been removed.

At this point, inspect the deck to determine whether it is sound. Then, as a result of this inspection, prepare the deck as follows. (See also Fig. 7-18.)

- Make repairs to the existing roof framing where required to level and true it to provide adequate strength.

- Remove all rotted or warped sheathing and delaminated plywood. Replace with new sheathing.

- Fill in all the spaces between the boards with securely nailed wooden strips of the same thickness as the old deck or remove existing sheathing and resheath the deck.

- Pull out all protruding nails and renail the sheathing firmly at the new nail locations.

- If the deck consists of spaced sheathing, fill in all spaces with new boards of the same thickness as the old deck or cover the

✓	Item
	Make repairs to the existing roof framing where required to level and true it to provide adequate strength.
	Remove all rotted or warped sheathing and delaminated plywood. Replace with new sheathing.
	Fill in all the spaces between the boards with securely nailed wooden strips of the same thickness as the old deck or remove existing sheathing and resheath the deck.
	Pull out all protruding nails and renail the sheathing firmly at the new nail locations.
	If the deck consists of spaced sheathing, fill in all spaces with new boards of the same thickness as the old deck or cover the entire area with decking of the type and thickness required by local codes or the shingle manufacturer.
	Cover all large cracks, slivers, knotholes, loose knots, pitchy knots, and excessively resinous areas with sheet metal securely nailed to the sheathing.

FIGURE 7-18 Removing old roofing.

entire area with decking of the type and thickness required by local codes or the shingle manufacturer.

■ Cover all large cracks, slivers, knotholes, loose knots, pitchy knots, and excessively resinous areas with sheet metal securely nailed to the sheathing.

Sweep the deck clean again. Then follow the underlayment application directions described earlier in this chapter for new construction.

Reroofing Operations

Preparatory procedures depend on the type of existing roofing. Five situations are generally encountered: asphalt shingles over old asphalt shingles, over wooden shingles, over roll roofing, over slate, and over BUR. Some shingles have a greater tendency to reveal the unevenness of the surface over which they are applied. This phenomenon is known as *telegraphing*, and can influence the choice of roofing material. If telegraphing might be a problem, consult individual manufacturers for advice before proceeding with the installation.

Take measurements at both rakes. It is not unusual to find a difference of several inches between rakes on the same roof. Knowing this before you apply new shingles enables you to compensate for the difference in small increments over a series of courses.

If the asphalt shingles with a 5-inch exposure are to be recovered with metric shingles, either a *bridging* or a *nesting* procedure is acceptable. Either provides good results when properly applied. The bridging method, shown in Fig. 7-19A, requires fewer shingles than the nesting procedure, illustrated in Fig. 7-19B.

Telegraphing of the underlaying material can result, however, when the bridging method is used. Nesting, which minimizes an uneven appearance, is recommended when applying metric size shingles over an existing roof (Fig. 7-19C). For specific recommendations about bridging or nesting, contact the shingle manufacturer.

REROOFING OVER OLD ASPHALT SHINGLES

If the old asphalt shingles are to remain in place, remove any curled or lifted shingles and all loose and protruding nails. Remove all badly

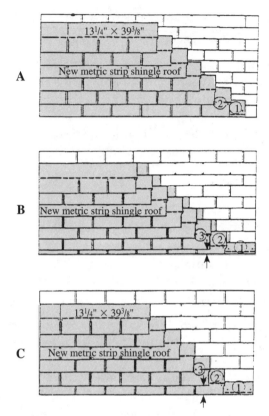

FIGURE 7-19 (*A*) Bridging method; (*B*) nesting procedure; (*C*) nesting metric size shingles.

worn edging strips and replace with the new ones. Just before applying new roofing materials, sweep the old surface clear of all loose debris. If the old asphalt shingles are square-butt, and they are to be covered with new self-sealing, square-butt strip shingles, follow the application procedures in this section for applying shingles over an old roof.

If the old asphalt shingles are locked down, stapled down, or badly curled strip shingles, the size, shape, and unevenness of the shingles can produce an uneven roof surface. New shingles tend to conform to the uneven surface. If a smoother surface is desired, remove the old shingles and prepare the deck using double underlayment.

REROOFING OVER OLD WOODEN SHINGLES

If an inspection of the roof indicates that the old wooden shingles can remain, carefully prepare the surface of the roof to receive the new roofing.

- Remove all loose or protruding nails, and renail the shingles in a new location.
- Nail down all loose shingles.
- Split all badly curled or warped old shingles and nail down the segments.
- Replace missing shingles with new ones.

When the roof is subject to unusually high winds, cut back the shingles at the eaves and rakes far enough to allow the application at these points of 1-inch-thick wooden strips that are 4 to 6 inches wide. Nail the strips firmly in place and allow their outside edges to project beyond the edge of the deck the same distance as did the wood shingles.

To provide a smooth deck, apply a backer board over the wooden shingles or use beveled wooden feathering strips along the butts of each course of old shingles.

REROOFING OVER OLD ROLL ROOFING

When asphalt shingles are laid over old roll roofing, prepare the deck as follows.

- Slit all buckles and nail segments down smoothly.
- Remove all loose and protruding nails.
- If some of the old roofing has been torn away and sections of the deck are exposed, examine these areas to note any loose or pitchy knots and excessively resinous areas.
- Sweep the deck clean before applying new shingles.

REROOFING OVER OLD BUR

BUR on a slope between 2 and 4 inches per 12 inches can be reroofed with asphalt shingles provided no insulation exists between the deck and the felts. Remove any old slag, gravel, or other material and leave the surface of the underlying felts smooth and clean. If a smooth, clean surface cannot be obtained, remove all of the old roofing and follow the procedures for reroofing when existing roofing is removed. Apply the asphalt shingles directly over the felts in the same manner as that for new construction on low slopes.

If rigid insulation exists under the BUR, remove the BUR and the insulation to expose the underlying deck. Then follow the low-slope application procedures outlined earlier in this chapter.

REROOFING OVER OLD SLATE ROOFS

It is generally recommended that old slate roofs be removed before new asphalt shingles are applied, although it is possible to apply asphalt shingles successfully over this type of substrate by following the general recommendations described below and those suggested by the individual shingle manufacturer. The owner's architect must be consulted.

The old roof slates must not be thicker than 3/16 inch. The roof deck must be solidly sheathed and in good condition. The slates must not shatter when nails are driven into them, regardless of placement. After years of exposure, most slates become quite soft, and needle-point nails can be used with very little breakage or spalling.

Begin by nailing down or removing any loose or projecting slates to provide a smooth surface for the new roofing material. Nails must be long enough to pass through the slates and penetrate the roof deck.

Note: Failure to follow these general application recommendations, especially those of the shingle manufacturer, can result in a poor roofing installation. Refer to local building codes for applicability and requirements regarding asphalt shingle applications over old slate roofs.

Laying Asphalt Shingles

Shingle application depends on whether the existing roofing has been removed or left in place and on the type of roofing material left in place. If the old roofing has been removed or if reroofing is over old roll roofing or BUR, the shingle-application procedures are the same as those for new construction. Wooden shingle roofs that are modified with feathering strips also are shingled in the same manner as new construction.

The situation that differs from new construction is the application of new strip shingles over existing strip shingles. The nesting procedure, described earlier, minimizes any unevenness that might result from the shingles bridging over the butts of the old shingles. It also ensures that the new horizontal fastening pattern is 2 inches below the old one. The nesting procedure is based on the assumptions that the new roofing has a 5-inch exposure and the existing roofing was installed with a 5-inch exposure and properly aligned.

If new eaves flashing has been added, snap chalklines on it to guide the installation of the new shingles until the courses butt against the existing courses.

For the starter course, remove the tabs plus 2 inches or more from the top of the starter strip shingles so that the remaining portion is equal in width to the exposure of the old shingles, which is normally 5 inches. Apply the starter strip so that it is even with the existing roof at the eave. If self-sealing shingles are used for the strip, locate the factory-applied adhesive along the eaves.

Be sure the existing shingles overhang the eaves far enough to carry water into the gutter. If they do not, cut the starter strip to a width that will do so. Do not overlap the existing course above. Remove 3 inches from the rake end of the first starter strip shingle to ensure that joints between adjacent starter strip shingles are covered when the first course is applied.

For the first course, cut 2 inches or more from the butts of the first course of shingles so that the shingles fit between the butts of the existing third course and the eave edge of the new starter strip. Start at the rake with a full-length shingle. Use four fasteners per shingle, locating them in the same positions as in new construction. Do not fasten into or above the factory-applied adhesive.

Use full-width shingles for the second and all successive courses. Remove 6 inches from the rake end of the first shingle in each progressive course, through the sixth. Repeat the cycle by starting the seventh course with a full length shingle. Place the top edge of the new shingles against the butt edge of the old shingles in the course above.

Using the full width on the second course reduces the exposure of the first course to 3 inches, but this area is usually concealed by gutters and the appearance should not be objectionable. For the remaining courses, the 5-inch exposure is automatic and coincides with that of the existing shingles. As in new construction, apply the shingles across the diagonal as you proceed up the roof.

Flashing and Reroofing

Once the old roofing has been removed, flashing details generally follow those for new construction. If the existing flashings are still serviceable, however, they can be left in place and reused. If the old roofing is left in place, some flashing application procedures can differ from those used in new construction.

VERTICAL WALLS

If existing flashings are in good condition, reroof as follows. At a sidewall, trim ends of new shingles to within ¼ inch of the wall. Embed 3 inches of each shingle adjacent to the wall in plastic asphalt roofing cement. Finally, with a caulking gun, apply a thin line of roofing cement between the cut edge of the shingle and the wall.

At a frontwall, remove nailed tabs adjacent to the wall. Embed them in roofing cement and nail on the butt edge opposite the wall. Be careful not to nail within 3 inches of the vertical wall. Finish with a thin line of roofing cement between the cut edge of the shingle and the wall.

VALLEY FLASHING

If the existing roof has an open valley, build up the exposed area of the valley with a mineral-surfaced roll roofing to a level flush with the existing roofing. Then install new open-valley flashing, in the same manner as for new construction, and overlap the existing shingles. The preferred treatment, however, is to construct a woven or closed-cut valley with the new shingles crossing over the valley filler strip.

VENT PIPES

Carefully examine old metal flashing around these protrusions. If it has deteriorated, remove it and apply new flashing. If the metal flashing is in good condition, proceed as follows.

- Lift the lower part of the flange and apply the shingles underneath it up to the pipe.

- Replace the flange and embed it in asphalt roof cement.

- Protect the junction of the metal sleeve and the flange with an application of roof cement.

- Apply shingles around the pipe and up the roof.

DORMERS

A gable-type dormer is reroofed by using vertical sidewall and front-wall methods, where applicable, and the closed-cut valley technique at the valleys. The roof of a shed dormer is shingled as a continuation of the main roof slope.

CHIMNEYS

Reuse serviceable metal chimney flashing, but, if the old flashing has deteriorated, remove and replace it. To do this, apply a strip of roll roofing approximately 8 inches wide on the reconditioned roof surface at the front and sides of the chimney. Lay it so that it abuts the chimney on all sides and secure it to the old roof with a row of nails along each edge.

At the junction where it meets the chimney, apply a heavy coating of plastic asphalt cement to each course of shingles. Hold the lower edge slightly back from the exposed edge of the covering shingle, bend it up against the masonry, and secure it with a suitable asphalt cement. Drive nails through the lower edge of the flashing into the roof deck. Cover these nails with the plastic cement used to secure the end shingle to the horizontal portion of the flashing and also the shingle itself. Repeat the operation for each course.

The flashing units are wide enough to lap each other at least 3 inches, the upper one overlaying the lower one each time. Clean the masonry with a wire brush for a distance of 6 or 8 inches above the deck, and apply a suitable asphalt primer to the masonry surface.

After the primer is applied, trowel plastic asphalt cement over the shingles for approximately 2 inches and up the chimney against the masonry surface for 4 to 6 inches. Press a strip of mineral-surfaced roll roofing wide enough to cover the cement into the cement. Return the side pieces around and over the ends of the front piece and around the back of the chimney for a distance of about 6 inches.

If the original construction did not provide a cricket behind the chimney, the base flashing at this point should consist of a 36-inch-wide strip of mineral-surfaced roll roofing, applied to lie 24 inches up the roof, over the old roofing, and 12 inches up the rear face of the chimney. Embed the strip in plastic asphalt cement applied over the old roofing and against the primed masonry surface. Trowel the cement into all irregularities between the roof deck and the masonry. Secure the upper edge by nailing into a mortar joint.

Applying Individual Shingles

Individual shingles are manufactured in three basic types: hexagonal, giant, and interlocking. Selection depends on many factors, including slope, wind resistance, coverage, aesthetics, and economics. Installation details vary according to the type of shingle used. The preparatory procedures are the same as those for applying strip shingles.

Laying Hexagonal Shingles

Two types of hexagonal shingles are available: those that lock together with a clip and those that have a built-in locking tab. Both the clip and lock-down shingles are relatively lightweight and intended primarily for placement over old roofing. They also can be used at times for new construction. For either application, the slope should be 4 inches per 12 inches or greater. Consult the roofing manufacturer for specific application instructions.

Laying Giant Shingles

This type of shingle can be used for new construction or reroofing depending on the method of application. The dutch lap method is intended primarily for reroofing over old roofing when a smooth surface and adequate anchorage is provided for nailing. It also can be used to cover new decks when single coverage provides the intended protection. For either application, the slope should be 4 inches per 12 inches or greater. Consult the roofing manufacturer for specific application instructions.

The American method can be used for new construction or reroofing. In either case, the slope should be 4 inches per 12 inches or greater. Consult the roofing manufacturer for specific application instructions.

Laying Interlocking Shingles

Interlocking shingles are manufactured with an integral locking device that provides immediate wind resistance. These shingles can be used for reroofing over existing roofing on slopes recommended by the shingle manufacturer. They can also be used for new construction depending on whether single or double coverage is required. In general, single-coverage interlocking shingles are not recommended for new construction. Check local building codes before installing them on new roofs.

The location of the shingle fasteners is essential to the proper performance of the mechanical interlock. For best results, follow the shingle manufacturer's specifications concerning fastener placement.

Follow the manufacturer's directions concerning application of the starter, first, and succeeding courses. Although interlocking shingles are self-aligning, they are flexible enough to allow limited adjustment. It is especially important to snap horizontal and vertical chalklines to keep the work in alignment.

The integral locking tabs are manufactured within close tolerances to ensure a definite space relationship between adjacent shingles. Be sure, therefore, to engage the locking devices carefully and correctly. Figure 7-20 illustrates two common locking devices used in interlocking shingles.

During installation, locking tabs on shingles along the rakes and eaves might have to be removed in part or entirely. To prevent wind damage, shingles that have their locking tabs removed should be either cemented down or fastened in place according to the individual manufacturer's recommendations.

Applying Roll Roofing

Asphalt roll roofing is used as both a primary roof covering and a flashing material. Asphalt roll roofing materials can be applied by two methods, depending on slope and other considerations. Asphalt roll roofing materials can be applied parallel to the

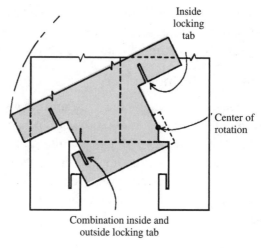

Inside locking tab

Center of rotation

Combination inside and outside locking tab

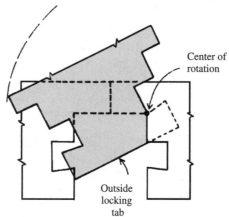

Center of rotation

Outside locking tab

FIGURE 7-20 Two common locking devices.

- Rake using the exposed-nail method on roof decks having a slope of 4 inches per 12 inches or more

- Rake using the concealed-nail method on roof decks having a slope of 3 inches per 12 inches or more

- Eaves using the exposed-nail method on roof decks having a slope of 2 inches per 12 inches or more

- Eaves using the concealed-nail method on roof decks having a slope of 1 inch per 12 inches or more

When maximum service life is an important consideration, use the concealed-nail method of application. As a primary roof covering, roll roofing is used on slopes as gradual as 2 inches in 12 inches. The roofing is applied either parallel to the eaves or parallel to the rakes. Nails should fit the application and have a length sufficient to penetrate ¾ inch into the deck or through the deck panel.

Store the material on end in a warm place until ready for use, especially during the colder seasons of the year. It is not good practice to apply roll roofing when the temperature is below 45°F. If rolls do have to be handled below this temperature and have not been stored as suggested, warm them before unrolling them to avoid cracking the coating. Then cut the rolls into 12- to 18-foot lengths and spread them in a pile on a smooth surface until they flatten out.

Before applying roll roofing, prepare the deck and install the necessary flashing in the manner described earlier for strip shingles. Use open valleys and follow the appropriate valley flashing procedures. Because all roll roofing is applied with a certain amount of top and side lapping, the proper sealing of the laps is crucial.

Use only the lap cement or asphalt plastic cement recommended by the roofing manufacturer. Store the cement in a warm place until ready to use. The plastic cement is asphalt-based and contains solvents. Take proper safety precautions. Never heat asphalt cement directly over a flame. Do not attempt to thin the cement by diluting it with solvent.

Always apply cement in a continuous but not excessive layer over the full width of the lap. Press the lower edge of the upper course firmly into the cement until a small bead appears along the edge of the sheet. Use a roller to apply pressure uniformly over the entire cemented area. Unless otherwise noted by the roofing manufacturer, apply asphalt plastic cement at the rate specified by the coating manufacturer.

Caution: Excessive amounts of cement can cause blistering. During application allow sufficient time for volatiles to flash off.

Applying Parallel to Eaves (Concealed-Nail Method)

When using this method, place narrow edging strips along the eaves and rakes before applying the roofing material. Figure 7-21 illustrates

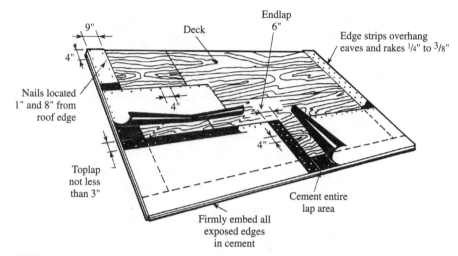

FIGURE 7-21 Application parallel to eaves using concealed-nail method.

the general installation procedure, including lapping, cementing, and nailing.

EDGE STRIPS

Place 9-inch-wide strips of roll roofing along the eaves and rakes and position them to overhang the deck ¼ to ⅜ inch. Fasten the strips with rows of nails located 1 inch and 8 inches from the roof edge and spaced 4 inches on center in each row.

FIRST COURSE

Position a full-width strip of roll roofing so that its lower edge and ends are flush with the edge strips at the eaves and rakes. Fasten the upper edge with nails 4 inches on center and slightly staggered. Locate the nails so that the next course overlaps them a minimum of 1 inch.

Lift the lower edge of the first course and cover the edge strips with cement according to the manufacturer's specifications. In cold weather, turn the course back carefully to avoid damaging the roofing material. Press the lower edge and rake ends of the first course firmly into the cement-covered edge strips. Work from one side of the sheet to the other to avoid wrinkling or bubbling.

Endlaps should be 6 inches wide and cemented over the full lap area with the recommended cement. Nail the underlying sheet in rows 1 inch and 5 inches from the end of the sheet with the nails spaced 4 inches on center and slightly staggered. Endlaps in succeeding courses cannot line up with one another.

SECOND AND SUCCESSIVE COURSES

Position the second course so that it overlaps the first course by at least 3 inches or as specified by the roofing manufacturer. Fasten the upper edge to the deck, cement the laps, and finish installing the sheet in the same manner as the first course. Follow the same procedure for each successive course. Do not apply nails within 18 inches of the rake until cement has been applied to the edge strip and the overlying strip has been pressed down.

HIPS AND RIDGES

Trim, butt, and nail the sheets as they meet at a hip or ridge. Next, cut 12-x-36-inch strips from the roll roofing and bend them lengthwise to lay 6 inches on each side of the joint. Do not bend the strips in cold weather without first warming them. Use these as shingles to cover the joint, each one overlapping the other by 6 inches, as shown in Fig. 7-22.

Start the hips from the direction opposite the prevailing winds. To guide the installation, snap a chalkline 5½ inches from and parallel to the joint on both sides. Apply asphalt plastic cement evenly over the entire area between the chalklines, from one side of the joint to the other. Fit the first folded strip over the joint and press it firmly into the cement. Drive two nails 5½ inches from the edge of the end that is to be lapped.

Cover the 6-inch lap on this strip with lap cement. Then place the next strip over it. Nail and cement in the same manner as the first strip. Continue the same procedure until the hip or ridge is finished.

Applying Parallel to Eaves (Exposed-Nail Method)

Figure 7-23 illustrates the general installation procedures, including lapping, cementing, and nails.

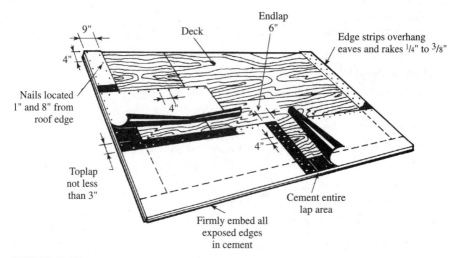

FIGURE 7-22 Concealed-nail method for hips and ridges.

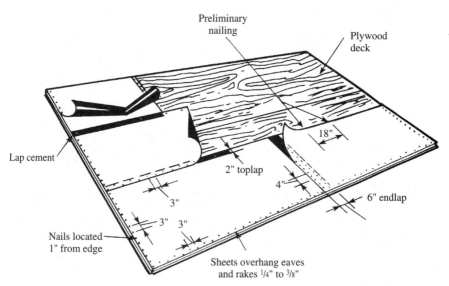

FIGURE 7-23 Application parallel to eaves using exposed-nail method.

FIRST COURSE

Position a full-width sheet so that its lower edge and ends overhang the eaves and rakes between ¼ and ⅜ inch. Nail along a line ½ to ¾ inch parallel to the top edge of the sheet. Space the nails 18 to 20 inches apart. This top nailing holds the sheet steady until the second course is placed over it and fastened. Nail the eaves and rakes on a line 1 inch parallel to the edges of the roofing with the nails spaced 3 inches on center and staggered a bit along the eaves to avoid splitting the deck.

If two or more sheets must be used to continue the course, lap them 6 inches. Apply lap cement to the underlying edge over the full lap width. Embed the overlapping sheet into it and fasten the overlap with two rows of nails 4 inches apart and 4 inches on center within each row. Stagger the rows so that the spacing is 2 inches between successive nails from row to row.

SECOND AND SUCCEEDING COURSES

Position the second course so that it overlaps the first course by 2 inches. Fasten the second course along the top edge following the same nailing directions as the first course. Lift the lower edge of the overlapping sheet and apply lap cement evenly over the upper 2 inches of the first course. Then embed the overlapping sheet into it.

Fasten the lap with nails spaced 3 inches on center and staggered slightly. Place the nails not less than $\frac{3}{4}$ inch from the edge of the sheet. Nail the rake edges in the same manner as for the first course. Follow the same procedure for each successive course. Endlaps should be 6 inches wide and cemented and nailed in the same manner as for the first course. Stagger endlaps so that an endlap in one course is never positioned over the endlap in the preceding course.

HIPS AND RIDGES

Trim, butt, and nail the roofing as it meets at a hip or ridge. Snap a chalkline on each side of the hip or ridge, located $5\frac{1}{2}$ inches from the joint and parallel to it. Starting at the chalklines and working toward the joint, spread a 2-inch-wide band of asphalt lap cement on each side of the hip or ridge (Fig. 7-24).

Cut strips of roll roofing 12 inches wide and bend them lengthwise along the centerline so that they lay 6 inches on each side of the hip or ridge. In cold weather, warm the material before bending it. Lay the bent strip over the joint and embed it in the cement. Fasten the strip to the deck with two rows of nails, one on each side of the hip or ridge.

Locate the rows $\frac{3}{4}$ inch from the edges of the strip and space the nails 3 inches on center. Be sure the nails penetrate the cemented area

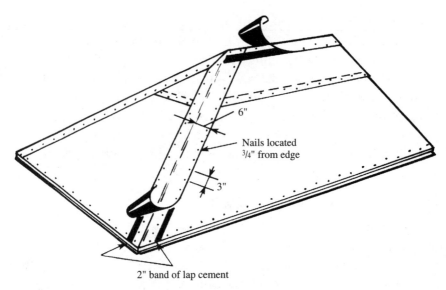

6"

Nails located
3/4" from edge

3"

2" band of lap cement

FIGURE 7-24 Hips and ridges using exposed-nail method.

underneath in order to seal the nail hole with asphalt. Endlaps should be 6 inches and cemented the full lap distance. Avoid excessive use of cement, as it can cause blistering.

Applying Parallel to Rake (Concealed-Nail Method)

With this method, illustrated in Fig. 7-25, sheets are applied vertically beginning at the eaves. Lay the sheets out and let them warm in the sun until they lie smoothly on a flat surface. If the sheets are nailed before they have time to relax, wrinkling can occur.

EDGE STRIPS

Place 9-inch-wide strips of roll roofing along the eaves and rakes. Placement and nailing of these strips is identical to that for the concealed-nail method for applying parallel to the eaves.

FIRST COURSE

Position the first sheet so that it is flush with the edge strips at the rake and eaves. Fasten the upper edge with nails located 1 inch and

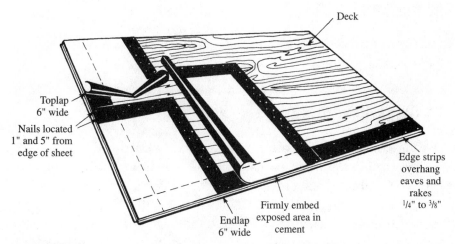

FIGURE 7-25   Applying parallel to rake using concealed-nail method.

5 inches from the top edge of the sheet. Space the nails 4 inches on center. Carefully lift the first sheet back and cover both edge strips with an even layer of plastic cement. Always remember that excessive amounts of plastic cement can cause blistering. Press the sheet firmly into the cement. Work from the top of the sheet down to avoid wrinkling or bubbling.

Endlaps should be a minimum of 6 inches wide. Fasten the length of the endlap with nails 4 inches on center and slightly staggered. Locate the nails so that the next sheet overlaps them a minimum of 1 inch. Cement endlaps over the full lap area with recommended lap cement.

Toplaps should be a minimum of 6 inches wide. Stagger the toplaps of adjoining sheets to prevent a buildup where the sheets intersect. Cement endlaps over the full lap area with the recommended lap cement.

SECOND AND SUCCESSIVE COURSES

Position the second course so that it overlaps the first course at least 6 inches. Fasten the upper edge to the deck, cement the laps, and finish installing the sheet in the same manner as the first course. Do not apply nails within 18 inches of the rake until cement has been applied to the edge strip and the overlying strip is pressed down.

HIPS AND RIDGES

Finish the roof at these joints in the same manner as in the concealed-nail method of application parallel to the eaves.

Applying Double-Coverage Roll Roofing

Double-coverage roll roofing is a 36-inch-wide sheet, of which 17 inches are intended for exposure and 19 inches for a selvage edge. It provides double coverage for the roof and can be used on slopes down to 1 inch per 12 inches. The 17-inch exposed portion is covered with granules, while the 19-inch selvage portion is finished in various manners depending on the manufacturer. Some vendors saturate the selvage portion with asphalt; some saturate it and coat it as well.

Cement the selvage edge and all endlaps according to the manufacturer's recommendations. It is important to know the requirements of the particular product being used and to follow the roofing manufacturer's directions concerning the type and quantity of adhesive. Unless otherwise noted by the roofing material manufacturer, apply asphalt plastic cement at the rate of 2 gallons per 100 square feet of covered area or as recommended by the coating manufacturer.

Make certain there is adequate roof drainage to eliminate the possibility of water standing in puddles. This is especially important on low slopes on which double-coverage roofing is commonly used. Choose the correct type and length of nail to fit the application. The fastener should be able to penetrate the deck ¾ inch or through the deck panel.

Application of double-coverage roll roofing can be parallel to the eaves or to the rake. Although 19-inch selvage roll roofing is discussed here, any roll roofing can be applied in the same manner to obtain double coverage if the lapped portion of the sheet is 2 inches wider than the exposed portion. Before applying the roofing, prepare the deck and install flashings in the manner described earlier for strip shingles. Use open valleys and follow the appropriate valley flashing procedures.

Applying Parallel to Eaves

Remove the 17-inch granule-surfaced portion from a sheet of double-coverage roll roofing. Place the remaining 19-inch selvage portion parallel to the eaves so that it overhangs the drip edge $\frac{1}{4}$ to $\frac{3}{8}$ inch at both the eaves and rakes. Fasten it to the deck with two rows of nails, one on a line $4\frac{3}{4}$ inches from the top edge of the strip, the other on a line 1 inch above the lower edge. Space the nails 12 inches on center and slightly stagger them in each row (Fig. 7-26).

FIRST COURSE

Cover the entire starter strip with asphalt plastic cement. Avoid excessive use of cement, as it can cause blistering. Then position a full-width sheet over it. Place the sheet so that the side and lower edge of the granule-surfaced portion are flush with the rake and eave edges of the starter strip. Fasten it to the deck with two rows of nails in the selvage portion. Locate the first row $4\frac{3}{4}$ inches below the upper edge and the second row $8\frac{1}{2}$ inches below the first with the nails spaced 12 inches on center and staggered.

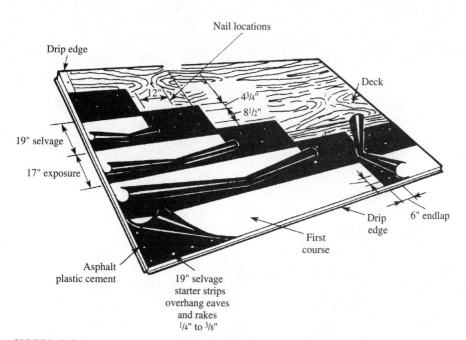

FIGURE 7-26 Applying parallel to eaves using starter strip.

SUCCESSIVE COURSES

Position each successive course to overlap the full 19-inch selvage width of the course below and nail the selvage portion in the same manner as the first course. Turn the sheet back and apply cement to the full selvage portion of the underlying sheet. Turn the sheet back carefully to avoid damaging it. Spread the cement to within ¼ inch of the edge of the exposed portion. Press the overlying sheet firmly into the cement. Apply pressure over the entire lap using a broom or light roller to ensure complete adhesion between the sheets. It is important to apply the cement so that it flows to the edge of the overlying sheet under the application pressure.

ENDLAPS

All endlaps should be 6 inches wide. Fasten the underlying granule-surfaced portion of the lap to the deck with a row of nails 1 inch from the edge. Space the nails 4 inches on center. Then spread asphalt plastic cement evenly over the lap area. Embed the overlying sheet in the cement and secure the selvage portion of the sheet to the deck with nails on 4-inch centers in a line 1 inch from the edge of the lap. Stagger all endlaps so that those in successive courses do not line up with one another.

Caution: Never cement roll roofing directly to the deck. This ensures that the sheets do not split due to deck movement. To make certain that roll roofing is not cemented to the deck when hot application is allowed, nail down a base sheet.

Applying Parallel to Rake

With this method, the sheets are applied vertically from the ridge down. Begin by applying starter strips to both rakes using the procedures for horizontal application. Cover the starter strip with asphalt plastic and apply a full-width sheet over it as the first course. Position all endlaps so that the upper sheet overlaps the lower one, thereby carrying drainage over the joint rather than into it. The remainder of the application is the same as that for applying parallel to the eaves. Figure 7-27 shows the general arrangement for application parallel to the rake.

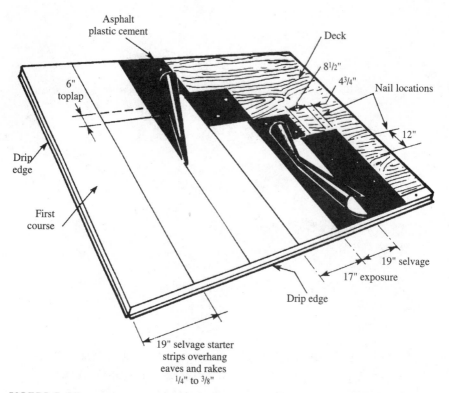

FIGURE 7-27 Application of double-coverage roll roofing parallel to rake.

HIPS AND RIDGES

For both hip and ridge applications, trim, butt, and nail the roofing sheets as they meet at a hip or ridge. Snap chalklines 5½ inches from and parallel to the joint on each side to guide the installation. Next, cut 12-×-36-inch strips of roll roofing that include the selvage portion. Bend the strips lengthwise to lie 6 inches on either side of the joint. In cold weather, be sure to warm the strips before bending. Start applying the strips at the lower end of the hip or at the end of the ridge opposite the direction of prevailing winds.

Cut the selvage portion from one strip to use as a starter. Fasten this strip in place by driving nails 1 inch from each edge and 4 inches on center over the full length. Cover it completely with asphalt plastic cement. Fit the next folded strip over the starter and press it firmly

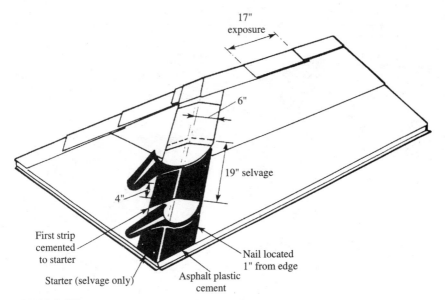

FIGURE 7-28 Application of double-coverage roll roofing.

into the cement, nailing it in the same manner as the starter, but only in the selvage portion. Continue the process until the hip or ridge is completed. Figure 7-28 illustrates the procedure.

Double-coverage roll roofing is frequently used on sheds that contain no hips or ridges. To finish this type of roof trim, nail the selvage portion of the last course to the edge of the roof. Then trim the exposed, granule-surfaced portion that had been cut from the starter strip to fit over the final selvage portion and cement it in place. Finally, overlay the entire edge with metal flashing and cement it in place.

Inspecting the Job

A quality installation is dependent on partial roof inspections during application and a final roof inspection upon completion. Ongoing review offers the applicator a number of opportunities to make certain that the materials are properly placed, properly fastened, and in good condition.

Upon final inspection, give the roof a thorough cleaning. Remove any loose shingles, cuttings, nails, wood shavings, boards, or other debris that has been left lying on the roof. Remove any debris from valleys and gutters to ensure that they are unobstructed and can carry water away quickly and efficiently. Damaged shingles cause leaks and detract from the appearance of the finished job.

Be sure to inspect the finished roof for any broken, torn, or damaged shingles. Remove damaged shingles by drawing the fasteners from the damaged shingle and the one immediately above it. Carefully fit a new shingle into place, fasten, and hand seal both shingles.

Inspect all areas on which asphalt cement has been used to ensure that the materials have adhered properly and that there are no bubbles in the roofing. If recementing is required to obtain a good bond, carefully lift the overlapping material to avoid damaging it, apply the cement, and work it thoroughly into the inplace material. Replace the overlying layer and press it firmly into the cement.

Regardless of the type of asphalt roofing material used, check for proper fastening. Fasteners should not be exposed unless the installation procedures specifically call for exposure. No fastener should protrude from, or be driven into, the surface of the roofing material. Seat all protruding fasteners if they are placed improperly. A protruding fastener must be driven properly or replaced. Remove any bent or incorrectly placed fasteners as well as any that are driven too far into the surface of the asphalt roofing. Then pack the hole with asphalt plastic cement and drive a new fastener nearby.

Before leaving the roof, examine the entire roof area for overall appearance and any defects that might have been overlooked at closer range. Thoroughly clean the site of roofing materials, cuttings, scraps, wood, and any other debris remaining from the roofing work. If possible, before leaving the job have the owner make a final inspection of the grounds and obtain his or her approval of the work.

Resolving Problem Areas

Two problem areas that can result in callbacks are shading and algae discoloration. Neither of these problems is the result of poor workmanship.

Shading

As a completed asphalt shingle roof is viewed from different angles, certain areas can appear darker or lighter. This difference in appearance is called shading. Shading also depends on the position of the sun and the overall intensity of light. For example, slanting sun rays emphasize shading, while direct, overhead rays cause color shading to disappear.

Shading is a visual phenomenon that in no way affects the performance of the shingles. It occurs primarily as a result of normal manufacturing operations that produce slight differences in surface texture that cannot be detected during the production process. These unavoidable variations in texture simply affect the way the surface reflects light. Shading is usually more noticeable on black or dark-colored shingles because they reflect only a small part of the light shining on them. This magnifies the slight differences in surface texture.

White and light-colored shingles reflect a great amount of light, which diminishes observable shading differences. Blends made from a variety of colors actually tend to camouflage shading, with lighter-colored blends reducing the effect of shading more than darker blends.

Racking or straight-up application accentuates shading. For this reason, be sure shingles are applied across and diagonally up the roof. This blends the shingles from one bundle into the next and minimizes any shade variation from one bundle to the next.

Shading can also occur when the backing material used to keep shingles from sticking together in the bundle rubs off onto the exposed portion of a shingle. Staining can also appear in shingles that have been stacked too high or for too long a period of time. This type of shading develops because of minor staining from the oils in the asphalt coating. In either of these cases, weathering usually eliminates the problem over a period of time.

Algae Discoloration

Roof discoloration caused by algae, commonly referred to as fungus growth, is a frequent problem throughout the country. It is often mistaken for soot, dirt, moss, or tree droppings. The algae that cause this discoloration do not feed on the roofing material and, therefore, do not affect the service life of the roofing. The natural pigments in the algae, however, can gradually turn a white or light roof dark brown or black over a period of years.

Algae discoloration is difficult to remove from roofing surfaces, but the coloration can be lightened with a diluted cleaning solution. Contact the appropriate manufacturer for details. Gently sponge the solution onto the roofing material. Scrubbing loosens and removes granules, so apply the solution carefully to avoid damaging other parts of the building and the surrounding landscape. If possible, work from a ladder or walkboards to avoid walking directly on the roof. Observe safety precautions whenever working on or near the roof. After sponging, rinse the solution off the roof with a hose.

Caution: Observe extra caution, as this process makes the roof slippery and potentially hazardous during treatment.

The effectiveness of such cleaning is only temporary, and the discoloration can recur. Several types of algae-resistant roofing have been developed and are now commercially available. These asphalt roofing products are specifically designed to inhibit most algae growth for extended periods of time. When undertaking an asphalt roofing job in an area where algae discoloration is a problem, advise the property owner of available algae-resistant asphalt roofing products.

Wooden Shingles and Shakes

Wooden shingles and shakes are cut from logs of red cedar, redwood, cypress, or pine. Western red cedar is by far the most common shingle lumber used in the United States. The cellular composition of Western red cedar, with millions of tiny, air-filled cells per cubic inch, provides a high degree of thermal insulation on both roof and wall applications. The wood also possesses outstanding rigidity in hurricane winds and demonstrates great resilience under the pounding of hail. These natural advantages are amplified by the structural strength derived from the overlapping method typically used for application.

Most cedar roofs are not chemically treated. Solutions are used, however, to get the best service in climatic areas that experience a combination of heat and humidity. Check local building codes for conformity with these recommendations.

To make shingles and shakes, selected logs are cut to specified lengths and the resulting blocks or *bolts* are trimmed to remove bark and sapwood. Then shingles are sawn from the bolts, while shakes can be either hand-split or sawn (Fig. 8-1).

Grading Shingles and Shakes

Like lumber and plywood, wooden shingles are graded and labeled to indicate the quality of the product. The Cedar Shake and Shingle

Shake

Shingle

FIGURE 8·1 The physical difference between wooden shingles and shakes.

Bureau grades red cedar shingles. The thickness at the butt end, referred to in the following grades, shows the height of the number of butts stacked on top of each other. For example, 5/2 means that 5 butts stacked on top of each other are 2 inches thick.

The premium grade of shingle is No. 1, or blue label. These shingles are cut tapered and smooth sawn on both faces for a neat, tailored appearance. They are cut to one of three lengths:

- 16-inch shingles with a total thickness of 5/2 inches

- 18-inch shingles with a total thickness of 5/2¼ inches

- 24-inch shingles with a total thickness of 4/2 inches

No. 1 shingles are clear heartwood that have no defects and are 100 percent edge grain. On 4/12 and steeper pitched roofs, a No. 1 shingle is applied at 5-inch exposure for a 16-inch shingle, 5½ inches for an 18-inch shingle, and 7½ inches for a 24-inch shingle.

A No. 2, or red label, shingle is cut in the same lengths and thicknesses as the No. 1 shingle. The face must be 10 inches clear on the 16-inch shingles, 11 inches clear on the 18-inch shingle, and 16 inches on the 24-inch shingle. Limited sapwood and flatgrain are allowed. Flatgrain can easily be detected by the grain designs on the face of the shingle. Limited knots and defects are allowed above the clear portion. On a 4/12 or steeper pitched roof, No. 2 shingles are applied at

4 inches for 16-inch shingles, 4½ inches for 18-inch shingles, and 6½ inches for 24-inch shingles. These shingles are most often used as a starter course and for reroofing.

No. 3, or black label, shingles are a utility grade for economy applications and secondary buildings. A No. 3-grade shingle is cut in the same lengths and thicknesses as the No. 1 and No. 2 shingles. The face must be 6 inches clear on the 16- and 18-inch shingles and 10 inches clear on 24-inch shingles. Sapwood and flatgrain are allowed. Limited knots and defects are permitted above the clear portion. On a 4/12 or steeper pitched roof, No. 3 shingles are applied at a 3½-inch exposure for 16-inch shingles, 4 inches for 18-inch shingles, and 5½ inches for 24-inch shingles.

The No. 4 grade is a utility grade for starter courses and undercoursing.

Manufacturers of shakes produce four distinct types, based on texture and method of manufacture. The *tapersawn* shake is available in three types. On the No. 1 type, both faces are sawn like that of a shingle, but it is cut thicker. This shake presents the naturally tailored look of a wooden shingle roof, but the thicker butt creates the complimentary sharp shadowline of a shake.

The No. 1-grade shake is cut from clear heartwood and is 100 percent edge grain with no defects. This shake is ⅝ inches thick at the butt and is cut in two lengths: 18 and 24 inches. On a 4/12 or steeper pitched roof, this product is exposed at 7½ inches for 16-inch shakes and 10 inches for 24-inch shakes.

The No. 1 grooved tapersawn shake is machine-grooved to resemble a hand-split or resawn shake. Being a tapersawn type, the grooved tapersawn shake is applied at the same exposure as the tapersawn. As with the tapersawn shake, a ¾-inch-thick butt can be custom cut for an even sharper shadowline. The exposure application is the same as that for the ⅝-inch-thick tapersawn shake.

The No. 2 tapersawn shake is cut in the same lengths and thicknesses as the No. 1 tapersawn shake. The lower half of the shake is clear. Flat and cross grains are allowed. The top half of the shake is allowed tight knots and other limited defects. The application exposure is dropped from 7½ inches to 5½ inches for the 18-inch shake and from 10 inches to 7½ inches for the 24-inch shake. This product is used for reroofing residences and for wall shake applications, starter coursing, sheds, etc.

The hand-split and resawn shake is usually available in two types. The heavy hand-split and resawn shake has a sawn back face. The face is hand-split so that the natural grain of the wood gives it a highly textured look. It is also heavier than any other shake, which gives the roof a rugged appearance.

This shake is cut from clear heartwood and 100 percent edge grain, with no defects. It is 1½ inches thick at the butt and cut in two lengths: 18 and 24 inches. On a 4/12 or steeper pitched roof, it is applied at 7½ inches for 18-inch shakes and 10 inches for 24-inch shakes.

The heavy hand-split and resawn shake is cut in the same manner as the medium hand-split but is not cut as thick. It still gives a very textured appearance, as it is made from clear heartwood and 100 percent edge wood, with no defects. This shake has a ½-inch-thick butt end and is cut in two lengths: 18 and 24 inches. It is applied with the same exposure as the heavy shakes.

The tapersplit shake is produced by reversing the block, end for end, with each split. The best part of the tree is required for this naturally split shake, which is largely produced by hand with a mallet and froe. A tapersplit roof gives a less rustic, rippled shadowline. It is cut from clear heartwood, 100 percent edge grain, and has no defects. The tapersplit shake is cut ½ inch thick and 24 inches in length and is applied at 10-inch exposure on a 4/12 or steeper pitched roof.

Straightsplit shakes, also known as barn or ranch shakes, are commonly mistaken for a product exclusively used on barns. They are produced mainly by machine, but also can be split with a mallet and froe like tapersplit shakes. The difference between the straightsplit and the tapersplit is that the straightsplit is split from the same end of the block, which produces the same thickness throughout.

The straightsplit creates a rippled shadowline on the roof. It is cut from clear heartwood and is 100 percent edge grain, with no defects. This shake is ⅜ to ½ inch thick and is cut in two lengths: 18 and 24 inches.

No. 1, or blue label, shingles and No. 1 split shakes are available pressure-impregnated with fire retardants that meet testing standards developed by Underwriters' Laboratories, Inc. (UL) and adopted by the

National Fire Protection Association (NFPA). These shingles and shakes also are warranted for 30 years by the treater and/or chemical company when they are .40 chromium copper alummate (CCA) pressure treated. Treated products are ideal in areas of high humidity where premature decay can occur. Specify the *Certi-Last* treating label for this extra protection.

Estimating Quantities

To calculate the material needed for reduced exposures, divide the square footage by the reduced coverage.

Table 8-1 shows that a typical four-bundle square of 16-inch (No. 1 blue label or No. 2 red label) shingles, at 4-inch exposure, covers 80 square feet. For 3200 square feet of surface, as the formula indicates, 40 squares of material (3200 ÷ 80 = 40) are needed. To estimate shake coverage, use the formula and information in Table 8-2.

Applying General Design Details

Use wooden shingles when the roof's slope or pitch is sufficient to ensure good drainage. The minimum recommended pitch is 1/6 or 4-in-12 (4-inch vertical rise for each 12-inch horizontal run), although there have been satisfactory installations on lesser slopes. Climatic conditions, the skill of the installer, and the application techniques are modifying factors when evaluating a proposed installation. For

TABLE 8-1 Shingle Coverage

	Weather exposure							
	$3\frac{1}{2}''$	$4''$	$5''$	$5\frac{1}{2}''$	$6''$	$6\frac{1}{2}''$	$7''$	$7\frac{1}{2}''$
Length × thickness	Approximate coverage (in sq. ft.) of 1 square (4 bundles) of shingles							
$16'' \times 5/2$	70	80	90	100	—	—	—	—	—
$18'' \times 5/2\frac{1}{4}$	—	$72\frac{1}{2}$	$81\frac{1}{2}$	$90\frac{1}{2}$	100*	—	—	—	—
$24'' \times 4/2$	—	—	—	—	$73\frac{1}{2}$	80	$86\frac{1}{2}$	93	100*

*Maximum exposure recommended for roofs.

TABLE 8-2 Shake Coverage

Length × thickness, shake type	Weather exposure				
	5"	5¹/₂"	7¹/₂"	8¹/₂"	10"
	Approximate coverage (in sq. ft.) of 1 square of shakes[a]				
18" × ¹/₂", hand-split and resawn mediums[b]	—	55[c]	75[d]	—	—
18" × ³/₄", hand-split and resawn heavies[b]	—	55[c]	75[d]	—	—
18" × ⁵/₈", tapersawn	—	55[c]	75[d]	—	—
24" × ³/₈", hand-split	50[e]	—	75[c]	—	—
24" × ¹/₂", hand-split and resawn mediums	—	—	75[c]	85	100[d]
24" × ³/₄", hand-split and resawn heavies	—	—	75[c]	85	100[d]
24" × ⁵/₈", tapersawn	—	—	75[c]	85	100[d]
24" × ¹/₂", tapersawn	—	—	75[c]	85	100[d]
18" × ³/₈", straightsplit	—	65[c]	90[d]	—	—
24" × ³/₈", straightsplit	—	—	75[c]	85	100[d]

15" starter-finish course: Use as supplemental with shakes applied not to exceed 10" of weather exposure.
[a]All coverage based on an average ¹/₂" spacing between shakes.
[b]5 bundles cover 100 sq. ft. roof area when used as starter-finish course at 10" weather exposure; 7 bundles cover 100 sq. ft. roof area at 7¹/₂" weather exposure. See a.
[c]Maximum recommended weather exposure for 3-ply roof construction.
[d]Maximum recommended weather exposure for 2-ply roof construction.
[e]Maximum recommended weather exposure.

shakes, the recommended minimum slope is 3 inches vertical rise for every vertical foot.

Shingles and shakes can be applied over spaced sheathing. Spaced sheathing is usually 1-×-4- or 1-×-6-inch softwood boards. Solid sheathing is acceptable and might be required in seismic regions or

TABLE 8-3 Shingle Exposure

Pitch	No. 1, blue label			No. 2, red label			No. 3, black label		
	Length (in.)								
	16	18	24	16	18	24	16	18	24
3/12–4/12	$3^3/_4$	$4^1/_4$	$5^3/_4$	$3^1/_2$	4	$5^1/_2$	3	$3^1/_2$	5
4/12 and steeper	5	$5^1/_2$	$7^1/_2$	4	$4^1/_2$	$6^1/_2$	$3^1/_2$	4	$5^1/_2$

under treated shakes and shingles. It is recommended for shake applications in areas where wind-driven snow is common. Solid sheathing is usually strand-board panels or plywood, which provides a smooth base for roofing.

Shingle Sheathing

There are two acceptable application methods for spaced sheathing. One is to spread 1-×-4-inch boards to coincide with the weather exposure (Table 8-3) of the shingles (Fig. 8-2). In other words, if the shingles are to be laid at 5½ inches to the weather, the sheathing boards also should be spaced at 5½ inches on center. In this method of application, each shingle is nailed to the center of the sheathing board.

In the second method of application, which uses 1-×-6-inch boards, two courses of shingles are nailed to each 1-×-6-inch board (Fig. 8-3), up to and including 5½ inches of weather exposure. With 7½ inches of weather exposure, the center of the sheathing board equals the distance of the weather exposure. Although not commonly used, a breather-type underlayment, such as roofing felt, can be applied over either solid or spaced sheathing.

Shake Sheathing

In shake applications, spaced sheathing is usually 1-×-6-inch boards spaced on centers equal to the weather exposure (Fig. 8-4) at which the shakes are to be laid. The spacing should never be more than 7½ inches for 18-inch shakes and 10 inches for 24-inch shakes on roof installations. When 1-×-4-inch sheathing is installed at 10 inches on center, install

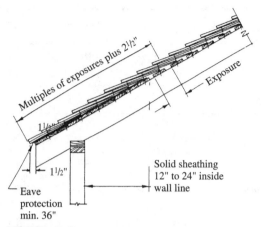

FIGURE 8·2 Shingles applied over 1-X-4-inch boards.

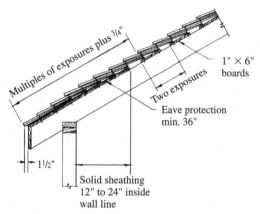

FIGURE 8·3 Shingles applied over 1-X-6-inch boards.

additional 1-x-4-inch boards between the sheathing boards.

A solid deck is recommended in areas where wind-driven snow is a factor. Roofing felt interlay between the shake courses is required whether the sheathing is spaced or solid. The felt interlay acts as a baffle that prevents wind-driven snow or other foreign material from entering the attic cavity during extreme weather conditions. The interlays also increase the roof's insulating value.

Take special care when installing the felt interlays over spaced

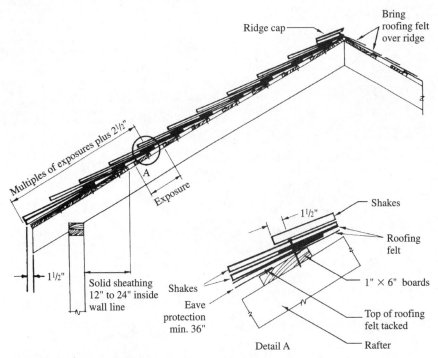

FIGURE 8-4 Shakes applied over 1-X-6-inch boards.

sheathing to ensure the formation of an effective baffle. Apply the felt over the top portion of the shakes and extend it onto the spaced sheathing. Position the bottom edge of the felt at a distance above the butt end that is equal to twice the weather exposure. To be an effective baffle, the top of the felt must rest on the sheathing. Place the shakes so that the nails are driven through the upper portion of the sheathing board and attach the tip of the roofing felt to the lower portion.

APPLYING SHINGLES

The following details (Fig. 8-5) must be observed on all applications. (See also Fig. 8-6.)

- Shingles must be doubled or tripled at all eaves.
- Project butts of first course shingles 1½ inches beyond the fascia.

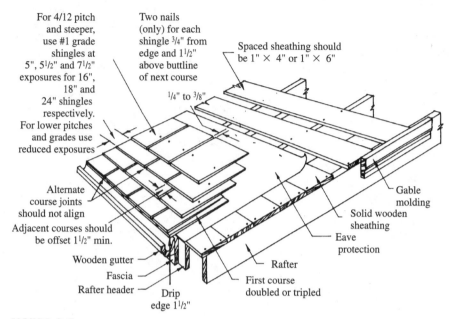

For 4/12 pitch and steeper, use #1 grade shingles at 5", 5½" and 7½" exposures for 16", 18" and 24" shingles respectively. For lower pitches and grades use reduced exposures

Two nails (only) for each shingle ¾" from edge and 1½" above buttline of next course

Spaced sheathing should be 1" × 4" or 1" × 6"

¼" to ³⁄₈"

Alternate course joints should not align

Adjacent courses should be offset 1½" min.

Wooden gutter

Fascia

Rafter header

Drip edge 1½"

Rafter

First course doubled or tripled

Eave protection

Solid wooden sheathing

Gable molding

FIGURE 8-5 Shingle application.

- Spacing between adjacent shingles (joints) should be a minimum of ¼ inch and a maximum of ³⁄₈ inch.

- Separate joints in any one course by not less than 1½ inches from joints in adjacent courses. In any three courses, no two joints should be in direct alignment.

- In lesser-grade shingles that contain both flat and vertical grain, do not align joints with the centerline of heart.

- Split flat grain shingles that are wider than 8 inches in two before nailing. Treat knots and similar defects as the edge of the shingle, and place the joint in the course above 1½ inches from the edge of the defect.

APPLYING SHAKES

Shakes, like shingles, are normally applied in straight, single courses. As with shingles, follow the application details (Fig. 8-7).

The starter course can be one or two layers of shingles or shakes overlaid with the desired shake. A 15-inch shake is made expressly for

✓	Item
	Shingles must be doubled or tripled at all eaves.
	Project butts of first course shingles 1½ inches beyond the fascia.
	Spacing between adjacent shingles (joints) should be a minimum of ¼ inch and a maximum of ⅜ inch.
	Separate joints in any one course by not less than 1½ inches from joints in adjacent courses. In any three courses, no two joints should be in direct alignment.
	In lesser-grade shingles that contain both flat and vertical grain, do not align joints with the centerline of heart.
	Split flat grain shingles that are wider than 8 inches in two before nailing. Treat knots and similar defects as the edge of the shingle, and place the joint in the course above 1½ inches from the edge of the defect.

FIGURE 8-6 Applying shingles.

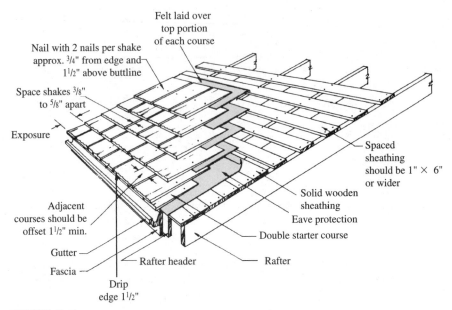

Nail with 2 nails per shake approx. 3/4" from edge and 1 1/2" above buttline

Felt laid over top portion of each course

Space shakes 3/8" to 5/8" apart

Exposure

Adjacent courses should be offset 1 1/2" min.

Gutter

Fascia

Drip edge 1 1/2"

Rafter header

Spaced sheathing should be 1" × 6" or wider

Solid wooden sheathing

Eave protection

Double starter course

Rafter

FIGURE 8-7 Shake application.

starter and finish courses. Project butts of first course shakes 1½ inches beyond the fascia.

Lay an 18-inch-wide strip of No. 30 roofing felt, or No. 15 felt depending on code requirements, over the top portion of the shakes and extend it onto the sheathing. Position the bottom edge of the felt so that it is above the butt of the shake by a distance equal to twice the weather exposure. For example, apply the felt 20 inches above the butt for 24-inch shakes laid with 10 inches of exposure. The felt covers the top 4 inches of the shakes and extends 14 inches onto the sheathing. Note that the top edge of the felt must rest on the spaced sheathing.

Spacing between adjacent shakes should be a minimum of ⅜ inch and a maximum of ⅝ inch. Offset joints between shakes 1½ inches over adjacent courses. Lay straightsplit shakes with the smoother end from which the shake was split, or *froe end*, toward the ridge.

Detailing Low-Slope Roofs

The minimum roof slope recommended for shakes is 1/3 and for shingles 1/4. It is possible, however, to apply shingles or shakes successfully to solid-sheathed roofs with lower slopes following a special

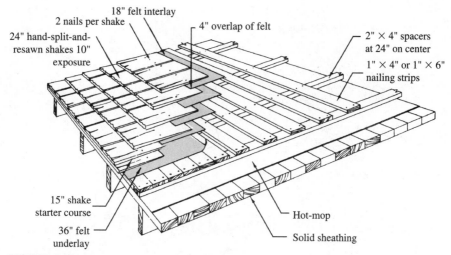

18" felt interlay
2 nails per shake
24" hand-split-and-
resawn shakes 10"
exposure
4" overlap of felt
2" × 4" spacers
at 24" on center
1" × 4" or 1" × 6"
nailing strips
15" shake
starter course
36" felt
underlay
Hot-mop
Solid sheathing

FIGURE 8-8 Application of shakes to low-slope roof.

application method (Fig. 8-8). The special method provides a double roof on which the shingles or shakes are applied to a lattice-like framework that is embedded in a bituminous surface coating.

Apply a conventional hot-mop or suitable roll-type asphalt roof over the roof deck (see Chap. 5). With the final hot-mop application, embed 2-×-4-inch spacers of Western red cedar or preservative-treated lumber in the bituminous coating. Install these spacers over the rafters at 24 inches on center and extend them from eave to ridge.

Next, nail 1-×-4-inch or 1-×-6-inch nailing strips, spaced according to the weather exposure selected for the shingles or shakes, across the spacers to form a lattice-like nailing base. For example, if 24-inch shakes are to be installed at a weather exposure of 10 inches, the nailing strips also should be spaced at 10 inches on center.

Finally, apply the shingles or shakes in the normal manner with a starter course at the eave and felt interlays between each course.

Detailing Hips and Ridges

Cap intersecting roof surfaces at the hip and ridge areas to ensure a weathertight joint. For the final course at the ridge line, as well as for hips, select uniform shakes. Apply a strip of 15-pound roofing felt, at least 8 inches wide, over the crown of all hips and ridges. Always

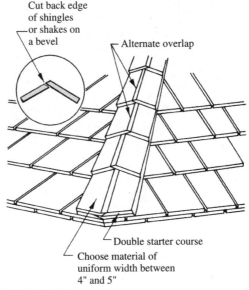

Cut back edge
of shingles
or shakes on
a bevel

Alternate overlap

Double starter course

Choose material of
uniform width between
4" and 5"

FIGURE 8-9 Hip and ridge application.

fabricate hips and ridges with shakes to create a harmonious appearance. Use prefabricated hip-and-ridge units, or fabricate the hips and ridges onsite.

To fabricate hips onsite, sort out shakes that are approximately 6 inches wide. Tack two wooden straightedges onto the roof. Place one on each side, 4 to 5 inches from the centerline of the hip (Fig. 8-9). Double the starting course of the shakes. Place the first shake on the hip with one edge resting against the guidestrip. Cut the edge of the shake projecting over the center of the hip on a bevel. Then apply the shake on the opposite side and cut back the projecting edge to fit.

Shakes in following courses are applied alternately in reverse order. Ridges are constructed in a similar manner. Weather exposure should be the same as that given the shakes on the roof. It is important that shakes on hips and ridges be attached with nails of sufficient length to penetrate the underlying sheathing by at least ½ inch.

When laying a cap along an unbroken ridge that terminates in a gable end, start laying the cap at each end so that it meets in the middle of the ridge. At that point, nail a small saddle of shake butts to splice the two lines. Always double the first course of capping at each end of the ridge. Conceal nails with the overlapping hip and ridge unit, as in the field of the roof.

Protecting Eaves

Extend eave protection on all shingle and shake roofs from the edge of the roof to a line up the roof slope not less than 12 inches inside the inner face of the wall. This protection is not necessary if ice dams are not likely to form along the eaves and cause a backup of water. Eave protection is not required over unheated garages, carports, and porches, where the roof overhang exceeds 36 inches as measured along the roof slope, or where low-slope shingles are used.

Detailing Roof Junctures

The correct construction of roof junctures ensures weathertightness. When metal flashing is employed, use nothing less than 26-gauge galvanized steel or an acceptable equivalent. Paint the metal on both sides with a good metal or bituminous paint. Paint flashing materials after bending; this helps maintain the integrity of the coating.

For convex junctures (Fig. 8-10), install metal flashings to cover the top 4 inches of the wall and the bottom 8 inches of the roof slope before the final course of shingles or shakes is nailed to the top of the wall. Apply a narrow, horizontal band of shingles or shakes, or a strip of wooden molding, after the final wall course is installed. Apply a double or triple starter course at the eave, with a 1½-inch overhang of the wall surface. Then complete the roof in the normal manner.

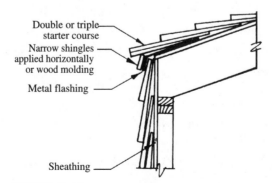

FIGURE 8-10 Convex roof juncture.

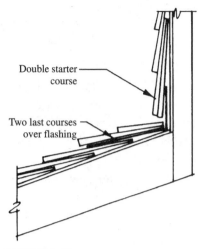
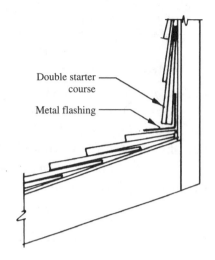

Double starter
course

Two last courses
over flashing

Double starter
course

Metal flashing

FIGURE 8-11 Concave roof juncture.

Metal flashings for concave junctures (Fig. 8-11) are similar to those for convex junctures. Install the metal flashings to cover the top of the roof slope and the bottom 4 inches of the wall before applying the final course of shingles or shakes. Install the final roof course so that the tips fit as snugly as possible against the wall at the juncture. Apply a double starter course at the base of the wall surface and then complete the remaining wall courses in the recommended manner.

For apex junctures (Fig. 8-12*A*), cover the top 8 inches of the roof and the top 4 inches of the wall with the metal flashing. Install the flashing before applying the final course of shingles or shakes to the wall. The recommended application sequence is to apply shingles or shakes first to the wall and then to the roof. Trim the overhanging roof material flush with the wall. Finally, apply specially prepared ridge units over the wall-roof juncture so that the roof piece overlaps the wall piece in each matching pair.

When shingles or shakes are to be applied to a swept or bell eave (Fig. 8-12*B*) with an excessive curvature, it might be necessary to soak the shingles or shakes, usually overnight, or steam them prior to installation. Employ a double starter course in the usual manner. Exposure is determined by the slope of the roof and the type of shingle or shake selected.

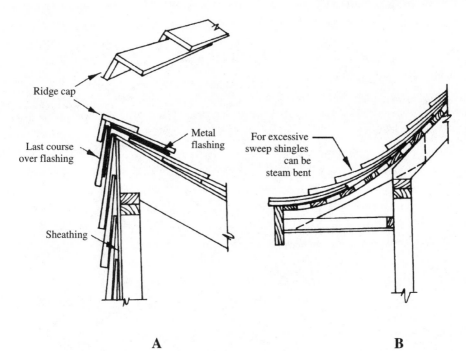

Ridge cap

Last course
over flashing

Metal
flashing

For excessive
sweep shingles
can be
steam bent

Sheathing

A

B

FIGURE 8-12 (*A*) Apex roof juncture; (*B*) swept or bell eave.

Constructing Valleys

Most roof leaks occur where water is channeled off the roof or where the roof abuts a vertical wall or chimney. At these points, metal valleys and flashings help keep the structure sound and dry.

Flash structural members that protrude through a roof at all intersecting angles to prevent leaks (Fig. 8-13). Extend step flashing under the shingles or shakes and up the vertical surface. Cover it with a second layer of flashing, or *counterflashing*.

Different flashing metals are available for different climatic requirements. Some metals have baked-on enamel coatings. It is a good practice to use metals that have proven their reliability under the specific conditions that are likely to be encountered once the roof is applied. Use metal flashings that have the same longevity as the shingles or shakes. Typical saddle flashing is shown in Fig. 8-14.

For roofs with slopes of 1/1 or greater, extend valley flashing not less than 7 inches on each side of the valley centerline. For roof slopes

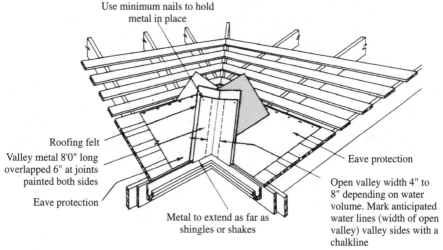

Use minimum nails to hold metal in place

Roofing felt

Valley metal 8'0" long overlapped 6" at joints painted both sides

Eave protection

Eave protection

Metal to extend as far as shingles or shakes

Open valley width 4" to 8" depending on water volume. Mark anticipated water lines (width of open valley) valley sides with a chalkline

FIGURE 8-13 Typical roof valley.

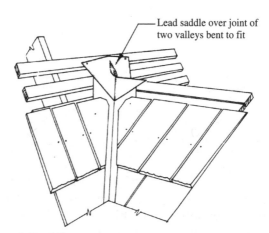

Lead saddle over joint of two valleys bent to fit

FIGURE 8-14 Typical saddle flashing.

less then 1/1, extend the flashing not less than 10 inches on each side. Valley flashing should be center-crimped and painted, and made from galvanized steel, copper, or aluminum.

Underlay valley metal with at least No. 15 roofing felt. Do not apply shingles with their grain parallel to the valley centerline. Cut shingles extending into the valley at the correct angle (Fig. 8-15). Joints between shingles must not break into the valley.

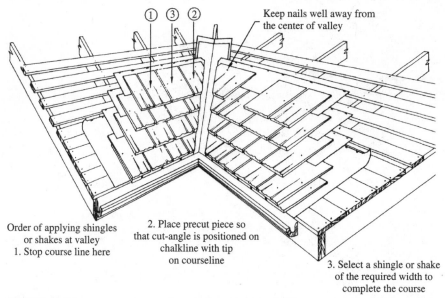

Keep nails well away from
the center of valley

Order of applying shingles
or shakes at valley
1. Stop course line here

2. Place precut piece so
that cut-angle is positioned on
chalkline with tip
on courseline

3. Select a shingle or shake
of the required width to
complete the course

FIGURE 8-15   Flashing details for shingle and shake valleys.

For shake roofs, follow the guidelines above for shingle valleys. Make the metal valley have a minimum total width of 20 inches. In some areas, however, flashing width requirements can differ. Consult local building codes.

Recommended application details for metal flashings around typical roof projections, such as chimney and vent pipes, are shown in Fig. 8-16. Application is accomplished in the manner described for asphalt shingles in Chap. 7.

Apply each shingle or shake with two corrosion-resistant fasteners, such as stainless steel-type 304 or 316, hot-dipped zinc-coated, or aluminum nails. If preservative-treated shingles or shakes are installed, pay close attention to the preservative manufacturer's recommendations about the compatibility of the preservative chemicals with the fastener. Minimum nail lengths are shown in Table 8-4.

Use aluminum or stainless steel-type 304 or 316 16-gauge staples. Drive two staples per shingle or shake with the staple crowns horizontal to the shingle or shake butt. The staples should be at least $7/16$ inch long or long enough to penetrate the sheathing at least $\frac{1}{2}$ inch. Drive staples in

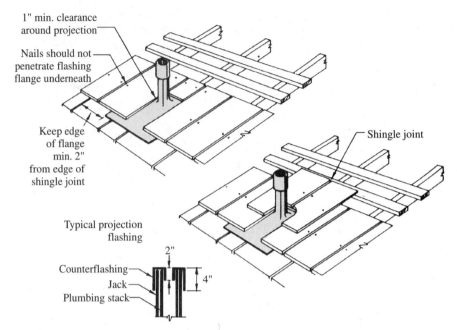

1" min. clearance around projection

Nails should not penetrate flashing flange underneath

Keep edge of flange min. 2" from edge of shingle joint

Shingle joint

Typical projection flashing

2"

Counterflashing

Jack

Plumbing stack

4"

FIGURE 8-16 Flashing details for typical roof projections.

the same location as nails relative to the sides and overlapping buttline, and flush with the surface of the shingle or shake. Preservative reaction with these metals should be checked with the manufacturer.

Detailing Vents

The importance of good attic ventilation beneath the roof cannot be overemphasized. Such air movement prevents or inhibits moisture condensation on the undersurface of the shingles or shakes, and on the roof deck and rafters.

Provide vents at the eaves, or *soffits*, as well as at gable ends. If cross-ventilation is desired, place vents at the ridgelines. Screen the vents to prevent the ingress of insects. For adequate ventilation, the rule of thumb is that the ratio of total net free ventilation area to the area of the attic should be not less than 1/150, with compensation made for screens over vent apertures. Attic fans can be beneficial, since these supply additional air movement in attic spaces. Several examples of construction techniques that provide roof ventilation are shown in Fig. 8-17.

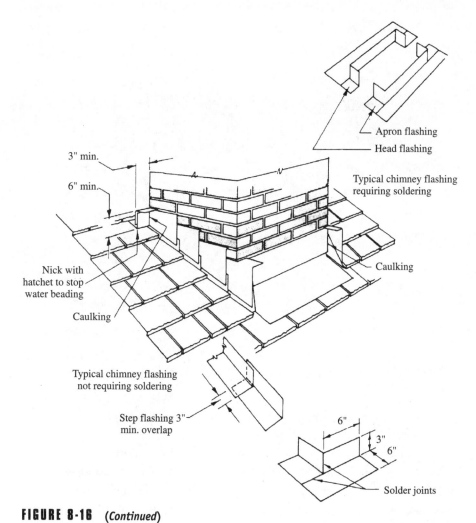

Apron flashing

Head flashing

Typical chimney flashing requiring soldering

3" min.

6" min.

Caulking

Nick with hatchet to stop water beading

Caulking

Typical chimney flashing not requiring soldering

Step flashing 3" min. overlap

6"

3"

6"

Solder joints

FIGURE 8-16 *(Continued)*

Designing Vapor Barriers

The decision to use a separate vapor barrier must be made by the designer. Factors to be considered include the type of building, its end use, and its geographic location.

When a vapor barrier is used, take care to ensure that the dew point is well to the outside of the vapor barrier in order to prevent condensation on the deck. Ideally, the vapor barrier should be as close as possible to the warm side of the roof. The thickness of the insulation

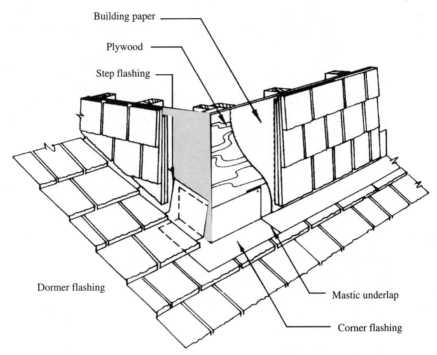

Building paper

Plywood

Step flashing

Dormer flashing

Mastic underlap

Corner flashing

FIGURE 8-16 *(Continued)*

should be increased as the deck thickness increases to maintain the correct location of the dew point.

In unevenly heated buildings, such as churches and halls, or buildings that generate an unusually high level of moisture, such as swimming pools, the excess humidity may have to be removed by mechanical means to prevent condensation on the deck. In air-conditioned buildings, a cold-weather roof system allows a constant flow of air between the insulation and the roofing. This helps reduce the energy required for cooling. Full details on the cold-weather roof system are given later in this chapter.

A new product, *Cedar Breather*, maximizes the performance and life of cedar roofs. It allows for a continuous air space beneath cedar shakes and shingles when they are installed on plywood decks. This reduces excess moisture and minimizes the potential for shingles to rot and warp. Made of patented three-dimensional nylon matrix, Cedar Breather enables the entire underside of the shingles to dry uni-

TABLE 8-4 Fasteners

Shingle or shake type	Nail type	Min. length (in.)
Shingles—new roof		
16" and 18" shingles	3d box	$1\frac{1}{3}$
24" shingles	4d box	$1\frac{1}{2}$
Shakes—new roof		
18" straightsplit	5d box	$1\frac{3}{4}$
18" and 24" hand-split and resawn	6d box	2
24" tapersplit	5d box	$1\frac{3}{4}$
18" and 24" tapersawn	6d box	2

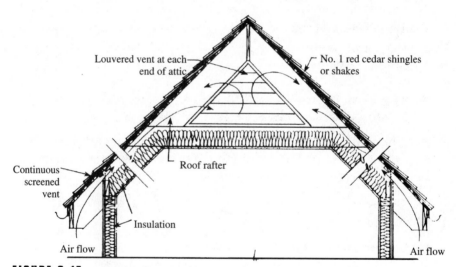

FIGURE 8-17 Examples of construction trechniques for roof ventilation.

formly. This prevents premature deck and shingle failure and eliminates the need for lathe or strapping.

Preparing for Reroofing Jobs

Generally, wooden shakes and shingles can be applied directly over existing roofing materials, with the exception of tile and slate. The following are important procedures that should be followed when you prepare to reroof with wooden shakes and shingles. (See also Fig. 8-18.)

- At the eaves and gable ends, cut back the old roofing material approximately 4 to 6 inches and replace it with a wooden filler of the same thickness.

- Remove the existing ridge and hip shingles before applying new roofing materials.

- When necessary, install new valleys, larger wall flashings, and counterflashings.

- Install a wooden strip in the existing sheet metal valleys to separate old metal from new metal.

If an inspection of the old roof indicates that the old wooden shingles can remain, carefully prepare the surface of the roof to receive the new material as follows.

- Apply wooden shakes with a felt interlayment, as in new-construction methods.

- When applying wooden shingles over wooden shingles, eliminate all obstructions and voids, such as curling shingles or missing shingles.

- Remove all loose or protruding nails and renail the shingles in a new location.

- Nail down all loose shingles.

- Split all badly curled or warped old shingles and nail down the segments.

- Replace missing shingles with new ones.

- If necessary, apply strip sheathing over existing wooden shingles.

- When necessary, remove or cut existing siding to allow for the installation of new and larger step flashings or tin shingles.

- Use nails that are long enough to penetrate the original roof sheathing by ½ inch.

✓	Item
	At the eaves and gable ends, cut back the old roofing material approximately 4 to 6 inches and replace it with a wooden filler of the same thickness.
	Remove the existing ridge and hip shingles before applying new roofing materials.
	When necessary, install new valleys, larger wall flashings, and counterflashings.
	Install a wooden strip in the exiting sheet metal valleys to separate old metal from new metal.
	If Old Shingles Are to Be Left in Place
	Apply wooden shakes with a felt interlayment, as in new-construction methods.
	When applying wooden shingles over wooden shingles, eliminate all obstructions and voids, such as curling shingles or missing shingles.
	Remove all loose or protruding nails and renail the shingles in a new location.
	Nail down all loose shingles.
	Split all badly curled or warped old shingles and nail down the segments.
	Replace missing shingles with new ones.
	If necessary, apply strip sheathing over existing wooden shingles.
	When necessary, remove or cut existing siding to allow for the installation of new and larger step flashings or tin shingles.
	Use nails that are long enough to penetrate the original roof sheathing by $1/2$ inch.

FIGURE 8-18 Preparing for reroofing jobs.

When the roof is subject to the impact of unusually high winds, cut back the shingles at the eaves and rakes and install 1-inch-thick wooden strips that are 4 to 6 inches wide. Nail the boards firmly and allow their outside edges to project beyond the edge of the deck by the same distance as the old wooden shingles. To provide a smooth deck to receive asphalt roofing, apply a *backer board* over the wooden shingles or use beveled, wooden *feathering strips* along the butts of each course of old shingles.

If old asphalt shingles are to remain in place, nail down or cut away all loose, curled, or lifted shingles. Remove all loose and protruding nails and all badly worn edging strips. Replace the edging strips with new ones. Just before applying the new roofing, sweep the surface clear of all loose debris.

Detailing Mansard Roofs

The mansard roof style is particularly well suited to renovation work on pitched-roof houses. The upper story can be enlarged without adding extra height to the structure. The conversion of a pitched-roof bungalow to a mansard provides a floor area on the upper floor that can be identical to the main floor area.

The low downward slope of the mansard roofline visually reduces the scale of the building and helps eliminate a boxy appearance. This technique is used frequently on large commercial projects, particularly those located near residential neighborhoods. It is also a common way to avoid a monotonous appearance on flat-roofed, framed apartment buildings. Properly used, a mansard roof can strengthen the building design without substantially increasing construction costs.

If raised up above the level of a built-up roof (BUR), the mansard can screen out roof penetrations or mechanical equipment. The variety of mansard roofs is practically infinite. One of the most widely used and misused roof designs, the mansard's proportions and scale are very important. Take care to avoid a mansard roofline that is either too skimpy or too generous.

Two of the most widely used roofing materials on the mansard roof are cedar shingles and shakes. Cedar shakes, with their heavier texture and solid appearance, are perhaps more frequently specified for mansards, although shingles are used when a lighter scale is desired.

The light weight of shingles and shakes and the ease with which they can be applied contribute substantially to economical construction. They can be installed over light framing, such as spaced battens, which affords a considerable savings in both materials and labor when compared to cladding, which requires a solid base. This cost-saving factor, combined with their excellent insulating qualities and attractive appearance, contributes to the increasing popularity of mansard roofs finished with shingles or shakes.

Preparing Special Roof Decks

Wooden decks are an ideal base over which to apply shingles or shakes because they can be attached in the conventional manner. The problem of how to fasten shingles or shakes is created when a layer of insulation, normally one of the rigid types, is included. Driving abnormally long nails through the shingles, the insulation, and into the deck is generally unsatisfactory.

Horizontal strapping is required to overcome the fastening difficulties. Strapping requires that fewer nails penetrate through the insulation and into the deck and achieves greater thermal efficiency by reducing the number of conductors. In addition, the nail lengths can be chosen to prevent the points from protruding through the deck where they might mar the inside face.

If ice dams are a potential problem, or if reverse condensation is likely to occur, such as might be encountered in an ice arena, use the cold-weather roof system in conjunction with horizontal strapping. Provide ventilation at the eaves and at the peak. In buildings such as ski cabins, which can be subjected to heavy snow loads, fasten wooden members, typically 2-x-4-inch boards on edge, from ridge to eave on the roof deck and place rigid insulation between the members. Then apply strapping across the top of these members to provide a ventilated air space and avoid compressing the insulation.

Often, you can eliminate the need for strapping by using a false plywood deck, to which the shingles or shakes are directly fastened, immediately over the insulation. Exterior-grade sheathing panels are ideal for this purpose, since they provide a strong, smooth surface.

Under certain conditions of pitch and loading, however, there might be a tendency for the entire roof above the decking to creep downward, which bends the nail fastenings, compresses the insulation, and generally reduces efficiency. In such cases, it often is desirable to install vertical members as previously described.

If the shingles or shakes are nailed directly through rigid insulation, a number of problems may be encountered. For instance, longer nails have thicker shanks that tend to split the shingles or shakes. Movement by the shingle or shake, which is caused by natural expansion and contraction cycles due to wet and dry conditions, tends to enlarge the holes in the insulation and reduce its efficiency. For this reason, the use of strapping or a false plywood deck is recommended.

Numerous types of rigid insulation are now in use. Rigid insulation can be made from expanded polystyrene beads, rigid urethane laminate, low-density fiberboard, or fast-setting liquids poured onsite. Its thickness can be more than 2 inches, and lengths and widths vary depending on the manufacturer. All these types of rigid insulation are efficient insulators and are usually of sufficient density to hold the weight of a normal roof covering without a lumber bridging.

Covering metal decks with shingles or shakes presents a rather unique problem. These decks are often used for economic reasons, but they generally require a finish roofing capable of withstanding weather conditions. In addition, aesthetic considerations often require that the deck itself be covered with a material such as red cedar shingles or shakes to provide a pleasing finish.

When a metal deck is to be covered, give consideration to the use and placement of vapor barriers. If insulation is to be placed on top of a metal deck, take into account the entire roofing system. For example, wooden members must not be sandwiched between two vapor barriers. If this is unavoidable, use preservative-treated wood. In some cases, the seams in a metal deck can be sealed to create an effective vapor barrier. This design should be passed by the design anchor or building owner.

Use boards or a panel deck as a nailing base for the shingles or shakes. Support it with vertical lumber members fixed to the deck. This can be achieved in a number of ways.

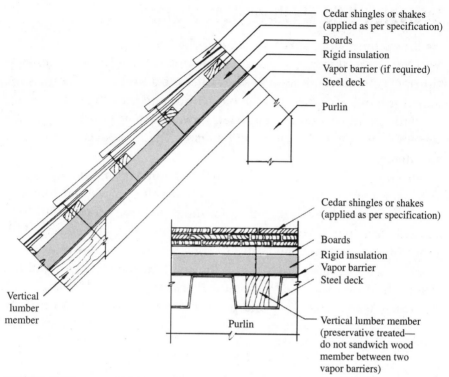

Cedar shingles or shakes
(applied as per specification)
Boards
Rigid insulation
Vapor barrier (if required)
Steel deck

Purlin

Cedar shingles or shakes
(applied as per specification)

Boards
Rigid insulation
Vapor barrier
Steel deck

Vertical
lumber
member

Purlin

Vertical lumber member
(preservative treated—
do not sandwich wood
member between two
vapor barriers)

FIGURE 8-19 Cedar shingles or shakes over steel roof.

Corrugated decks. On a corrugated deck, fasten vertical lengths of lumber to the deck and horizontal boards, or apply panels across the vertical pieces. If insulation is required, place it on top of the vertical members and hold it in place with the nails fastening the boards or panels to the members (Fig. 8-19).

Sheet decks. On a sheet deck, or where the corrugations are very shallow, it might be necessary to use angle clips to attach the vertical members to the deck. Nail the clips to the lumber and bolt or screw it to the deck. Then apply boards or panels as before.

If there is a likelihood of excessive moisture buildup, as might be encountered in icy areas, employ the cold-weather roof principle and supplement it with a mechanically produced air flow if necessary.

Designing Cold-Weather Roof Systems

Cedar shingles and shakes are an excellent roofing material for cold-weather areas that experience heavy snowfall and severe temperatures. A natural wood product, they offer the advantages of durability, superior wind resistance, and good thermal and acoustical properties. As with any other roofing material, however, their best performance depends on proper design, sound construction practices, and correct installation.

In cold-weather areas and particularly in mountainous regions that experience heavy snowfall, the cold-weather or vented roof system is recommended (Fig. 8-20). This system allows a constant flow of cold air above the insulation, but below the roofing material. With other roofing systems, ice buildup along the eaves can be a problem. Heat escapes from the insulation and melts the snow; the resulting water runs down the roof to the cold overhangs, where it freezes. This can cause water to back up and penetrate the roof system. A properly

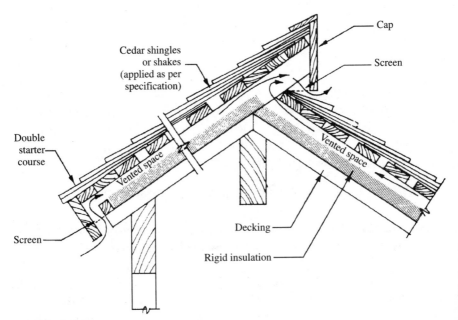

FIGURE 8-20 Details of cold-weather roof systems.

installed, vented cold-weather roof eliminates this problem. Venting space should be sufficient to allow a free flow of air from eave to rooftop.

There are a number of important considerations that influence roof performance in areas that experience heavy snowfall, particularly mountainous regions. Design, of course, is very important. The steeper the roof, the better the performance. Locate chimneys at the ridge or gable ends away from possible snow pressure on the slopes. Locate plumbing pipes on inside walls, extend them between the rafters, and vent them at the ridge. If this is not practical, then use galvanized iron for the plumbing vent pipes and anchor them inside the roof. Plastic vent pipes that extend through the roof can be dislodged by sliding snow.

Avoid wide overhangs at the eaves, as they provide large cold areas for snow and ice buildups. A strip of metal along the eaves helps shed ice quickly. Sliding ice and snow are constant hazards and should be given primary consideration in the total building design. Do not locate outside doors at the bottom of a roof slope. Entrances and all pedestrian traffic areas are better situated beneath the gable ends of the roof. Make certain that all design issues are agreed upon in contract documents.

In regions of heavy snowfall, lay shingles and shakes as a three-ply roof. Use three-layer starter courses for shingled roofs. For shake roofs, superior construction at the eaves is achieved by using two layers of shingles and one of shakes. Proper nailing is very important. All nails should be stainless steel-type 304 or 316 hot-dipped, zinc-coated, galvanized or equivalent and placed ¾ inch from the edge and 1½ inches above the buttline.

Take care to facilitate proper nailing when applying the sheathing boards. Increase shingle or shake sidelap to 2 inches. Lay the entire roof with the same precautions you would take with any other type of wooden shingle or shake roof. Include eave protection and an interlay felt between the shakes.

Properly installed and designed cold-weather roofs result in a sound roof system that gives many years of service during severe extremes of winter temperatures and snowfall.

Adding Novel and Distinctive Effects

One of the most unusual applications of cedar roofing ever used in this country is found in an architectural style that flourished in the 1920s. The roofs of this style, known curiously as English cottage thatch, thatch effect, and shingle thatch, tap back into 16th-century rural England and Ireland, where thatched-roof peasant cottages sprung up in the verdant countryside.

Possibly the most striking characteristics of the country-cottage roof are its rolled eaves and rolled gables. These rolled edges can be achieved through preassembled fascia boards with attached radiused blocking. They are usually sold in 10-foot sections and are nailed to existing roof framing members the same way a conventional fascia board is installed. Any method of gable construction is acceptable, as the preassembled gable sections are nailed on like a conventional fascia board. The radius of the blocking is designed to work with preformed shingles.

The No. 1 cedar shingles come preformed and precut for installation. A designed framing-component system has precut and some preassembled pieces. This enables builders basically to frame roofs with conventional procedures and then attach the manufactured component system to the existing framework.

The prebent shingles, which are used on all rolled sections, are of two different types. Eave-bent shingles are bent across the grain of the wood. Gable-bent shingles are bent with the grain of the wood. Both types of preformed shingles are available with either a straight buttline or a *thatched* buttline.

As the name implies, the eave-bent shingles, which are bent at approximately a 20-inch radius, are used from the rolled eaves to where the roll merges with the roofline. The starting course at the eave is doubled. Either straight-butt or thatched-butt shingles can be used for the starter course. Eave-bent shingles also are used at the start of eyebrow dormers and rolled roof vents.

Gable-bent shingles are used on the gables and can be bent to a 10-inch radius. They are more or less flexible and can be used on almost any radius. Gable-bent shingles can also be used on hips and in valleys. Enough mildly flexible shingles usually can be found in the bundles of straight shingles, however, to cover these areas. Gable-bent shingles come slightly overbent so that the piece can be pressed into place, which causes the shingle to hug the contour of the roof more tightly. On the rounded surface of the eaves it might be necessary to occasionally nail down the butt end of a shingle.

Next to the rolled eaves and gables there is no other feature of the country cottage more integral to the overall beauty of the finished product than the wave coursing. The artistic textures of the various styles of wave coursing create a dimension that cannot be found on any other type of roof. Wave coursing is achieved using a thatched-butt shingle. Wave coursing fits basically into two design categories. Patterned designs are shown in Fig. 8-21 and random designs are shown in Fig. 8-22.

In either style, the waves are achieved in a similar fashion. The long side of each shingle is laid on the short side of the previous one until the maximum exposure is reached. To bring the course down again to the minimum exposure, the procedure is reversed. The short side of each shingle is laid on the long side of the previous one.

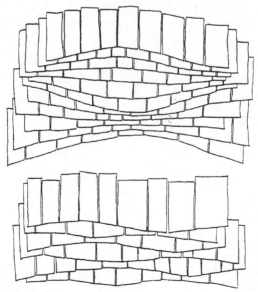

FIGURE 8-21 Pattern designs.

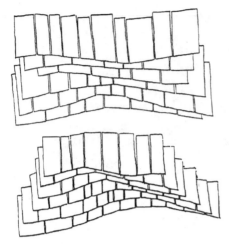

FIGURE 8-22   Random designs.

The following are other applications of distinctive shake effects.

Graduated Exposure: Among the many variations that can be achieved with hand-split shakes is one that is produced by reducing the exposure of each course from eaves to ridge. This requires shakes of several lengths.

A very serviceable graduated exposure roof can be built by starting at the eaves with 24-inch shakes, laid 10 inches to the weather. Lay one-half of each roof area with shakes of this length. Reduce the exposure gradually to about 8½ inches. Complete the area with 18-inch shakes, starting with an 8-inch exposure and diminishing to a 5-inch exposure. The resulting appearance exaggerates the actual distance from eaves to ridge.

Tilted Strip at Gables: While shake roofs are less prone to drip rainfall or snowmelt from the gables than are roofs of smoother materials, gable-drip and icicles can be eliminated by inserting a single strip of cedar beveled siding the full length of each gable end with the thick edge flush with the sheathing edge. The resulting inward pitch of the roof surface keeps moisture away from the gable edge. It is also a pleasing accent to the gableline.

Staggered Lines: To secure irregular and random roof patterns, shakes can be laid with butts placed slightly below or above the horizontal lines that govern each course. If an extremely irregular pattern is desired, longer shakes can be interspersed in the roof with their butts several inches lower than the course lines.

Mixtures of very rough shakes and fairly smooth shakes or wooden shingles can produce a more rugged appearance than a roof composed entirely of rough shakes.

Treating Wooden Roofs

To prevent owner dissatisfaction and complaints, it is in the best interest of the roofing contractor to explain to the customer the effects of weathering on a shake or shingle roof before contracts are signed. The surface of any untreated wood product changes when exposed to the effects of sunlight and precipitation. This change is partly physical and partly chemical. It is cumulatively referred to as weathering.

The first noticeable change is the color. The initial red-brown color tends to fade, and a graying process begins. This change to a silvery gray results when ultraviolet radiation from the sun strips the surface layer (less than 0.01 inch deep) of certain cell-wall materials. This first change in color is rather rapid, occurring within the first year of exposure or, under more severe conditions, within several months. Gradually, the silver gray changes to a darker, more graphite gray, which indicates the colonization of microfungi on the surface and completes the initial phase of the weathering process.

Good housekeeping for a wooden roof requires the removal of all leaf litter, pine needles, and debris that accumulate over time between the shingles and shakes and in the valleys of the actual roof structure. It is best to do this housekeeping before the wet season begins. In the Pacific Northwest, this is before the autumn rains begin. Along the Gulf Coast or in areas that are conducive to the growth of molds, mildews, and fungi, this task might be required several times each year.

The accumulated debris has two detrimental effects on the roof. First and foremost, debris tends to retard the shedding of water. Second, and more destructively, debris retains moisture that allows wood-destroying fungi to grow. Wood-rotting fungi require water as part of their life cycle. If you can reduce the amount of water so that the minimum level is never reached, the organisms cannot grow.

To promote drying, it is important to remove overhanging tree branches that excessively shade roof sections and retard the drying of the roof surface. Never allow tree branches to touch and rub against the surface of the roof. The mechanical action of this rubbing can literally wear sweeping grooves into the surface of the shingles or loosen the fasteners that hold them.

Finally, in excessively woody areas, branches can retard the flow of air so that surface drying occurs very slowly. In all cases, use sound judgment about pruning, trimming, or removing problem-causing trees. Excessive growth of lichens and mosses is a sure indicator that a roof is not drying properly and that enough moisture is available to support wood-rotting organisms.

Removing the accumulated debris is relatively simple. In most cases, anyone who is willing to climb on the roof can wash off most of the material with a garden hose. A stiff broom is handy for removing all the leaf and pine-needle litter.

It is easy to remove litter from around chimneys and in valleys, but all those areas between the individual shingles and shakes, or *keyways*, can be more difficult to clean. Wet sweeping the surface in a careful manner also removes the initial growths of moss and lichens. Professional services sometimes use high-pressure washers. These are effective, but they must be used with extreme care because cedar is a soft, low-density wood. Excessive or imprudent use of high-pressure systems can detach shingles or inflict many years of wear in moments, even though the roof initially looks like new. Many residential roofing contractors now provide a maintenance service, at an extra charge, for homeowners who do not want to climb on their roofs.

Most cedar roofs remain untreated and are known to provide excellent service. A surface chemical treatment is desirable, however, under conditions that can be conducive to premature deterioration, such as areas whose climate combines heat and humidity for considerable portions of the year, roofs of low pitch or slope, roofs beneath overhanging trees, etc. There are naturally resilient oils present in cedar, so it is more advantageous to let the roof weather for approximately one year before applying a treatment. Allowing the oils to leach out gives the wood a greater porosity to absorb the preservatives.

Fungicidal chemicals inhibit moss, fungus, and mildew and contribute to a roof's life service. When applying one of these chemicals, keep in mind that they are more effective when the moss is actively growing than when it is in the dormant phase. In most parts of our country, moss grows vigorously during rainy seasons.

If chemicals are applied during the dry season, it is recommended that you wet the moss first to promote chemical uptake. To control runoff and increase effectiveness, apply when rain is not likely. It may be several days before the growth dies and can be effectively removed. Remember to use care when handling and disposing of chemically treated moss and litter to avoid accidental exposure.

The corrosive nature of these chemicals means that unnecessary damage must be avoided by thoroughly rinsing metal tools, gutters, and flashings that come in contact with the solutions. Exercise care when applying some chemicals due to their toxicity. Follow the manufacturer's directions.

In areas of low humidity, where moss, fungus, and mildew are not a potential problem, commercial oil-based preservatives can prevent excessive dryness and prolong the natural resilience of the cedar. In such areas, fungicidal chemicals can be used to good effect, but are not required. Remember that these surface treatments are temporary and should be applied approximately every five years, depending on the products used.

When applying preservative chemicals to shakes and shingles, consider several important issues. First, the coating should be as uniform as possible. Avoid drips and runs. It is better to apply several light coats and obtain good absorption than to try to force the chemicals all in at once. The important point is to reach a certain level of preservative retention. A low retention level results in an ineffective treatment.

Second, since much of the decay occurs in the butt region, be conscientious when applying the solution so that the exposed butts are covered as well as the exposed shingle surface.

The three methods available to provide proper application are brushing, using a thick-napped roller, or spraying. Because of the irregular surface, a multigallon pump-type sprayer, set to produce a coarse spray, results in a more uniform coating. However, spraying requires a calm day, with a wind less than 4 miles per hour, to prevent overspray from drifting over plants or into neighboring yards. Brushes also are effective because they allow for the thorough treatment of the keyways and ends.

The Cedar Shake and Shingle Bureau recommends CCA pressure-treated cedar shakes and shingles in Southeastern states with a climate index of 65 or greater as identified by the U.S. Department of Agriculture's Forest Service. The pressure treatment is permanent and extends the service life of the roof. Most CCA-treated shingles and shakes are available with a 30-year warranty.

The services referred to in the paragraphs above are very important and can be translated into an excellent cash flow. Set up service contract sales while developing the original roofing contract. By doing so, you are working smart rather than hard—being paid to monitor your warranties.

Slate Shingles

The principal difference between slate and other stones is the natural cleavage in the slate. This cleavage permits it to be split in one direction, forming sheets of various thicknesses. Slate is cut and trimmed, mostly by hand, into the required sizes and thicknesses. A slate roof is typically a custom product and its use gives character to a building unlike that of any other covering. Slate quarried for roofing is of dense, sound rock and is exceedingly tough and durable.

The minerals that constitute slate were deposited in bodies of water many years ago. Erosion and deposition account for the different compositions of successive beds. Other materials then were deposited over the clays, and the pressure of the superimposed material gradually united the clays into shale.

Many of these shale beds were then subjected to intense pressure and high temperatures from the crumpling and folding of the earth's crust. This process turned the shale beds into slate. Slate's cleavage is the result of mineral grains being parallel to the extreme pressure forces.

The slate industry in the United States started in the mid to late nineteenth century. The first mines, or quarries, were developed in the Mettowee Valley, which lies between the Adirondack Mountains of New York and the Green Mountains of Vermont, by immigrants from Wales.

Huge holes are dug into the ground until the first beds of slate are found. Often, the top layers of slate, which usually are brown or gold in color, are very soft. These slates are not recommended for roofing use because they do not afford the same quality offered by the deeper beds. Only the best quality slate can be used to make roofing slate. From hot summer temperatures to the most brutal of cold winters, wind, water, ice, and snow take their toll on any slate that is not made from the best quality rock.

Today, blocks of slate are removed from quarry walls and driven to mills. Outside the mill, jackhammers cut these huge blocks into smaller, more manageable blocks. Conveyor systems then move the blocks inside the mill to saws with diamond blades. The diamond blades, one of the few advances in the slate roofing industry since its inception in the 1800s, can cut through the thickest of blocks. The process cuts the blocks to a size that is approximately two inches longer and an inch wider than needed.

Then the most impressive production step occurs. A craftsperson, known as a *splitter*, using a hammer and chisel and the measurement of his or her own eye, lays the chisel along the natural cleave of the stone and taps the end of the chisel with a hammer. This splits the slate evenly and cleanly. This is truly an art that must be mastered.

Those who have the feel for the stone and the natural ability to judge the thickness of each piece can quickly split the slates to the desired thickness. It is rather mind-boggling to think that every piece of roofing slate made today is split by hand. There is simply no machine available that can eye the slate, spot the cleavage, and gently, but firmly, evenly split the pieces of slate.

Once the pieces of slate are split for thickness, they are then trimmed to the ordered size. Trimming machines have a revolving blade that chops the slate to the desired size. The process leaves a *chamferred* edge, which is a rough edge that allows the texture of the slates to be seen.

The final step is to punch nail holes into the slate pieces. This step, done one piece at a time, can be a time-consuming process. It also acts as a type of quality control. As the craftsperson puts these slates through the punching process, any piece of slate that is cracked or broken, or not of the desired quality, is discarded. After the slate pieces are punched for nail holes, they are loaded onto pallets, and prepared for shipment.

The production of roofing slate is truly an art. Because these slates are a natural stone product, each piece has its own unique qualities. Roofing slate comes in thicknesses from ¼ to 1 inch and offers roof appearances that range from smooth and uniform to rough and textured. Slate roofs have a service life of 75 to 100 years.

Examining Slate Characteristics

As previously mentioned, slate has a natural cleavage that permits it to be more easily split in one direction than in others. A second direction of fracture or *scallop*, which is usually at right angles to the cleavage, is called the *grain*. Roofing slates are commonly split so that the length of the slate runs with the grain.

Slate quarried for roofing stock is of dense, sound rock that is exceedingly tough and durable. Slate, like any other stone, becomes harder and tougher upon exposure than it is when first quarried. It is practically nonabsorptive. Tests show that typical slate has a porosity of 0.15 to 0.4 percent.

The nature of the slate surface after splitting is dependent on the character of the rock from which it is quarried. Many slates split to a smooth, practically even and uniform surface, while others are somewhat rough and uneven. As a result, a wide range of surface effects is available for the finished roof.

Slate from certain localities contains comparatively narrow bands of rock differing to various degrees in chemical composition and color from the main body of the stone. These bands are called *ribbons*. Ribbons that do not contain injurious constituents and are of desirable color are not objectionable. Slates of this type, when trimmed so that the ribbons are eliminated, are known as *clear* slate. Slates that contain some ribbons are sold as *ribbon stock*.

Picking Colors

The color of slate is determined by its chemical and mineral composition. Since these factors differ in various localities, it is possible to obtain roofing slates in a variety of colors and shades.

It is truly remarkable to find a natural product that possesses, in addition to its other qualities, such unlimited color possibilities. Surface colors can be uniform, or have contrasting hues. Moreover, if the design of the building requires a roof of one general color, it can be graded up or down the slope from dark to light as desired.

To relieve the monotony of a flat, uniform body color, various shades of the same color can be used to provide interesting variations up and down, across, or interspersed throughout the roof. A low-eaved, prominent roof surface in a quiet or contrasting blend of autumnal colors can cause the structure to blend with its surroundings. Slate color can be *unfading*, or, by selecting *weathering* slate, the roof color can mellow with age and weather. These extremes and all the steps between are available to roof designers.

To take advantage of the various available effects, the source of the slate, as well as its ultimate color characteristics, should be known. For instance, Pennsylvania slate colors are blue-grey, blue-black, and black. Buckingham and other Virginia slates are generally blue-grey to dark grey with micaceous spots on the surface that produce an unusual luster. Vermont slates can be light grey, grey-black, unfading and weathering green, unfading purple, and variegated mottled purple and green. Unfading red slates are found only in Washington County, New York. These are the most costly.

For the purpose of classifying the basic natural colors of roofing slate now available in large quantities for general use, the Division of Simplified Practice of the U.S. Department of Commerce recommends the color nomenclature for slate materials shown in Table 9-1.

Estimating Quantities

In the United States, slate is sold by the *square*. A square of roofing slate is defined by the U.S. Department of Commerce, Bureau of Standards, in Simplified Practice Recommendation No. 14, as follows:

TABLE 9-1 Basic Slate Colors

Black	Grey	Purple	Green
Blue black	Blue grey	Mottled purple and green	Red

A Square of Roofing Slate: A square of roofing slate means a sufficient number of slate shingles of any size to cover 100 square feet of plain roofing surface, when laid with approved or customary standard lap of 3 inches. Slates for surfacing flat roofs are usually laid tile fashion, without lap, in which case a square of slate would cover an area greater than 100 square feet.

The quantity per square varies from 686 pieces for the 10-x-6-inch size to 98 for the 24-x-14-inch size, which includes the allowance for a 3-inch headlap. Note that for roofs of comparatively little slope, where a 4-inch lap is required, an additional quantity must be provided. For steep roofs or siding, where a 2-inch lap is sufficient, fewer slates are necessary. However, slate is always sold on the basis of quantity required for a 3-inch lap even for flat roofs.

Table 9-2 shows the sizes for standard $\frac{3}{16}$-inch-thick slate, the minimum number of slates required per square, the respective exposures for the slates listed, and the weight (per square) of the nails used to secure each size of slate.

Each roofing contractor has a method of compensating for waste, breakage, projections through roofs, dormers, hips, ridges, valleys, and other factors occasionally encountered in roofing work. The method is usually based on experience, labor skill, local conditions or practice, and items peculiar to a locality.

The following suggestions are therefore offered mainly for architects and others not actively engaged in selling and laying slate roofs. Use it as a guide to the many factors that should be taken into consideration. For rough estimates, a good slate roof costs from 6 to 8 percent of the total cost of the average building or home.

From the time work is first started at the quarry until the material is laid on the building, there are certain costs that must be taken into consideration. These can be listed as follows. (See also Fig. 9-1.)

Cost of slate (punched) on cars at the quarry

Freight from quarry to destination

Loading and hauling to storage yard

Unloading, piling, and waste at storage yard

Loading and hauling to job

TABLE 9-2 Schedule for Standard $^3/_{16}$-inch-thick Slate

Size of slate (in.)	Slates per square	Exposure with 3" lap (in.)	Nails per square (lb, oz)	
26 × 14	89	$11\frac{1}{2}$	1	0
24 × 16	86	$10\frac{1}{2}$	1	0
24 × 14	98	$10\frac{1}{2}$	1	2
24 × 13	106	$10\frac{1}{2}$	1	3
24 × 11	125	$10\frac{1}{2}$	1	7
24 × 12	114	$10\frac{1}{2}$	1	5
22 × 14	108	$9\frac{1}{2}$	1	4
22 × 13	117	$9\frac{1}{2}$	1	5
22 × 12	126	$9\frac{1}{2}$	1	7
22 × 11	138	$9\frac{1}{2}$	1	9
22 × 10	152	$9\frac{1}{2}$	1	12
20 × 14	121	$8\frac{1}{2}$	1	6
20 × 13	132	$8\frac{1}{2}$	1	8
20 × 12	141	$8\frac{1}{2}$	1	10
20 × 11	154	$8\frac{1}{2}$	1	12
20 × 10	170	$8\frac{1}{2}$	1	15
20 × 9	189	$8\frac{1}{2}$	2	3
18 × 14	137	$7\frac{1}{2}$	1	9
18 × 13	148	$7\frac{1}{2}$	1	11
18 × 12	160	$7\frac{1}{2}$	1	13
18 × 11	175	$7\frac{1}{2}$	2	0
18 × 10	192	$7\frac{1}{2}$	2	3
18 × 9	213	$7\frac{1}{2}$	2	7

TABLE 9-2 Schedule for Standard $^3/_{16}$-inch-thick Slate (*Continued*)

Size of slate (in.)	Slates per square	Exposure with 3" lap (in.)	Nails per square (lb, oz)	
16 × 14	160	6$^1/_2$	1	13
16 × 12	184	6$^1/_2$	2	2
16 × 11	201	6$^1/_2$	2	5
16 × 10	222	6$^1/_2$	2	8
16 × 9	246	6$^1/_2$	2	13
16 × 8	277	6$^1/_2$	3	2
14 × 12	218	5$^1/_2$	2	8
14 × 11	238	5$^1/_2$	2	11
14 × 10	261	5$^1/_2$	3	3
14 × 9	291	5$^1/_2$	3	5
14 × 8	327	5$^1/_2$	3	12
14 × 7	374	5$^1/_2$	4	4
12 × 10	320	4$^1/_2$	3	10
12 × 9	355	4$^1/_2$	4	1
12 × 8	400	4$^1/_2$	4	9
12 × 7	457	4$^1/_2$	5	3
12 × 6	533	4$^1/_2$	6	1
11 × 8	450	4	5	2
11 × 7	515	4	5	14
10 × 8	515	3$^1/_2$	5	14
10 × 7	588	3$^1/_2$	7	4
10 × 6	686	3$^1/_2$	7	13

✓	Item
	Cost of slate (punched) on cars at the quarry
	Freight from quarry to destination
	Loading and hauling to storage yard
	Unloading, piling, and waste at storage yard
	Loading and hauling to job
	Unloading and piling at job
	Placing on roof and laying
	Roofing felt
	Nails
	Elastic or plastic cement
	Snow guard or snow rails
	Sheet metal
	Labor, including compensation insurance
	Waste in handling, cutting, and fitting
	Contractor's overhead on organization and equipment
	Cost of guarantee or bond
	Contractor's profit

FIGURE 9-1 Choosing slate.

Unloading and piling at job

Placing on roof and laying

- Roofing felt

- Nails

- Elastic or plastic cement

- Snow guard or snow rails

- Sheet metal

- Labor, including compensation insurance

- Waste in handling, cutting, and fitting

Contractor's overhead on organization and equipment

Cost of guarantee or bond

Contractor's profit

This list is available to print for the field. The file is Cost Considerations for Slate on the CD included with this book. While it might seem a comparatively simple problem to estimate the net quantity, it is not as easy to allow for the additional material required for slate around chimneys, dormers, hips, valleys, etc. These allowances depend largely on the judgment and experience of the estimator and the roof design.

Estimating Weights

The weight of a square of slate varies from 650 to 8000 pounds, depending on the thickness of each slate. Slates of commercial standard thickness (approximately $^3/_{16}$ inch) weigh from 650 to 750 pounds per square. The dead-load weight of slate per square, however, can be estimated at a maximum of 800 pounds, or 8 pounds per square foot, when the weight of the slate, felt, and nails is combined.

The weight of slate shingles depends on the size, the color, and the source quarry. Table 9-3 shows the weights per square of slate products of different thicknesses for both sloped and flat roofs. The actual weight of slate products can vary from 10 percent above to 15 percent

TABLE 9-3 Average Weight of Slate Per Square

Slate thickness (in.)	Sloping roof with 3" lap (lb per square)	Flat roof without lap (lb per square)
$3/16$	700	240
$3/16$	750	250
$1/4$	1000	335
$3/8$	1500	500
$1/2$	2000	675
$3/4$	3000	1000
1	4000	1330
$1\frac{1}{4}$	5000	1670
$1\frac{1}{2}$	6000	2000
$1\frac{3}{4}$	7000	
2	8000	

below the weights shown in the table.

Installing Slate Roofs

As with all other roof installation methods, there are certain important considerations that must be kept in mind before you start to lay a slate roof.

Using Slater's Tools

The tools commonly used by the slater are the punch, hammer, ripper, and stake. If much punching is done at the yard or on the job, use a punching machine to punch the nail holes and cut the slates. It is adjustable to any size or shape, and it cuts and punches a countersunk hole in one operation. The hand or mawl punch is forged from fine tool steel that has been hardened and ground. It is about 4½ inches long with one tapered end. The butt end is struck with a mawl to punch the nail hole.

Slater's tools are all drop-forged. An approved hammer is forged solid, all in one piece, from crucible cast steel, with an unbreakable leather handle to avoid slipping and blistering the hands. One end terminates in a sharp point for punching slate, the other in the hammer head. There is a claw in the center for drawing nails, and on each side of the shank there is a shear edge for cutting slate. The head, point, and cutting edges are tempered to withstand heavy work. The slater's stock size hammer has a 12-inch handle.

The ripper is about 24 inches long and is forged from crucible cast steel. It is used to remove broken slate and make repairs. A hook on the end can be used to cut and remove slating nails. The blade is drawn very thin, and the hook end is correctly tempered for hard wear.

The stake is about 18 inches long and T-shaped. The long edge is used as a rest upon which to cut and punch slate or as a straight edge to mark the slate when cutting and fitting around chimneys, hips, valleys, etc. The short arm is tapered and pointed for driving into a plank or scaffold.

These tools, as well as 24-inch stakes, shorter rippers, left-handed hammers, special hammers, and other tools, can be obtained from any slate producer or tool manufacturer. A nail pouch, tinner's snips, rule, and chalkline complete the slater's equipment.

Laying Underlayment Felts

A watertight, standard slate roof can be laid on open lath boards instead of on felt, as is often done on buildings that do not require heat. If felt is used, the thickness of the felt has a relatively minor effect on the watertight integrity of the roof. If felt is installed as soon as the roof is sheathed, it protects the building from moisture until the slates are laid.

Felt also forms a cushion for the slate. This cushion is especially important when thick slates are used. For standard slate roofs, apply a No. 30 felt. For graduated slate roofs, apply two layers of felt and stagger the felt joints and laps. As with tile, whenever slate is laid on a roof with a minimum slope of 4 inches per foot or less, a double layer of felt set in either mastic or hot asphalt is recommended to ensure a watertight roof.

Felts for use in slate roofing should be asphalt-saturated felts and No. 30 or heavier. The following are some general guidelines for applying felts for slate roofing purposes.

- Lay felts in horizontal layers with joints lapped toward the eaves and at the ends.

- Employ a lap of at least 3 inches, and secure the edges of the felts to the surface over which the felts are laid.

- Extend the felt over all hips and ridges at least 12 inches to form a double thickness.

- Employ a lap of not less than 2 inches over the metal lining of valleys and gutters.

All felt should be preserved unbroken, tight, and whole. Many roofers drive the slater's stake into the roof boards. To avoid damaging the roofing felt, use a plank for this purpose or drive the stake into the scaffold only. In locations where the January mean temperature is 30°F or lower, apply two plies of No. 15 felt or one ply of No. 50 felt as the underlayment to serve as an ice shield. Work from the eaves to a point 24 inches inside the inside wall line of the building. Set the felt in hot asphalt or mastic, or an adhered bitumen membrane.

Nailing Slate

Like any other construction unit, a slate roof can only be as strong and enduring as its weakest part. Most slate roof failures can be attributed to the punching of the nail holes, the nailing of the slates, or the nails themselves.

Each piece of slate should have at least two nail holes. The standard practice is to machine punch, at the quarry, two holes in all architectural roofing slate ¼ inch and thicker. Four holes should be used for slates that are more than 20 inches in length and ¾ inch or more in thickness. Holes are punched ¼ to ⅓ the length of the slate from the upper end, and 1¼ to 2 inches from the edge (Fig. 9-2). When four holes are used, it is customary to locate the two additional holes about

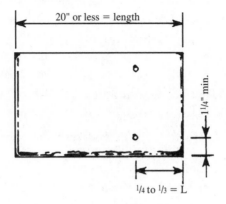

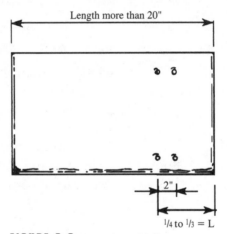

FIGURE 9-2 Location of holes.

2 inches above the regular holes.

Machine punching is preferable to hand punching. The term *hand punching* usually refers to the use of the double-headed slater's hammer. Machine punching can be done either at the quarry or on the job. Hand punching is necessary to fit hips and other complete joints.

The heads of slating nails should just touch the slate. Do not drive the slate "home" or draw the slate. The nailhead should clear the slate so that the slate hangs on the nail. Slate that is held too rigidly in place can shatter around the nailhole and eventually ride up and over the nail and be blown off in a heavy wind. The blame is placed on the material, whereas the real reason was to the nailing method. All nails should penetrate the sheathing and not the joints between boards. This is especially important near the ridge of the roof.

Slating nails should be 10-gauge, large-head hard copper wire nails, cut brass, or cut yellow metal type. Standard thickness slate usually can be applied with 1½-inch nails. Use 2-inch nails for hips and ridges. Thicker slate requires longer and heavier-gauge nails.

For all practical purposes, the ordinary diamond-point and smooth-shaft nail are sufficient. The needle-point nail is seldom necessary or advantageous. The nail shaft, since it supports a greater weight and must resist a small shearing stress, should be larger than that of the shingle nail. To prevent the slate from being lifted up and over the nail after it is laid, the diameter of the nail head should be greater than that of shingle nails.

Applying Elastic or Plastic Cement

Elastic or plastic cement is used under slates at hip and ridge locations to help secure those slates that are usually smaller than regular roofing slates and that cannot be nailed easily. Plastic cement is also used for pointing the peaks of hips and ridges.

Elastic or plastic cement must be waterproof. It must have a high melting point to prevent the slates from slipping under the heat of the sun, and a low freezing point so that it does not become brittle and crack in cold weather. The cement should be oily and sticky so that it adheres thoroughly to the roof.

Handling Materials

When piling slate, pay attention to the foundations of each pile, how you start the piles, how you arrange the piles and individual slates, and how you separate the tiers. The tiers of slate can be kept level only if the foundation is level and free from settlement. The earthen foundation, upon which the slates are to be piled, should be level, dry, and solid. A layer of 2-inch-thick wooden plank material keeps the slate off the ground, distributes the load, and helps maintain straight and even piles.

When stacking slate, start the first tier by stacking one pile of slate flat and to a height equal to the width of the slate. For example, when 20-×-12-inch slate is used, the flat pile of the first tier should be 12 inches high. The slates piled on top of the first tier should be placed in an upright position, on edge lengthwise, and kept as straight and vertical as possible. During inclement weather, cover piled slate with a tarp.

Designing a Slate Roof

In general, slate roofs are classified in three architectural designs: the standard slate roof, the textured slate roof, and the graduated slate roof.

Standard Slate Roofs

Standard slate roofs are made with standard commercial slate that is approximately $\frac{3}{16}$ inch thick and of a standard length and width, with square tails or butts laid to a line. Slate of this type is commonly obtainable in the basic slate colors.

Standard roofs are suitable for any building design that calls for permanent roofing material at a minimum cost. It differs from other slate roofs only in characteristics that affect the texture or appearance of the roof, such as the shape and thickness of the individual pieces. If desired, the butts or corners can be trimmed to give a hexagonal, diamond, or gothic pattern for all or part of the roof, as, for instance, on a church spire. Standard roofs are sometimes varied by laying two or more sizes (lengths and widths) of standard commercial slate on the same area.

Textured Slate Roofs

The term *textured* is used to designate those slates usually of rougher texture, with uneven tails or butts, and with variations in thickness or size (Fig. 9-3). In general, this term is not applied to slate more than $\frac{3}{8}$ inch thick. Varying shades are frequently used to enhance the color effect, which, with the characteristics just mentioned, adds interest in line and texture to the roof

FIGURE 9-3 Textured slate roof.

design. In addition to the basic colors of the commercial grades, accidental colorings of bronze, orange, etc., can also be used in limited quantities.

Graduated Slate Roofs

The graduated roof combines the artistic features of the textured slate roof with additional variations in thickness, size, and exposure. The slates are arranged so that the thickest and longest occur at the eaves. The pieces then gradually diminish in size and thickness to the ridges. Slates for roofs of this type can be requested in any combination of thicknesses from $\frac{3}{16}$ to $1\frac{1}{2}$ inches and heavier.

The graduated slate roof presents many opportunities for variation and offers excellent possibilities for interesting treatments. It is referred to as a custom roof because it often is designed to harmonize with the general character of the building. Many producers and distributors maintain special design staffs to help architects secure the most suitable and satisfactory graduated slate roofs. These services are freely offered to designers and architects in the interest of a better and more harmonious slate roof.

There are three other methods of laying slate that, when properly executed, have proven satisfactory for certain effects and locations. Dutch lap is placed with regular slate on shingle lath or tight sheathing (Fig. 9-4). Open slating is especially suitable for barns and other buildings or where ventilation is desirable (Fig. 9-5).

The French, diagonal, or hexagon slating method is less expensive than the standard American method (Fig. 9-6). When using this method to lay new roofs, if heat conservation is a factor, apply 30-pound (instead of 14-pound) asphalt-saturated felt on a solid roof deck of sheathing or roof boards. This is a popular method for both new construction and reroofing jobs in New England.

Laying Slates

Workmanship is as essential as the proper selection of the material. The more enduring the material, the more important intensive, skilled labor becomes. With slate, the most lasting roofing material known, it is important that application guidelines be followed. Slate application should be

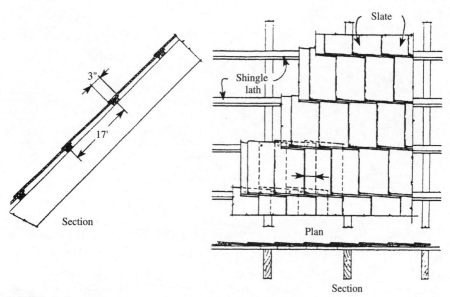

FIGURE 9-4 Dutch lap.

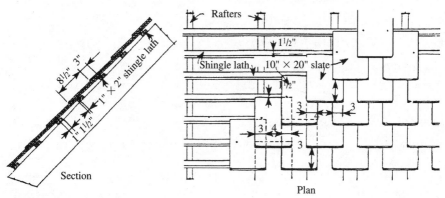

FIGURE 9-5 Open slating.

handled by a knowledgeable contractor who employs well-trained, experienced workers.

Through joints from the roof surface to the felt should not occur. Separate the joints in each slate course from those below. Otherwise, water can migrate through the joints and cause felts to disintegrate and leaks to develop in the roof.

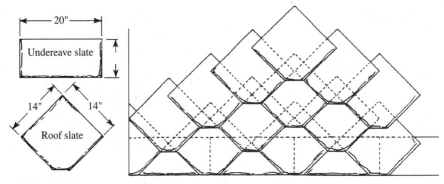

FIGURE 9-6 French method.

When slates of random widths are used, the overlapping slate should be joined as near the center of the underlying slate as possible and not less than 3 inches from any underlying joint. When all slates are of one width, start every other course with a half-slate or, where available and practical, with a slate that is one-and-a-half times the width of the other slates, as shown in Fig. 9-7.

Determining Slate Exposures

The exposure of a slate is the portion not covered by the next course of slate above. This portion is the length of the unit exposed to the weather. The standard lap of the alternate courses used on sloping roofs is 3 inches; this is the basis on which all roofing slate is sold and the quantity computed.

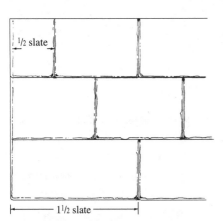

FIGURE 9-7 Starting slate.

To find the proper exposure for individual jobs, deduct 3 inches from the length of the slate and then divide by 2. For instance, the exposure for a 24-inch slate is 24 inches minus 3 inches equals 21 inches. Divide the total by 2 and the result is an exposure of 10½ inches. Use Table 9-4 to obtain the proper exposure.

Construct sloping roofs that have a rise of 8 to 20 inches per foot of horizontal run with the 3-inch lap. Buildings located in the southernmost parts of the country, or on the Pacific slope, can be safely roofed with a lap of 2 inches provided that a high standard of workmanship is maintained.

For steeper roofs, such as the mansard and others that are nearly vertical in plane, a 2-inch lap is usually sufficient. In some sections of the country, it is customary to increase the lap to 4 inches when the slope is from 4 to 8 inches per foot, while in other parts the 3-inch lap is considered entirely adequate. Employing a headlap of less than 3 inches reduces the amount of material over which water can be blown and increases the possibility of leaks.

TABLE 9-4 Exposure for Sloping Roofs

Length of slate (in.)	Exposure (in.)*
24	10½
22	9½
20	8½
18	7½
16	6½
14	5½
12	4½
10	3½

*Slope 8 to 20 inches per foot, 3-inch lap.

Finishing Ridges

There are two common methods to finish the roof ridge. These are usually known as the *saddle* ridge and *combing* ridge, but each might have other names and different laying procedures according to local practice.

SADDLE RIDGES

For the saddle ridge method, extend regular slates to the ridge so that pieces of slate on the opposite side of the roof butt flush. On top of the last regular course of roofing slate at the ridge, lay another course of slate, called combing slate, and butt flush pieces of slate on the opposite sides of the roof. Lay the combing slate with the grain horizontal. Use a width that maintains an approximately uniform exposure or gauge.

For example, if 20-x-12-inch slates are applied with an 8½-inch exposure, lay 12-x-8-inch slates horizontally on the ridge. Note that the combing slates should overlap and break joints with the underlying slate. In this way, all nails in the combing slate are covered by the succeeding

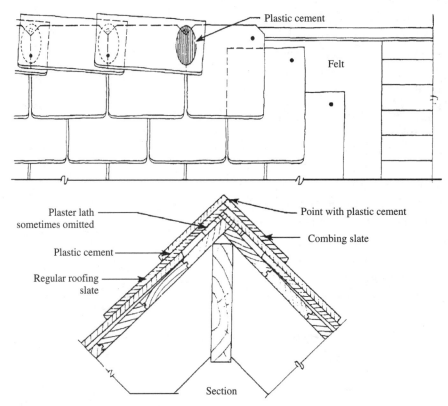

FIGURE 9-8 Saddle ridge.

slates, except for the nails in the last slate, which is called the finishing slate. Cover the nails in the finishing slate with plastic or elastic cement.

Fill the joints on top of the ridge formed by the butted edges of the combing slate with cement when these joints are subject to heavy rainfall (Fig. 9-8). Many architects prefer to keep the grain of the slate vertical, using combing slate that is the same width and exposure as the regular slate used on the roof. In such cases, the starting slate should be a slate-and-a-half wide rather than a half-slate wide.

A variation of the saddle ridge is known as the strip-saddle ridge (Fig. 9-9). Lay this type of ridge in a manner similar to the saddle ridge, but do not overlap the combing slates. Rather, the combing slates should butt flush and be attached with four nails. The combing slate for strip-saddle

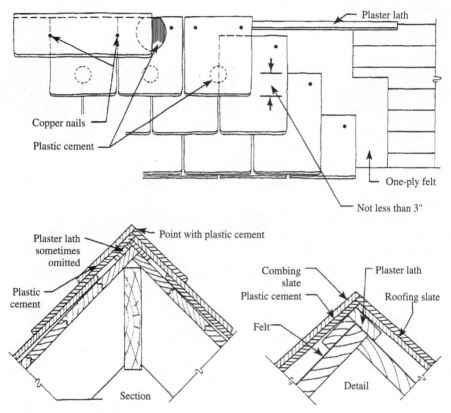

FIGURE 9-9 Strip-saddle ridge.

ridges can be the same width as the regular slate used on the roof, or narrower if the designer wishes. Cover the four nails with plastic cement, and set the edges of the combing slate in plastic or elastic cement. Clarify all design criteria with care and include them in contract documents.

COMBING RIDGES

The combing ridge is laid in the same manner as the saddle ridge, except that the combing slate of the north or east side extends beyond the ridgeline, as shown in Fig. 9-10. This extension should not be more than 1 inch. This type of ridge can be laid with the combing slate grain either vertical or horizontal. In either case, the edge of the slate should be set in elastic cement and the nails covered with elastic cement. If the

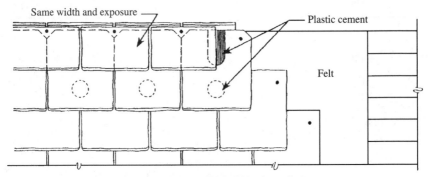

Combing slate laid with grain vertical

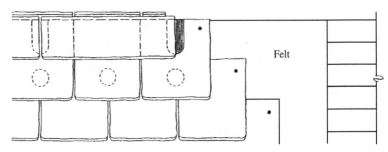

Combing slate laid with grain horizontal. Smaller
slate of proper size can be used to give same
exposure as rest of roof courses.

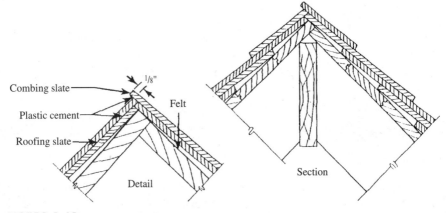

FIGURE 9-10 Combing ridge.

top or combing course projects $\frac{1}{16}$ to $\frac{1}{8}$ inch above the top courses, it makes a better finish and is more easily filled with elastic cement.

A variation of this type of ridge is known as the *coxcomb* ridge. The combing slate alternately projects on either side of the ridge.

Dealing with Hips

There are several ways to form hips on slate roofs. Descriptions of the most common follow.

SADDLE HIPS

The saddle hip can be formed by placing one or two plaster laths or a $3\frac{1}{2}$-inch cant strip on the sheathing. This forms the hip. Run the roofing slate up to this strip. On top of the cant strip and the slate, lay the hip slates, which are usually the same width as the exposure of the slates on the roof, although they can vary in width on different classes of work.

Drive the four nails used to fasten the hip slate to the roof into the cant strip. Do not nail between the joints of the slate. Cover the heads of these nails with elastic cement and embed the lower part of the next slate. Elastic or plastic cement is also placed on the joint between the roofing slate and the plaster lath and on the peak of the hip before the hip slates are laid.

A variation of the saddle hip is known as the strip-saddle hip. This method is used on less expensive work and can be made with narrower slates laid with butt joints that do not necessarily line up with the course of the slate on the roof.

MITRED HIPS

For this type of hip, the slates that form the roof courses and the hip are all in one plane. Cut the hip slates accurately to form tight joints and fill the joint with elastic cement. Place the nail holes so that they come under the succeeding hip slate.

It is sometimes recommended that metal or slip flashings be woven in with each course of mitred hips. This is usually unnecessary if proper care and workmanship are exercised when cutting, fitting, and embedding the hip slates. Some roofers do not use plastic or elastic cement on the hip slates and secure satisfactory results.

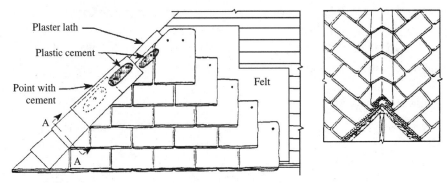

FIGURE 9-11   Boston hip.

A variation of the mitred hip is known as the *fantail* hip. It is laid in the same manner as the mitred hip, but the bottom edge of the hip slate is cut at an angle to form a fantail.

BOSTON HIPS

Another popular type of hip is known as the Boston hip. The slates are woven in with the regular courses of the roofing slates, as shown in Fig. 9-11. The nails then are covered with elastic or plastic cement and the lower part of the succeeding slate embedded therein.

Building Valleys

Of the two methods used to form valleys, the first—and without doubt the most satisfactory—is the open valley. The second, known as the closed valley, is considered by many to be more pleasing in appearance and often is used in high-grade work. Variations of the closed valley, frequently used in connection with graduated or textured roofs, are the *round* valley and the *canoe* valley.

OPEN VALLEYS

The open valley is formed by laying strips of sheet metal in the valley angle and lapping the slate over it on either side. This leaves a space between the slate edges that acts as a water channel down the valley angle. The width of the valley, or the amount of space between the slate edges, should increase uniformly toward the bottom. The amount of this increase, or taper, has been determined as 1 inch in 8 feet.

For example, in a valley 16 feet long, the distance between slates should be 2 inches greater at the bottom than at the top because the width increases at the rate of ½ inch in 8 feet on each side of the valley. This permits a uniform flashing width that is approximately two-thirds the slate width under the slate adjacent to the valley. The difference in width allows the slate to be laid closer to the valley at the upper end than at the lower end and takes care of the increased water flow at the base of the roof. This tapering of the valley also has the very practical effect of allowing any ice that might form to free itself and slide down as it melts.

Allow for this increase in valley width when placing the flashing strips. Valley flashings are generally laid in pieces up to 8 feet long. Theoretically, the best way to achieve the taper is to taper the sheets. Since this involves additional labor and material expenses, it often is more practical to use sheets that are shorter than 8 feet and increase the width of each sheet by an amount sufficient to take care of the taper. In this case, the increase in the width equals ⅛ inch per foot. The increase in the widths of succeeding sheets of various lengths is given in Table 9-5.

TABLE 9-5 Taper Sizes

Length of sheets (in.)	Increase in width (in.)	Length of sheets (in.)	Increase in width (in.)
24	¼	60	⅝
30	⁵⁄₁₆	66	¹¹⁄₁₆
36	⅜	72	¾
42	⁷⁄₁₆	84	⅞
48	½	96	1
54	⁹⁄₁₆		

EXAMPLE: A valley is 19 feet long. The sheet extends 5 inches under the slates and is fastened by cleats. What width of sheets can best be used? Because a 4-inch minimum under the slate is necessary, two 8-foot and one 3-foot length can be used. Starting at the top, the first sheet would be 2 + 2 + 5 + 5 + ½ + ½ = 15 inches wide. If the 3-foot sheet is used at the top, the first 8-foot sheet would be 15⅜ or 15½ inches wide and the second one 16⅜ or 16½ wide.

Start the slate 2 inches up each side of the valley center at the top and taper away from the center at the rate of ½ inch for every 8 linear feet. The metal flashing should be of sufficient width to extend up and under the slate not less than 4 inches (preferably 6 to 8 inches), and as far as possible without being punctured by the slating nails.

When the two roofs that form the valley have considerable differences in slope, or when the differing roof sizes create a large variation in the volumes of water delivered into the valley, crimp the metal or make a standing seam. This breaks the force of the water from the steeper or longer slope and prevents it from being driven up under the slate on the opposite side.

Condensation that forms on the underside of valley flashings, when not free to run off or evaporate, can attack the metal. It is recommended that the felt be omitted under the metal, unless copper is used. If felt is used under other metals, the metal should be well painted on the underside. Discuss flashing metal with the owners and design anchor before contract. Omitting copper on such expensive roofs is penny-wise and pound-foolish.

For inexpensive roofs, the copper for valleys is laid flat, without crimps or cleats. For high-grade work, secure the copper sheets to the roof boards and over the felt with metal cleats set 8 to 12 inches apart. Turn over the edge of the sheet ½ inch and hook it under the bent end of the cleat. Then nail the cleat to the roof boards with two nails and bend the cleat over to cover the nails. A high-quality method, sometimes used on wide valleys, is to fold the metal 4 or 5 inches from the valleyline and 3 inches from the cleat fold. This is known as foldover flashing.

CLOSED VALLEYS

The closed valley is formed with the slate worked tight to the valleyline and the pieces of metal placed under the slate. The size of the sheet is determined by the length of the slate and the slope of the adjoining roofs. Extend each sheet 2 inches above the top of the slate on which it rests so that it can be nailed along the upper edge of the roof sheathing without the nails penetrating the slate.

Each sheet should be long enough to lap the sheet below by at least 3 inches. Set back the sheet from the butt of the slate above so that it is not visible. Separate these sheets by a course of slate. Each sheet must be wide enough so that the vertical distance from the center of the valley to a line that connects the upper edges of the sheet is at least 4 inches. This dimension depends on the nailing of the slate, which should not penetrate the sheets.

Some roofers form the sheets with a center crimp, which stiffens them and forms a straight line on which to set the slates. This prevents water on one slope from forcing its way above the sheet and onto the other slope.

Another method of forming a closed valley is shown in Fig. 9-12. The sheets are laid in long pieces directly on the paper or felt that covers the roof sheathing before the slate is laid. The sheets can be of any desired length and should lap in the direction of the flow by at least 4 inches. Nail the sheets about every 18 inches along the outer edge. Take care to avoid penetrating the sheet when nailing the slate.

ROUND VALLEYS

The round valley forms a pleasing transition between two intersecting slopes when used in connection with the graduated or textured roof. If not properly laid out, however, it produces disaster and mars an otherwise beautiful roof. For this reason, only experienced workers should lay this type of valley. It requires the most careful workmanship and a veteran craftsperson's knowledge of the situation to produce a job that is both pleasing in appearance and watertight.

The round valley requires a suitable foundation to establish the general contour. The building of this foundation usually is considered a sheathing/deck function and might or might not be the job of the roofer. The construction and use of the foundation, however, is of great interest to the roofing contractor. Three methods can be used. The first, suitable for valleys of slight curvature, usually consists of a

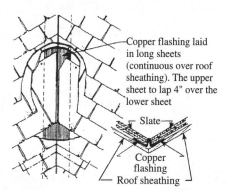

Copper flashing laid in long sheets (continuous over roof sheathing). The upper sheet to lap 4" over the lower sheet

Slate

Copper flashing

Roof sheathing

FIGURE 9-12 Method of forming closed valley.

12-inch-wide board, with tapering sides, that is nailed into the angle formed by the intersecting roofs.

In the second method, suitable for any curvature or radius desired, 3-inch blocks are cut to fit the valley angle and sawn to the proper radius. Then the blocks are nailed over the roof sheathing and spaced approximately the same as the exposure of the slates. The blocks form nailing strips under the slate. The size of the blocks varies due to the diminishing size of the valley as it approaches the ridge.

The third and usually the most satisfactory method is a combination of the first two. The 3-inch nailing blocks are spaced from 20 to 30 inches apart over the regular sheathing. Tapered strips $\frac{7}{8} \times 2$ or 3 inches wide are nailed over the blocks for the length of the valley. The important consideration in any of these methods is a solid, accurately formed foundation that supports the slates and establishes the desired shape of the valley.

As to the installation itself, the valley slates must be at least 4 inches longer than the slate used in the corresponding roof courses. Trim the sides of the slates to the proper radius and shoulder the tops to make the slates lie flat.

The round valley slates are sometimes embedded in elastic cement. If proper care is used when trimming and fitting the slate, no flashings should be necessary. Where the workmanship is not dependable, use flashings of metal or prepared roofing cut to the proper radius as a precautionary measure. But remember, when installing slate, the use of poorly trained mechanics should be avoided with diligence. Always use flashings wherever ice can form.

The radius of the round valley is at maximum size at the eaves and gradually diminishes to practically zero at the ridge. For appearance, as well as to facilitate laying the valley slates, the distance across the eaves should not be less than 26 inches. If the roof condition does not permit this, use the canoe valley.

CANOE VALLEY

The canoe valley is a variation of the round valley and is laid in the same manner, except that the radius at the eaves and ridge is practically zero. The radius gradually increases until it reaches maximum size halfway between the eaves and the ridge.

Eaves and Gables

Start the slate under the eaves on a cant strip of a thickness suitable to the thickness of the slate. This enables the second course of slate to be laid correctly. In the case of a cornice, the slate should project about 2 inches beyond the cant strip, sheathing, or finishing member. The length of the slate under the eaves is determined by adding 3 inches to the exposure being used on the regular slates. Thus, if 16-inch slates are used, the exposure is 6½ inches and the size of the required slate is 9½ inches.

Half slates are sometimes used, or roofing slates of the proper width can be laid horizontally. If the first course is ¾ inch thick, use ⅜-inch slates for the under-eave course or ¼-inch slates if the starters are ½ inch, although the under-eave and first course are sometimes made the same thickness.

The first course of slate is laid over the under-eave course with the butts of both courses flush and the joints broken.

When changing from a roof with a flat slope to one with a steeper slope, as in the case of a gambrel roof, project the slate of the upper and flatter roof 2 to 2½ inches beyond the steeper roof below. Use a cant strip to start the slate on the roof of lesser slope, the same as at the eaves.

At the gables, the slate should overhang the finishing member of the verge board by not more than ½ inch. This dimension can be increased when close-clipped gables are used or the construction is such that the gable slates have ample nailing. The projection ought not to be too great for good appearance. Also, there are many interesting ways to lay gable-end or barge slates under regular courses along gable ends where shadow effect is desired. The ways and means of using and securing all gable-end slates depend on the type of construction.

Applying Flashing

As in all steep-slope applications, flashing must be used at all intersections of vertical or projecting surfaces or those against which the roof abuts, such as walls, parapets, dormers, sides of chimneys, etc.

Flashings used over or under the roof covering and turned up on the vertical surface are known as *base* flashings. Metal built into the vertical surface and bent down over the base flashing is termed a *cap* flashing or counterflashing.

BASE FLASHING

Extend base flashings under the uppermost row the full depth of the slate, or at least 4 inches over the slate immediately below the metal. Turn up the vertical leg not less than 4 inches and preferably 8 inches on the abutting surface. Where a vertical surface butts against the roof slope, build in the base flashing as each course of slate is laid. Turn out 4 inches on the slate and at least 8 inches above the roof.

If the roof stops against a stuccoed wall, secure a wooden strip 4 inches wide, with a beveled top edge, to the wall. Then turn the base flashing out over the slate at least 4 inches and bend it up vertically at least 3 inches on the board. Except in unusual cases, it is satisfactory to turn the base flashing out 4 inches on the roof surface and up on the vertical surface from 6 to 8 inches for either sloping or flat roofs. Protect soil pipes and ventilators with base flashing. Vent pipes require base flashing in the form of either special sleeves or in one of the numerous patented flashing devices.

COUNTERFLASHING OR CAP FLASHING

When the base flashing is not covered by vertical slate, siding, etc., use a cap flashing. Build this member into the masonry joints not less than 2 inches, extend it down over the base flashing 4 inches, and bend the edge back and up ½ inch. Reglets in stone or concrete are usually about 1 inch wide and 1 inch deep.

Lay and form the flashing in the bottom of the cut and thoroughly caulk with molten lead on flat surfaces or lead wool on upright work. After caulking, fill the reglet to the surface with elastic or plastic cement. Use flashing hooks to secure stepped flashings. Make the vertical legs tight with roofers' cement colored the same as the masonry. On the best work, these flashings are soldered.

Many flashing installations are a combination of cap and base flashings. Locations where this might occur include

Chimneys. As shown in Fig. 9-13*A*, the chimney's base flashing is woven into the slate course and extends up under the counterflashing by at least 4 inches. The counterflashing for a chimney on the slope of a slate roof must lap at least 2 inches. Figure 9-13*B* illustrates the proper installation for a chimney on the ridge of a slate roof. The cap flashing laps the base by at least 2 inches.

Dormers. The built-in base for a dormer window is shown in Fig. 9-14. The flashings are woven into the slate course with each flashing sheet lapping the next lower one by 2 inches. Figure 9-15 details the flashing features of a recessed dormer window.

SADDLES OR CRICKETS

Where a chimney or other vertical surface breaks through the roof at a right angle to the slope, build a saddle or cricket to throw the

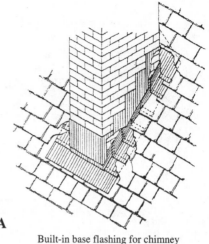

A

Built-in base flashing for chimney on slope of slate roof

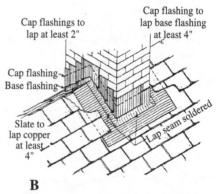

B

FIGURE 9-13 (*A*) Chimney base flashing; (*B*) proper installation for chimney on ridge of slate roof.

water away from the back of the vertical member. If the roof is constructed of wood, use light rafter construction covered with sheathing boards, paper, and sheet metal. If it is a large area that is exposed to prominent view, slate it the same as the other roof areas. The size of the saddle is largely determined by the roof condition. It is usually sufficient to make the slope of the saddle the same as that of the roof.

It is most important that the saddle or cricket be of adequate size, of ample slope, and well flashed.

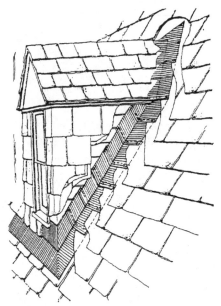

FIGURE 9-14 Built-in base for dormer window.

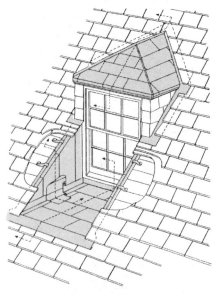

FIGURE 9-15 Flashing features of recessed dormer window.

Reroofing with Slate

Slate possesses qualities that make it suitable for reroofing any type of building. Its neutral or more vibrant colors quickly assume the characteristics of age, blend with the natural surroundings, and, with proper selection, conform to any desired color scheme. Its texture meets the demands of any design. Its cost is only slightly more than that of less permanent roofing materials.

Removing the old covering exposes the sheathing or roof lath and permits a thorough inspection. Replace any broken boards or lath with new whole material and securely nail all loose boards in place. Take out and replace original boards so that the joints are not broken over bearing rafters. As an alternative, cut in short rafters or blocking to act as a bearings. Fill and secure any low spots or loose areas. When possible to do so, inspect the sheathing or lath from the underside and repair all broken or loosened boards.

When saving heat is essential and old roof laths are not to be removed, use new boards to fill the space between the old laths. Both should be of the same thickness to provide a reasonably smooth surface.

TABLE 9-6 Lath Spacing and Slate Length

Spacing of lath (O.C.) (in.)	Length of slate (in.)
10½	24
9½	22
8½	20
7½	18

It is important to go over the roof and remove or drive home any projecting nails. Cut down any warped or raised edges or ends of sheathing or lath. Before laying the felt, thoroughly sweep off the sheathing to remove all chips, blocks, and loose nails.

Rafters adequate for wooden shingle roofing are of sufficient strength for slate of commercial standard (³⁄₁₆-inch) thickness. Sometimes the rafters of old buildings, and some not classed as old, are of ample strength to support the present roof covering, but were not designed to carry the additional snow or wind load recommended by present-day engineering practice or required by the local building code.

When the existing roof appears to sag or there are indications that existing supports are not adequate for the present roof covering, it is unwise to replace the old covering or to cover the roof with any new material before strengthening the roof supports.

When ordering nails, make allowances for the thickness of the old roofing and use a nail of sufficient length to secure penetration into the roof boarding. Note the width of the roof boards or lath spacing and order slate of proper length to secure a nailing that avoids the joints. Table 9-6 is useful in this regard.

Making Additions or Alterations to Slate Roofs

A properly laid slate roof requires little, and sometimes no, upkeep or care. However, houses are sometimes enlarged or remodeled. In such cases, it is often necessary to join new and old roofs or to remove and alter sections of the existing roof.

It is desirable and necessary that the altered or additional roof matches the existing one in both shade and texture. To obtain this result, attempt to secure slate of the same quality and color as the original slate. In many remodeling jobs, however, the slate has been on the roof for many years and no record is available of the quarry from which it came.

In this case, the best way to proceed is to remove small adjoining sections and relay the slate, mixing some new slate with the old. This prevents a clear line of demarcation where the new work adjoins the old.

In minor alterations, such as adding or removing a dormer, the old slate that was removed can be used again. In dormers and other projections, the lights and shadows differ from those on a roof expanse, which makes it easier to add new slate. For example, it would be best to use new slates on the new dormer cheeks and the old slates on other parts. Some roofers buy up a number of old roofs from buildings that are being torn down. They keep the old weathered slate in their yards so that they can readily match slates when minor alterations are made.

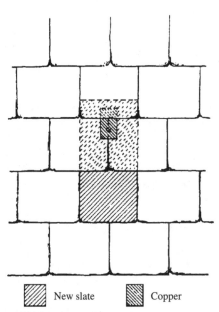

New slate   Copper

FIGURE 9-16 Proper method of inserting a new slate.

Due to unavoidable causes, however, slates are sometimes broken on the roof. The broken slate can be repaired by the slate roofer. The best method is to first remove the broken slate, cut the old nails with a ripper, and remove any remaining small pieces of slate (Fig. 9-16). Insert a new slate and nail this slate through the vertical joint of the slates in the overlying course approximately 5 inches from the head of the slate, or 2 inches below the tail of the second course of the slate above.

Over this nail, insert a piece of copper approximately 3 inches in width by 8 inches in length. Insert the piece of copper under the course above, lengthwise, so that it extends a couple of inches under the succeeding course. This ensures a proper lap and protects the exposed joint in which the nail is driven. Before installing, slightly bend this small piece of metal to ensure that it remains tightly in place.

Only responsible and experienced slate roofers should make alterations to a slate roof. If other workers are required to use ladders or scaffolds on slate roofs, use boards under the legs or uprights to distribute the pressure.

Overcoming Slate Problems

One of slate's major advantages is that it needs no ongoing maintenance: no painting, no preservative coatings, and no cleaning. Slate resists seasonal weather changes better than other roofing materials, though some slates have a greater porosity than others and eventually begin to spall due to freezing cycles. Any roof, however, should be checked and maintained periodically. Gutters and flashings are particularly prone to problems and may need occasional repairs.

Complete failure of a slate roof is almost always due to poor installation methods, bad flashing details, or inferior nails. The nails sometimes give way. The worst condition is when all the nails need replacing because false economy or ignorance led to the use of the wrong nails. If some slates are letting go because their nails have rusted through, this could mean that eventually all the slates will have to be relaid with the proper copper nails. Today's galvanized nails are not recommended. Old slates can be reused.

Leaks in slate roofs are usually caused by deteriorated flashing or missing slates. Flashings gradually erode from ice and atmosphere. Flashing repair often can be tackled by the homeowner, especially the flashings around chimneys and stacks and in open valleys. Replacing flashing in closed valleys, where the metal is covered by slates, is more complicated.

Without a doubt, it is more economical to keep up the repairs on a sound slate roof by replacing missing slates and deteriorated flashings than to replace the roof or cover it with a modern, less permanent material. If the majority of the slates are delaminating or crumbling, it is not possible to save the roof. Recognize that such a roof is probably many years old already and that the condition resulted from the original installation of inferior, less expensive slates.

Here, again, is an excellent opportunity to gain a maintenance business client. Owners with slate roofs typically care a great deal about their houses, welcome information, and can afford to maintain their homes.

If trees and vines are present, the leaves and branches can build up in valleys and dam water, which backs up under the slate. Maintaining drainage without breaking slates is an excellent service.

Clay and Concrete Tiles

Today's clay and cement roof products all started to develop thousands of years ago when man discovered the function and durability of roofs constructed of clay. First appearing in the Bronze Age, sun-baked tiles were found in Crete adorning the palace roofs of local rulers. Millenniums later, Greeks learned to fire-bake clay tiles and applied them to structures as majestic as their renowned temples.

Making Clay Tiles

Clay tile is still among the most popular roofing materials in Europe, where homes and centers of commerce are designed to last for generations. Their simplicity of form and shape makes traditional, tapered mission tiles ideal for funneling and shedding water from pitched roofs.

Modern extrusion and pressed-formed processes and high-tech gas-fired kilns have replaced the primitive method of shaping clay tiles over human thighs and then either baking them in the sun or using wood-fired beehive kilns. With these advances in manufacturing have come tremendous improvements in performance, quality, and product diversity.

Today, the manufacture of clay tiles starts with shale that is crushed to a fine powdery clay. The clay is mixed with water and kneaded, or pugged, to the consistency of cookie dough. Then, to produce simple clay tiles, this moist, plastic clay is extruded through a die, like dough through a cookie press, and sliced into lengths. More complicated shapes come from pressing the clay into molds. Some ornamental tiles are even sculpted by hand. The formed tiles dry in a room kept at a temperature of approximately 90 to 95°F. From there, they go to a kiln for firing.

For their trip through the kiln, the tiles are either stacked on a refractory, which is similar to a railroad car, and pulled through a tunnel kiln, or laid on ceramic rollers that convey individual tiles through a roller-hearth kiln.

Regardless of the method, the object is to raise the temperature of the clay to the point of vitrification, which is about 2000°F. At this point, the clay minerals lose their individual identity and fuse together.

Coloring Clay Tiles

Clay tiles are available in a wide range of colors. The more sophisticated clay tile manufacturers achieve colors through the careful blending and mixing of various clays into complex clay bodies. Hues from ivory and almond to deep reds and browns, apricot to peach and buff tones, and variegated accents are now available to designers. (See App. A for a vendor list.)

In addition, these colors can be enhanced by adding natural flash or variegated effects through the introduction of streams of natural gas during the firing process. By controlling the timing, frequency, and location of the kiln flashing, an infinite combination of randomly flashed tiles can be created for that truly custom, one-of-a-kind color blend.

Another method of coloring tiles is to spray a thin creamy layer of clay, called a slip, onto the tile before it is fired. The tile then takes on the color of the slip. The most dramatic, and most expensive, way to color tile is with a glaze. The metallic pigments in the glaze, when fired, melt to a glossy, vitreous, richly colored surface, much like that found on ceramic tile used indoors.

Premium clay roof tiles ensure protection from the elements and offer extended warranty periods. Some manufacturers offer lifetime warranties and even include fade coverage.

Clay tiles offer the homeowner and roofing contractor numerous advantages:

- Tile roofs typically last 50 years or longer and do not rust or otherwise deteriorate.

- The color and texture of most tiles is integral and of natural materials that do not fade.

- Tile roofs are more insulating. Clay and concrete tiles resist the passage of heat gain from summer sun and winter heat loss.

- The mass of tile roofing provides superior insulation from sound.

- Tile roofs are noncombustible and protect the structure from burning embers without suffering irreparable damage. Tile and cement roof products carry Class A fire ratings.

Designing Tile Shapes

The roofing industry generally separates roof tile designs into three categories: high profile, low profile, and flat. High-profile tiles are the familiar mission, barrel, S, or Spanish-influenced styles (Fig. 10-1). They are available in a variety of integral and applied colors.

According to legend, the curved shape of high-profile tile evolved in ancient times when craftspeople formed wet clay sections over their knee to provide added stiffness. These primitive tiles were probably baked in the sun or placed in wood-fueled fires.

Low-profile tiles are manufactured in numerous different styles by several manufacturers (Fig. 10-2). They are also available in a variety of colors that complement any architectural style.

Flat clay tiles have a shingle shape and are ribbed to simulate wooden shakes or colored to represent slate. Many of the flat clay tiles feature an interlocking system.

Also available are special clay tile shapes (Fig. 10-3). For instance, closed-ridge end tiles and gable-terminal tiles are designed for gable roofs. Hip-terminal tiles are intended for decorative purposes and are used where a ridge and two or more hips intersect. Graduated tiles of diminishing widths are used for round towers, circular bays, and porches. Tile manufacturers furnish graduated tiles in all popular

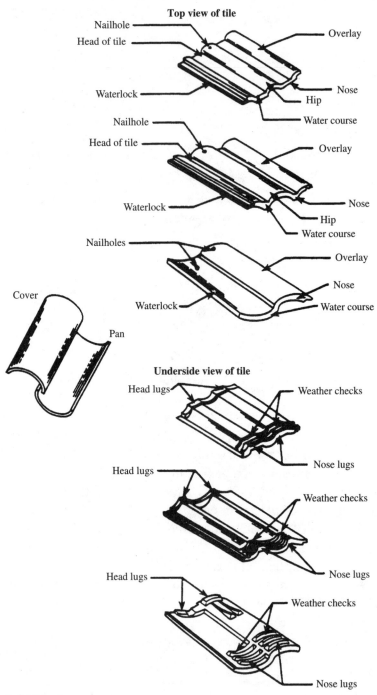

FIGURE 10-1 High-profile S tiles.

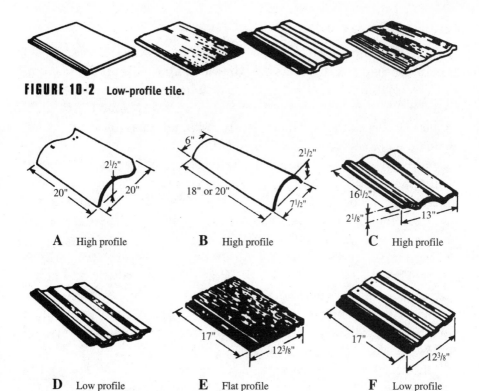

FIGURE 10-2 Low-profile tile.

A   High profile B   High profile C   High profile

D   Low profile E   Flat profile F   Low profile

FIGURE 10-3 Special clay tile shapes.

shapes. Some manufacturers also offer special valley tile, manufactured in angular or round form, and other special shapes for particular applications.

Making Concrete Tiles

Concrete tiles are relatively new compared to clay tiles. Although concrete tiles have been used in Europe and Australia since the mid-1800s, they have enjoyed widespread use in the United States only since the mid-1960s.

Concrete tile is composed of portland cement, sand, and water, mixed in varying proportions. These materials are mixed and extruded on individual molds under high pressure to form the tile product.

Coloring Concrete Tiles

Concrete tiles are colored by one of two methods. The first is to add iron or synthetic oxide pigment to the batch mix. This produces a uniform color all the way through the tile. A less expensive method is to coat the tile with a slurry of cement and iron-oxide pigment. This technique also allows the manufacturer to add highlights of a second color, which creates a shaded or variegated effect.

The particular coloring method used is dictated by several different requirements. The first of these relates to cosmetic and aesthetic appeal. The color-coated product gives wider color hues, while the body-colored product is less spectacular, is more subdued in appearance, and has a limited color range.

Application requirements for both product types relate primarily to atmospheric and climatic conditions, as well as aesthetic and architectural intent. Experience has shown that the surface-coated product is, generally speaking, more resistant to growth of, and discoloration from, moss and lichen, found in tropical areas with high humidity and in areas with large amounts of rainfall.

On the other hand, the color-coated product does not fare well in areas with extreme freeze and thaw conditions or where there is a large amount of industrial pollution that contributes to such phenomena as acid rain and deposits of atmospheric dirt and grime. The through-color product is more resistant to freeze and thaw conditions and more subject to discoloration and staining by moss, lichens, and atmospheric pollution.

With either coloring method, the tiles are usually sprayed with a clear acrylic sealer. The sealer helps the tiles cure properly. It also controls any efflorescence, which is the white powder consisting of free lime that surfaces as concrete ages. The acrylic sealer forces the lime out the underside of the tile, where it doesn't spoil the appearance.

As a side effect, the sealer gives the tile a slight gloss. This gloss wears off in a few years, and the color softens to its true matte finish.

Designing Tile Shapes

Concrete tile has three classifications: flat, roll, and graduated. Flat tiles vary in size and have the appearance of slate or wooden shakes.

Some of these tiles are made of fiber-reinforced cement and come in various colors. They can be installed on roofs up to 40 feet high in areas with wind speeds up to 80 miles per hour (mph). Some flat tiles are available with interlocking water locks.

Roll tile is pan and cover shaped. It is better known as barrel or mission tile. Graduated tiles of diminishing widths are used for round towers, circular bays, and porches. Some tile manufacturers furnish graduated tiles in all popular shapes.

Preparing the Roof

A new roof must receive the same preparation whether clay or concrete tiles are to be applied.

Matching Roof Slope and Underlayment

Roll or flat tile can be applied to roof decks with slopes of 4 inches per foot or more when a minimum of one layer of 30- or 43-pound felt is applied horizontally to serve as the underlayment, and the tiles are nailed or wired with a minimum 3-inch headlap.

Any style of clay or concrete tile can be applied on solid-sheathed roof decks with slopes less than 4 inches per foot when a minimum of two layers of 30- or 40-pound nonperforated, asphalt-saturated felt are set in hot asphalt or mastic to serve as the underlayment. One layer of a modified-bitumen-coated sheet, with laps either torched or heat-welded, is also acceptable.

Over the underlayment, install vertical lath stringers with horizontal battens fastened over the stringers. This creates a simulated surface over which the tile can be installed. The tile must be installed with a minimum 4-inch headlap.

Preparing Deck Surfaces

If plywood is used as the deck material, use exterior plywood thick enough to satisfy nailing requirements. Separate the plywood panels by at least $\frac{1}{16}$ inch to allow for expansion. If wooden planks are used for the roof deck, the boards should be a minimum of 1 × 6 inches and should span a maximum of 24 inches between trusses or rafters.

When the roof deck is made of concrete, a surface must be provided onto which the tiles can be applied. To create this surface, run 1-×-2-inch, beveled wooden nailing strips of treated lumber from the eaves to the ridge. Embed the strips in the concrete and space them 16 or 24 inches on center.

Nail the felt to these nailing strips. Then nail lengths of lath, applied vertically, directly into the beveled nailing strips through the felt. Finally, nail 1-×-2-inch battens, or stringers, spaced according to the type of tile to be used, horizontally across the lath. This simulates a wooden surface that can accommodate the application of tiles. Some concrete tile manufacturers also produce special wiring systems for securing concrete tile to concrete roof decks.

Working with Tiles

When working with tiles, keep the following precautions in mind.

Broken tiles. More tiles are broken in transit and on the ground than are broken on the roof. Therefore, take great care when unloading tile at the jobsite. Unload the tile as near as possible to the building, and distribute it so that delivery to the roof is convenient. Save tiles that have been broken either in transit to the roof or during application. Use these tiles when cut tiles are required.

Loading tiles. To prevent tiles from breaking or becoming soiled, stack them not more than six high. Keep nails, cement, and coloring material covered until needed.

Cutting tiles. When tiles must be cut, mark the desired break line on the tile. Then carefully cut along this line. If too much tile is cut at once, the tile can fracture. The correct cutting procedure requires that the tile be placed on a tile stake well back from the cutting line and tapped with a hammer. Further trimming should be done with a large pair of pincers. Tile saws or power saws also can be used to cut tiles. Diamond-tipped tile saw blades are best for this process.

Narrow tiles. Concrete tiles less than three-quarters of the width of a full tile are susceptible to wind damage when used on gable ends.

Drill an additional nail hole in the top of the tile and place a dab of roofer's mastic under the butt end. Mechanically fasten the tile through the newly drilled hole. Gable-end partial tiles installed in this manner can withstand the same windy conditions as the field tile.

Drains. To avoid choking drains with broken tile or tile trimmings, do not permanently install drains until tiling work is complete. Keep drain outlets covered when tiling to prevent any debris from clogging the leader pipes.

Making Allowances for Cold Weather

Freeze and thaw conditions are encountered in many areas of the United States. A freeze and thaw area is defined as one that experiences 30 cycles of freezing and thawing per year. One cycle encompasses a change in temperature from more than 32°F to less than 30°F that is accompanied by moisture that freezes into ice and thaws to a liquid state.

Freeze and thaw failure can occur in just about every product, manufactured or natural, that is exposed to such conditions. Dry, cold-weather temperature fluctuations above and below freezing have little or no effect on concrete tile.

To protect tile from freeze/thaw conditions, apply a minimum of one layer of 40-pound coated felt horizontally with a minimum 4-inch headlap and 6-inch sidelap. Slopes below 4 inches in 12 require a functional built-up roof (BUR), a modified bitumen roof (MBR), or a self-adhering, ice-and-water-shielding bitumen membrane underlayment system.

In addition to the 40-pound underlayment, the following is required as an ice shield, regardless of slope, on eaves and barges, or rakes. Starting from the eave and barge to a point 36 inches beyond the inside wall line of the structure, use one layer of 40-pound coated felt set in roofer's mastic or cold-process adhesive, or one layer of self-adhering, ice-and-water-shielding bitumen membrane.

Over the underlayments or decks, fasten a vertical counter-batten, at least 1 × 2 inches, at a minimum of 24 inches on center from the eaves to the ridge. Over the vertical counter-batten, fasten a second horizontally installed batten, at least 1 × 2 inches, spaced to ensure a minimum 3-inch tile headlap. Use only treated lumber.

This counter-batten system minimizes condensation by allowing air circulation and, with proper ventilation, helps prevent ice-dam buildups. A minimum 4-inch tile headlap is recommended in areas with heavy snowfalls.

Arranging Colored Tiles

Blending modern colored clay and concrete tiles can be a rather difficult procedure. To guard against a spotted colored roof, mix tiles in the correct color arrangement on the ground and then send them up to the roof in bundles along with strict application instructions. For example, if the color scheme calls for 10 percent of one color, 30 percent of another color, and 60 percent of a third color, send the tiles up to the roof in bundles of 10 tiles, with each bundle having one tile of the first color, three of the second color, and six of the third color.

In this way, the tiles can be applied in the order in which they were bundled, and no time is wasted selecting colors on the roof. Separate the 10-tile bundles into two stacks of five tiles each when loading them onto the roof deck.

After 75 to 100 tiles have been installed, visually inspect the applied tiles from ground level and at a distance from the building to ensure that the tile courses follow straight and true lines and that the colors of the tile blend well. Repeat this procedure at regular intervals during installation to ensure an attractive and acceptable roof. The blending of tile shades to avoid streaks or hot spots is particularly important. Preblended tiles can be obtained from some manufacturers.

Discuss quality control of material shading and uniformity of tile with the manufacturer, architect, and building owner prior to contract and prior to placing an order for the product.

Fastening Tiles

Use nails and screws to hold clay and concrete tiles to decks. Use $^3/_{16}$-inch compression spikes for concrete decks and No. 12 TEK screws for steel decks. On a plywood deck, use ring-shank nails of sufficient length for slight penetration through the underside of the deck. For board plank decks, use smooth-shank nails at least 1½ inches long that do not penetrate the underside of the deck. For gypsum plank and nailable concrete decks, use stainless steel or silicon-bronze screw-shank nails of a length sufficient to penetrate ½ to ¾ of their length into the deck. Do not penetrate the underside of the deck. If the deck is excessively hard, use smooth-shank nails.

Fasteners also can be used to hold tiles. Do not drive home fasteners or draw the tile. Drive fasteners to a point where the fastener head just clears the tile, so that the tile hangs on the fastener. When tiles are fastened too tightly, they lift up at the butt. This allows high winds to blow them off the roof or to blow water under them. On exposed overhangs, the fasteners should not penetrate the sheathing.

Drive all fasteners into the roof sheathing and not between sheathing joints. This is especially important near the top and sides of the roof. When battens are used, drive all fasteners into the batten boards.

Fasten tiles individually. Secure hip and ridge tiles with one nail in each tile and with a golf-ball-size dab of roofer's mastic under the tile at the headlap, recessed so that it does not show. Barge or verge tiles, when available, require two nails and roofer's mastic.

Table 10-1 gives the nailing procedures for attaching clay and concrete tiles to plywood and wooden sheathing. Battens should consist of a nominal 1-×-2-inch approved material spaced parallel with the eaves to achieve a minimum 3-inch tile headlap. Use battens over solid sheathing and an approved underlayment. Make provisions for drainage at a maximum of every 4 feet past or beneath the battens. Battens must be attached with four corrosion-resistant fasteners per batten.

Over concrete decks, use wire tying strips instead of battens to secure the tiles. Secure angular strips of wire, 1½ × ½ inches in size, to the concrete deck with expansion bolts. Perforate the wide flange of the wire strip with holes spaced at regular intervals suitable to the tile exposure. Then run 14-gauge tie wire through these holes and tie it around the holes in the tile covers. Turn up the wire and twist it under the lap of the succeeding tile.

TABLE 10-1 Attachment of Tiles to Sheathing

	Field tile nailing		Nailing for perimeter tile and tile on cantilevered areas[2]
Roof slope	Solid sheathing with battens	Solid sheathing without battens[1]	
3/12 to and including 5/12	Not required	Every tile	Every tile
Above 5/12 to less than 12/12	Every tile every other row	Every tile	Every tile
12/12 and over	Every tile	Every tile	Every tile

1. Battens are required for slopes exceeding 7/12.
2. Perimeter nailing areas including three tile courses but not less than 36 inches from either side of hips or ridges and edges of eaves and gable rakes. In special wind areas, as designated by the building official, additional fastenings might be required.

When tiles are attached directly to metal purlins, fasten them with No. 14-gauge, rust-resistant wire or self-tapping screws with a minimum $7/16$-inch-diameter head. If self-tapping screws are used, they should be capable of penetrating a minimum of $3/4$ inch into the purlins.

Either nail or wire each ridge and hip tile in place. If tiles are to be wired in place, lace No. 14-gauge rust-resistant wire through the nail holes in the tiles and securely tie it to the heads of nails driven into the ridge or hip boards (Fig. 10-4). A golf-ball-size dab of roofer's mastic is required at the tile headlap. Recess it so that it does not show.

On all roofs, over all roof slopes, and under all conditions, securely fasten all tiles installed on cantilevered sections of the roof, such as gables or eaves, and all tiles installed at the perimeter of the roof. Where tiles overlap sheet metal, secure them with appropriate tie-wire systems.

Nails for tile roofs should be made of No. 11-gauge, rust-resistant, aluminum, copper, yellow metal, galvanized, or stainless steel and be of sufficient length to penetrate either $3/4$ inch into the sheathing or

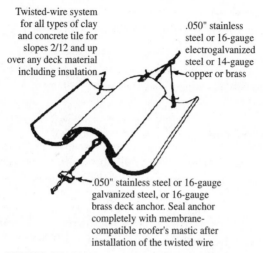

Twisted-wire system
for all types of clay
and concrete tile for
slopes 2/12 and up
over any deck material
including insulation

.050" stainless
steel or 16-gauge
electrogalvanized
steel or 14-gauge
copper or brass

.050" stainless steel or 16-gauge
galvanized steel, or 16-gauge
brass deck anchor. Seal anchor
completely with membrane-
compatible roofer's mastic after
installation of the twisted wire

FIGURE 10-4   No. 14-gauge rust-resistant wire.

through the thickness of the sheathing, whichever is less. On exposed overhangs, nails should not penetrate the sheathing.

When specifications require that all tiles be embedded in plastic cement, cover all lateral laps with cement. Use approximately 40 pounds of cement per square. Continuously embedding tiles in plastic cement throughout the roof restricts roof movement. Cracked tiles can result because of the expansion and contraction of the roofing during changes in temperature. This method of application is not recommended.

Where building officials have designated their localities high-wind hazard areas, special fasteners must be used. In these wind hazard areas, secure the nose end of all eaves-course tiles with hurricane clips (Fig. 10-5). Hurricane clips are available in different shapes to suit the type of roof sheathing used. Lay the tiles with a minimum 3-inch headlap. Nail each tile to the roof sheathing with one No. 11-gauge, rust-resistant nail with a minimum $^5/_{16}$-inch-diameter head. Apply a bead of roofer's mastic over the nailheads that fasten gable, barge, and ridge tiles.

Use one per
tile on sidelap

FIGURE 10-5   Hurricane clips.

On extremely steep or vertical roofs, wind currents can cause tiles to rattle. The recommended method for preventing rattling is to use hurricane clips. Another way to prevent tile rattling is to set the butt edge of each tile in a dab of roofer's mastic. Be careful not to stain the surface of the exposed tile.

Applying Flashing

Basic tile flashing is applied in much the same manner as for slate and wooden shake roofs, which is fully described in Chaps. 8 and 9. For valley flashing, use at least 28-gauge, corrosion-resistant metal and extend it at least 11 inches from the centerline of the valley each way. Form a splash diverter rib, as part of the flashing, not less than 1 inch high at the flowline. Overlap flashing at least 4 inches. Valley metal and flashing should be in place prior to tile application.

For other flashing, use at least 26-gauge, corrosion-resistant metal. At the sides of dormers, chimneys, and other walls, extend the flashing at least 6 inches up the vertical surface. Thoroughly counterflash and extend the flashing under the tile at least 4 inches. Turn the edge up 1½ inches.

Long runs of flashing material at parapet walls (Fig. 10-6) and copings, where roof tiles come to an abrupt termination, can be made of rigid materials, such as 26-gauge, galvanized sheet metal. If rigid materials are used, form them in such a way that they provide sufficient coverage and adequate drainage. Establish an acceptable windblock at longitudinal edges of flashings by grouting the longitudinal edges with portland cement mortar or by using alternative materials acceptable to local building officials. Flashing around roof penetrations

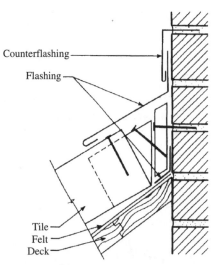

FIGURE 10-6 Flashing material at parapet walls.

should be in place prior to the application of tiles. Flashing metal should be at least 28-gauge, galvanized metal or an equivalent noncorrosive, nonstaining material.

Installing the Tile Roof

Examine the areas and conditions under which the tile is to be installed. Do not proceed until unsatisfactory conditions have been corrected. To avoid later disputes, report any such conditions to the contractor and other subcontractors in writing, and keep these letters on file. Verify that deck surfaces are clean and dry. Remove all foreign particles from the substrate to assure proper seating and to prevent water damage. Install the specified tiles in strict accordance with pertinent local code requirements.

On vertical applications, and on extremely steep pitches where wind currents can cause lift, set the butt of each tile in a bead of the specified plastic cement or sealant, or provide copper hurricane clips at intervals. Carefully use plastic cement and sealant. Avoid smearing the exposed tile surface.

Chalk horizontal and vertical guidelines on the membrane to assure watertightness and proper appearance. Space the chalklines by measuring the delivered tiles for average length and width exposures. Do not exceed an average exposure length of ¼ inch.

Applying Flat Tiles

Mark off the roof horizontally. Vertical lines, marked off randomly, help maintain a good vertical alignment. For roofs with pitches of 4/12 and above, install 9½-inch eave blocking, which is available from the manufacturer, eave metal, a bead of sealant, and 43-pound felt. Lay the felt parallel to the eave metal and extend it ¼ inch over the lower edge (Fig. 10-7). Note that some local building codes require an ice-and-water shield along the eaves.

For roofs with pitches below 4/12, install 12-inch eave blocking, eave metal, and a minimum 3-foot-wide strip of ice-and-water shield along all the eaves. Lap the felt 5 inches instead of the standard 3¾ inches and install battens that are notched 8 inches on center (Fig. 10-8).

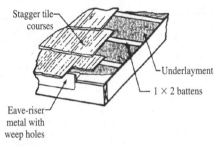

Stagger tile courses

Underlayment

1 × 2 battens

Eave-riser metal with weep holes

FIGURE 10-7 Flat-tile application, pitches 4/12 and above.

On reroofs, or when the eave fascia is not raised, install eave blocking with a cant strip and use new eave metal.

When laying out the roof, install a ridge nailer (vent). Next, strike lines that are centered on each hip, if applicable. Then strike the horizontal line for the top edge of the first batten 13¼ inches above the eave for a typical 15⅜-inch tile to ensure proper fit of the bottom row of tile. Next, strike the horizontal line for the top edge of the last bat-

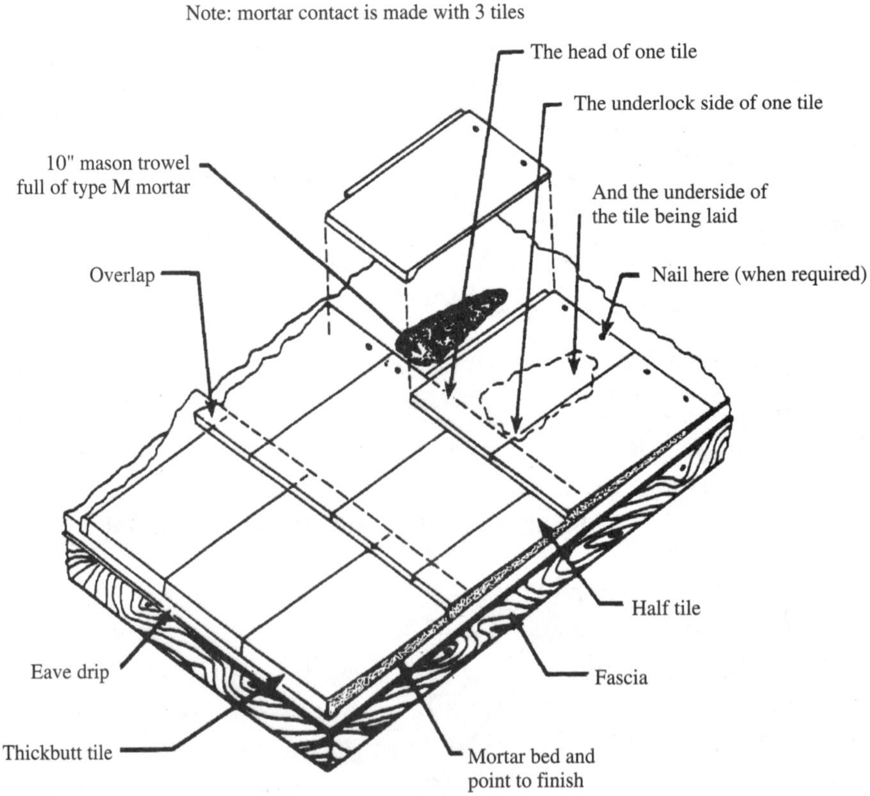

Note: mortar contact is made with 3 tiles

The head of one tile

The underlock side of one tile

10" mason trowel full of type M mortar

And the underside of the tile being laid

Overlap

Nail here (when required)

Eave drip

Half tile

Fascia

Thickbutt tile

Mortar bed and point to finish

FIGURE 10-8 Flat-tile application, pitches below 4/12.

ten so that the field tile butts the ridge nailer (vent), which is approximately 1 inch below.

Then divide the distance between these two lines into equal increments not to exceed 12 inches and strike lines for the top edges of the battens. If different eave lines do not allow for equal spacing, overlap the bottom row to allow the second row to match the rest of the equally spaced rows. If a short row is required, lower the height of the eave fascia board and use ½-x-16-inch battens with 1-inch spacing for the bottom row. Doing so allows less than a 10-inch exposure on the bottom tile without causing the second row to lay at a different pitch from the rest of the tile.

If different ridge heights do not allow for equal spacing, add a short row along the shortest ridge. If a short row is required, cut off the head of the tile, drill a new nail hole, and install a thicker batten tight to the ridge nailer so that the short row maintains the same pitch as the rest of the tile.

Install hip nailers to within 6 inches of the bottom corner and then install the horizontal, 1-x-2-inch batten strips, leaving 1-inch spaces between the ends. Use pressure-treated 1-x-2-x-8-inch battens with notches or ports 16 inches on center. On pitches 4/12 and above, install with 18 fasteners.

On 3/12 to 4/12 pitches and vented cold-roof applications, use pressure-treated 1-x-2-inch-x-6-foot battens with notches or ports every 8 inches; install with 14 fasteners. Use noncorrosive fasteners of sufficient length to fully penetrate the roof sheathing. Another approved procedure is to use 1-x-2-inch horizontal batten strips, without notches, installed over 1-x-2-inch pressure-treated vertical battens 16 inches on center or 1-x-4-inch horizontal battens installed over 1-x-2-inch, pressure-treated vertical battens 24 inches on center. When using vertical battens, remember to raise the eave and fascia metal to the additional thickness of the vertical batten.

For flat tile on gable roofs, put an X on the third batten above the eave, 35 inches on center, starting at the left gable edge. Randomly place stacks of four tiles above each X on every other batten. Then stack four more tiles randomly on top of the existing stacks to get a good color blend across the entire roof position. Put one barge on each batten next to the gable edge and install with gable tile or later from a ladder or staging after the field tiles are in place.

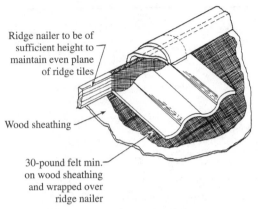

Ridge nailer to be of sufficient height to maintain even plane of ridge tiles

Wood sheathing

30-pound felt min. on wood sheathing and wrapped over ridge nailer

FIGURE 10-9 Installation of ridge trim.

To lay flat tile on hip roofs, place the Xs at 35 inches on center on the third batten above the eave and stack the tile above each X. Add extra stacks along the hip and valley to allow for 1½ tiles per row to be cut on each side.

Install ridge trim as the roof progresses. Lay the trim with approximately a 17-inch exposure and seal between the laps with mortar (Fig. 10-9). In fact, fill all hip, ridge, and other voids with American Society for Testing and Materials (ASTM) C-270 Type M cement mortar and neatly point it.

Applying Roll Tiles

When roll or mission tile is applied, an under-eaves course of tile does not need to be applied. Apply the first course of roll tile above an eaves strip. This metal eave-closure strip, better known as a *birdstop*, is specially formed to fit the underside of the tile. They are fastened 1¼ inches back and inside the cover of the tile. Apply the balance of the tile with the head of the tile aligned with the horizontal guidelines. Tile spacing may need to be adjusted to provide a uniform exposure.

The procedure for installing mission tile varies according to the tile design and size. Each manufacturer covers the specifics in its installation instructions. Basically, as with clay and concrete flat tiles, the roof is first felted and then horizontal and vertical guidelines are chalked to indicate the courses. Layout is critical because any deviations stand out against the pronounced vertical pattern. If the tile is designed with lugs to hang on battens, nail those battens next. Some manufacturers approve of hanging their tiles on spaced sheathing over heavy felt underlayment that is draped over the rafters. Load the tiles on the roof so that they are evenly distributed and within easy reach.

Extend the first row of tiles ½ to 1¼ inches over the eaves. For example, when a 13¾-inch tile is used, strike the first horizontal guideline 12 to 12¾ inches from the eavesline. Where chimneys or dormers project through the roof, loosely lay the first course of tile along the eaves. To minimize tile cutting, secure the tiles to the eaves only after making adjustments for the projections.

Cut or weave tiles installed down each side of a valley. Valleys can be open, mitered, swept, rounded, or closed with special tile. Special valley tile produces much the same effect as the rounded valley tile. It is longer than regular tile and fan-shaped. Because of its shape, valley tile need not be nailed, but should be cemented at the laps.

When valley tiles are used, first extend a row of tiles up the valley. Tiling then should proceed back toward the valley from the verge or gable. The last tile should be large and trimmed against the valley tile.

Fit tiles that converge along the hips of a roof close against the hip board. Make a joint by cementing the hip tile to the hip board with roofing cement or mortar. Color the cement to match the tile. Notch these cut tiles and either nail or wire them to the hip board. It is advisable to lay a golf-ball-size dab of roofer's mastic between the tiles.

Begin the hip roll with a hip starter, which is a hip roll with one end closed, or a hip stack, which is a stack of hip roll pieces equal in height to the hip stringer. Nail the hip-starter tile to the hip board with nails of appropriate length and follow with the regular hip roll, lapped either 3 inches or in accordance with the manufacturer's requirements. Cement between the laps. Do not fill the interior spaces of hip or ridge rolls with pointing material, as this material inhibits air circulation.

Cover ridges in much the same manner as hips. Fill the spaces between the tiles in the top row with special ridge fittings or with cement mortar colored to match the tile. When tile fixtures are used at the ridge, nail the diagonal half of a 2 × 4 on either side of the ridge board to provide a nailing surface for the tile fixtures. This is not required when portland cement mortar is used as fill material between the last course of tile at the ridge.

Some tile manufacturers make batten strips available for these layouts, while others provide detailed drawings with which contractors can make their own battens.

Spanish or S Tile Designs: Mark off the roof vertically and horizontally. Interlocking unlugged tile can be laid with a minimum 2½-inch headlap. Lugged tile should maintain a 3-inch design for mortar application. Check with the manufacturer of the particular tile.

Prefabricated Birdstops or Eave Closures: Prefabricated eave closure strips or mortar can be used to elevate the butt end of the first, or eave, tile to attain the proper slope. When using mortar, provide weep holes next to the deck to allow proper drainage of any moisture accumulation under the tiles. Place a full 10-inch mason's trowel of mortar under the pan section of each tile, beginning at the head of the tile in the preceding course. Press each tile into the interlocking position so that the cover rests firmly against the lock of the adjacent tile.

Installation details for eaves, ridges, gables, and so forth are given in Fig. 10-10.

Two-Piece Roll or Barrel Mission Layouts: Mark off the roof vertically and horizontally. Maintain a minimum 3½-inch headlap. Use mortar or prefabricated eave-closure strips or birdstops to elevate the butt end of the first, or eave, tile to attain the proper slope. Provide weep holes next to the deck to allow proper drainage of any moisture accumulation under the tiles.

For roof pitches of 3/12 and steeper, provide a minimum of one layer of 30-pound felt or upgraded material. Install a birdstop or a 1-×-2-inch wooden strip to boost the first course of tile. Then install the first row of tile leaving a 3-inch overhang. Use one corrosion-resistant nail not less than a No. 11-gauge, $\frac{5}{16}$-inch head per tile or a tile-tie system. Then lay a booster tile above the birdstop or eave strip, followed by the starter tiles. Other details necessary for laying a barrel/mission roof are given in Fig. 10-11.

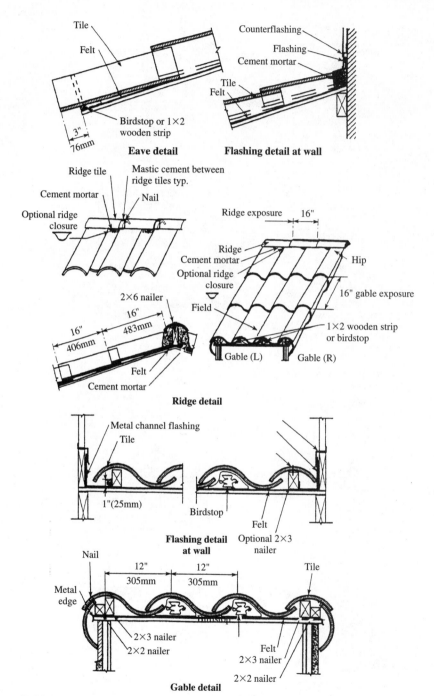

FIGURE 10-10 Installation details for eaves, ridges, gables.

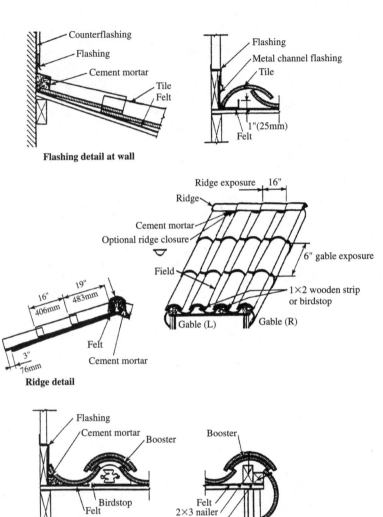

Flashing detail at wall

Ridge detail

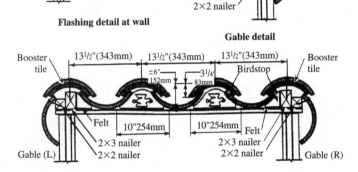

Flashing detail at wall

Gable detail

Spacing detail

FIGURE 10-11 Laying barrel/mission tile on roof.

Oriental Style: Figure 10-12 shows a typical oriental roll-style tile layout and installation details. Without the use of ornaments, this tile is used in western contemporary designs. While the most traditional color for a Japanese-style tea house or temple is black, these oriental-style tiles are available in natural red and glazed colors.

Turret Tile

True turret roof designs or fan-shaped applications are now possible without compromising design concepts. To determine the quantity of tile needed and specific installation guidelines for a given job, ask the manufacturer to provide a scale drawing or blueprint of the top and side views. All that the manufacturer needs is the diameter of the circle and roof pitch (Fig. 10-13).

The following are general instructions for installing turret tiles.

- At the first course, between vertical chalklines, install a clay birdstop and then place the pan tile on top of the vertical chalkline.

- Fasten each pan tile with a copper or other noncorrosive 11-gauge, large-headed nail, or use the wire-tie system.

- If the job site is located in a high-wind area, use mortar or other sealant to secure the pan tile.

- Once the birdstop and pan tiles are in place, install the booster and starter tile. Secure with copper wire or noncorrosive nails.

- For the rest of the courses, lay 16 inches to the weather. When the tile becomes crowded, adjust to the next smaller size and continue to the top of the roof.

- Follow the chalkline and use the turret worksheet provided by the manufacturer.

Prior to installing the last two or three courses, lay a mockup to assure proper fit. Do not use adhesives or nails to secure the tiles until the mockup is complete and satisfactory. Note that the final two or three courses normally lose one to two lines, or more, close to the top. Start installing the final two or three courses from the top down and secure each tile.

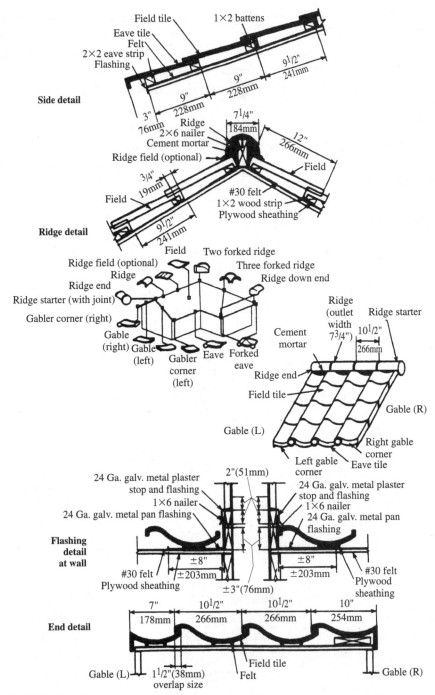

Field tile
1×2 battens
Eave tile
Felt
2×2 eave strip
Flashing
9¹/₂"
241mm
9"
228mm
Side detail
9"
228mm
3"
76mm
7¹/₄"
184mm
Ridge
2×6 nailer
Cement mortar
Ridge field (optional)
12"
266mm
Field
3/4"
19mm
Field
#30 felt
1×2 wood strip
Plywood sheathing
Ridge detail
9¹/₂"
241mm
Field
Field
Two forked ridge
Ridge field (optional)
Three forked ridge
Ridge
Ridge down end
Ridge end
Ridge starter (with joint)
Ridge
(outlet
width
7³/₄")
Ridge starter
Gabler corner (right)
Cement
mortar
10¹/₂"
266mm
Gable
(right)
Gable
(left)
Gabler
corner
(left)
Eave
Forked
eave
Ridge end
Field tile
Gable (R)
Gable (L)
Right gable
corner
Left gable
corner
Eave tile
24 Ga. galv. metal plaster
stop and flashing
1×6 nailer
24 Ga. galv. metal pan flashing
2"(51mm)
24 Ga. galv. metal plaster
stop and flashing
1×6 nailer
24 Ga. galv. metal pan
flashing
**Flashing
detail
at wall**
±8"
±203mm
±8"
±203mm
#30 felt
Plywood
sheathing
#30 felt
Plywood sheathing
±3"(76mm)
7"
178mm
10¹/₂"
266mm
10¹/₂"
266mm
10"
254mm
End detail
Gable (L)
1¹/₂"(38mm)
overlap size
Field tile
Felt
Gable (R)

FIGURE 10-12 Typical oriental roll-style tile layout and installation details.

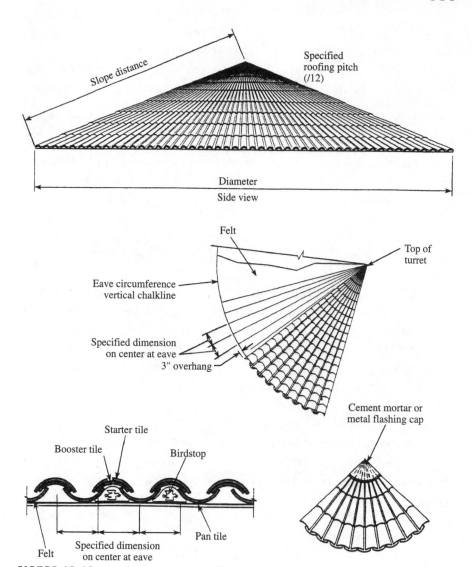

FIGURE 10-13 Turret tile diameter of the circle and roof pitch.

After any style clay or concrete tile roof is laid completely, do not allow traffic on the roof that might vibrate the framing or roof sheeting. At least 24 hours are needed to ensure a proper set. Prohibit roof traffic for at least 72 hours.

Installing Low-Slope, Mortar-Set Roofs

Roofing tiles have been installed in mortar for centuries. The practice of installing tiles with mortar over a built-up subroof evolved in high-wind and high-moisture areas of the southeastern United States. In this system, the built-up subroof provides the moisture barrier and the tiles, in addition to being aesthetically pleasing, protect the subroof from the sun's ultraviolet (UV) rays, high winds, and external damage. The system also allows the use of tile on lower-sloped roofs.

Apply the mortar over a solid sheathing of at least ⅝-inch plywood or 1-inch tongue and groove and mechanically fasten one layer of at least 30-pound organic felt underlayment to the sheathing. After applying the first layer, install metal eave flashings. Next, apply one layer of mineral-surfaced rolled roofing material to the underlayment with hot steep asphalt or mastic and then backnail. Other mineral-surfaced products can be used, such as MBR, although care should be taken to guard against roof slippage.

Mortared tile can be used on slopes a minimum of 2 inches in 12. On slopes between 5 inches in 12 and 7 inches in 12, additional mechanical fastening is required for the first three courses in areas subject to high winds. On slopes 7 inches in 12 and steeper, mechanically fasten all tiles. Mortar used to adhere tile to the subroof should be as specified in ASTM Specification C-270 Type M. Soaking the tile prior to installation, adding additives to the mortar, or both, might be required to achieve proper adhesion between mortar and tile.

In areas of the country subject to blowing sand or heavy rainfall, use mortar at ridge or hip intersections to provide a weatherblock. Use mortar sparingly and only to provide proper bedding for hip or ridge tiles. Specially designed metal weatherblocks are available from most manufacturers.

Reroofing Tile Roofs

Roofing contractors, architects, and specifiers, particularly those in the Sunbelt or Western states, sometimes encounter a reroofing project that involves a clay tile roof that is 70 years old or more. Their first impulse might be to draw up specifications for the job on the assumption that the existing tile needs to be entirely removed and replaced with new material. This thinking is understandable. UV radiation, heat, moisture, and exposure to the elements work together to limit the life span of most roofing materials to 25 years or less.

But first-quality clay roofing tiles are different. They last indefinitely if the roof is properly laid and maintained. There are many examples of old roofs that remain basically sound, with the clay tile intact and the underlayment in generally good repair. When these roofs begin to leak, it is often a result of problems with the underlayment. The tile often can be used again.

Failure to understand the long-lasting nature of the best clay tiles could be a costly mistake. A simple computation based on the money saved in new materials, plus the life-cycle cost benefits offered by tile roofing, normally results in a reroofing specification based on lifting and relaying the existing tile. Even if the existing tile is in good condition overall, however, there can be broken or damaged tiles on the roof.

The first step in assessing the condition of the roof can be conducted from the ground. Using a predetermined test area, count the number of broken or damaged tiles to get a percentage of probable breakage. The survey can be done with binoculars, and normally provides a reliable, rough estimate of how many tiles need to be replaced.

Deterioration in clay roofing tile is easy to spot. It is almost invariably the result of water absorption in tiles that were not manufactured properly. Most roofing tile failures can be traced to the use of inadequate raw materials or lack of proper time and care in the production process. The inferior tiles that result have a tendency to absorb moisture. The moisture then expands and contracts in response to the extremes of the freeze and thaw cycle in the north and to heating by the sun in warmer climates. This process causes the tile body to flake and spall.

The first sign of trouble is usually small chips of tile in gutters or around the foundation. Areas of discoloration, visible from a distance, might indicate that the internal body of the tile, which is lighter in color, is showing because the surface has chipped or flaked. The tile might appear fuzzy at the edges, and the shapes might be unclear. A closer inspection on the roof itself might reveal a crazed pattern of cracks in individual tile bodies. If such problems with spalling, cracking, etc., are widespread, the tile may in fact be deteriorated and in need of replacement.

With luck, the buildings's owner made the decision decades ago to use a tile designed to last the lifetime of the structure and damage is limited to isolated cases of broken or detached tiles. If there is any doubt, take representative samples and send them out to a reputable tile manufacturer who offers a testing service. The tests look at the major factors in tile condition: pore structure, compressive strength, and water absorption rates. Good results on these tests indicate that the tile is a strong candidate for additional decades of useful life.

Contractors reviewing the specifications for a job involving the removal and relaying of a tile roof may not be completely familiar with the procedure. There are many questions that should be answered before they go forward with a bid based on this approach. These questions can be answered by the tile manufacturer or the local representative.

An experienced roofer working from the manufacturer's installation manual normally has no more difficulty laying a tile roof than a shingle one. The major difference between the two is that tile is a fired material. It must be cut using special tools and is subject to breakage. More care is required when handling tiles.

When lifting and relaying tile roofs, weight can be a concern. If adequate scaffolding is available, or if the structure is deemed strong enough, the tiles can be lifted in sections and stacked near the work area. Common sense and experience are normally enough to let the roofer know whether this kind of loading might cause movement or possible collapse. When any doubt exists, consult with an engineer. If the roof or scaffolding is not strong enough, move the tile to the ground via a conveyor belt and develop a plan for restocking the roof with tile as needed.

The actual removal of existing tile is extremely easy. The tile is simply lifted and rotated, which normally pries out the nail. After the underlayment and deck are deemed satisfactory, the tiles can be relaid in the same manner as new tiles.

Installing Fiber Cement Shakes

A relative newcomer to concrete roofing is the fiber cement shake. This roofing product has the natural texture and tones of real cedar shakes with the added benefit of a Class A fire rating. Shakes resist the damaging effects of sun, water, humidity, rot, fungus, and termites. They contain no asbestos, formaldehyde, or resins.

For roof pitches 4/12 and greater, cement slates and shakes can be installed over spaced or solid sheathing (Fig. 10-14). Install 18-inch-wide, 30-pound underlayment, the starter course at a 9-inch exposure, and then succeeding field courses at a typical 10-inch exposure. Lap the underlayment over the ridge and hip to create a double layer.

For a 1-inch stagger, use a 9-inch underlayment exposure. Some regions require application over solid sheathing. Contact the local building department for sheathing code requirements.

For roof pitches 3/12 and less than 4/12, lay shingles over solid sheathing. For 4/12 and greater pitches, install a 36-inch, 15-pound underlayment described previously. For roof pitches under 3/12, install for appearance only over an approved sealed-membrane, low-slope roof system.

In snow areas, an approved 36-inch snow-and-ice moisture barrier is recommended at the eave.

To lay fiber-cement shingles follow these steps. (See also Fig. 10-15.)

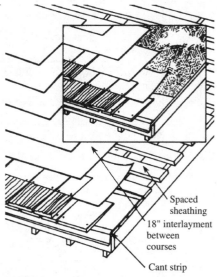

Spaced sheathing
18" interlayment between courses
Cant strip

FIGURE 10-14 Installing cement slate and shakes over spaced or solid sheathing on roofs with pitch of 4/12 and greater.

✓	Item
	Install a $\frac{1}{4}$-inch cant strip flush along the eave over the interlayment and beneath the starter course.
	Install the starter course face down, with up to $1\frac{1}{2}$ inches overhang at eave and $1\frac{1}{2}$ to 2 inches at rake.
	Fasten the starter course with $1\frac{3}{4}$-inch corrosion-resistant nails, staples, or screws, located within 1 inch of the eave line.
	Interlap field shakes with interlayment to create an approximate 2-inch headlap and a typical 10-inch exposure to the weather.
	Install shakes with an approximate $\frac{1}{2}$-inch keyway and $1\frac{1}{2}$-inch minimum sidelap. Applications with more narrow keyways require additional material, which changes the appearance of the roof.
	Alternate the sequence of shake widths every third to fifth course to avoid a stair-step pattern.
	Fasten the field shakes with either two 16-gauge, $\frac{7}{16}$-inch crown, $1\frac{3}{4}$-inch galvanized staples, or 13-gauge nails or screws. Locate the fasteners approximately $10\frac{1}{2}$ to $11\frac{1}{2}$ inches above the butt, and 1 inch in from each side of the shake.
	Do not walk on the smooth surface of installed field shakes.

FIGURE 10-15 Installing fiber cement shingles.

■ Install a ¼-inch cant strip flush along the eave over the interlayment and beneath the starter course.

■ Install the starter course face down, with up to 1½ inches overhang at eave and 1½ to 2 inches at rake.

■ Fasten the starter course with 1¾-inch corrosion-resistant nails, staples, or screws, located within 1 inch of the eave line.

■ Interlap field shakes with interlayment to create an approximate 2-inch headlap and a typical 10-inch exposure to the weather.

■ Install shakes with an approximate ½-inch keyway and 1½-inch minimum sidelap. Applications with more narrow keyways require additional material, which changes the appearance of the roof.

■ Alternate the sequence of shake widths every third to fifth course to avoid a stair-step pattern.

■ Fasten the field shakes with either two 16-gauge, ⁷⁄₁₆-inch crown, 1¾-inch galvanized staples, or 13-gauge nails or screws. Locate the fasteners approximately 10½ to 11½ inches above the butt, and 1 inch in from each side of the shake.

■ Fasteners must penetrate through the tail of the shake beneath and ¾ inch of the sheathing or its full thickness, whichever is less.

■ Do not walk on the smooth surface of installed field shakes.

The treatment at ridge and hip and in the valleys and the flashing at chimneys, vents, and so on, are handled in the same manner as that described for wooden shakes and slate in Chaps. 8 and 9.

Cement fiber shakes usually can be applied over one existing composition or wooden-shingle roof system, if the existing roof is relatively smooth and uniform. Wind-resistance performance applies only if fasteners penetrate the sheathing as specified. Structural evaluation and local building code approval is required. Longer fasteners are required.

Metal Roofing

The concept of using sheet metal as a roofing material dates back to the twelfth century. Craftspeople in Germany, Scandinavia, and other countries malletted together small copper sheets for some of the first such applications. Lead was similarly utilized in 15th-century Europe as artisans exploited its ductility and malleability. In the United States, samples of fine metal work can be seen on many buildings erected during the Revolutionary War period, including the copper domes found on some of the nation's state capitol buildings.

While the use of lead and copper can be traced to antiquity, the origins of contemporary metal roofing systems can probably best be attributed to a man named Henry Palmer, who in 1829 devised a unique form of corrugated wrought iron sheeting to cover warehouse buildings on the London docks.

With the development of commercial hot-dip galvanization techniques in the mid 1800s, metal roofs came into a period of widespread use on all sorts of buildings. They were chosen in the past for reasons that remain compelling today.

- Metal roofs are lightweight, yet extremely durable and cost effective over the long term.

- They have strong architectural appeal and can be adapted to suit a wide variety of historic and contemporary designs.

- They install easily and quickly, yet are flexible enough to overcome unique or difficult situations.

- They are readily available throughout the United States in a broad range of colors and configurations.

Many technical improvements have been made in the basic, venerable metal roof system. Advances in coatings, treatments, and forming technologies, as well as major improvements in systems for seaming and anchoring, have combined with a renewed interest in historic architectural styles to make metal roofing systems a leading choice today. The vast improvements in metal roofing, while offering many benefits, have not altered the need for great attention to detail and proper field practices.

Taking Advantage of Metal Roofing

There are a number of good reasons to use metal on commercial and residential buildings. First, metal is a predictable and stable product. Unlike other single-ply technologies like rubber, a metal panel is not affected by ultraviolet (UV) rays, which are one of the leading sources of roof degradation. Second, advances in metal coating and finishing technologies have significantly improved the life cycle of metal roofs, and the related cost benefits.

Metal roofing also is extremely puncture-resistant, which is a valuable benefit to today's building owner who is looking for long-term, low-maintenance solutions. Actually, durability is one reason for metal's growing popularity. Many standing-seam roof systems have earned a Class 90 wind-uplift rating (the highest in the industry) from Underwriters' Laboratories, Inc. (UL). Some systems also carry a Factory Mutual Class A fire rating as a result of their noncombustible surface. Both ratings can help the building owner reduce insurance rates substantially.

Metal roofing is extremely versatile. It is compatible with all types of building materials and can be incorporated into all design concepts. Metal roofing is an effective accent for masonry, wood, stucco, glass-curtain walls, marble, and granite. The various profiles make metal panels easily adaptable for large and small projects. Commercial structures typically are best suited to 18- to 24-inch-wide panel configurations, while 8- to 12-inch-wide profiles are most often seen on smaller structures, such as residential or smaller commercial buildings.

Colors and finishes are available to suit any application, from earth tones to exotic tropical hues, from bare, unfinished surfaces to special coatings designed to resist corrosive atmospheres. These colors and finishes have warranties for as long as 20 years. The premium architectural finishes are highly resistant to fading and chalking. In addition to the 20-year warranties available on some panel finishes, the metal of the panel itself also carries warranties.

Most manufacturers cover their panels against leaking due to penetration or rust for a period of 20 years. In the long run, for commercial, low-slope construction, metal is the most economical material available. It is not the least expensive product initially, but when true life-cycle cost comparisons are made, metal usually outperforms many other products that require periodic maintenance or replacement.

On high-pitch applications, architectural metal panels are favorably priced along with premium, fire-treated wooden shakes, clay tile, or slate. The use of metal roofing panels on a residence offers pride of ownership to the homeowner who wants to have something different from the neighbors. Metal panels are made to simulate clay tile, wooden shakes, and slate. Metal roofing has always had aesthetic appeal since the clean look is an attractive option for many applications.

Perhaps the most significant benefit of all, given today's social and economic environment, is the environmental friendliness of metal roofing, from installation through eventual tear-off. To begin with, many metal roofing products use recycled metal in their initial fabrication. Second, the building owner and contractor does not have to contend with external flames, hot kettles, or noxious fumes during installation. Finally, after the roof has served its useful life, most of its metal is 100 percent recyclable, which eliminates concerns about hazardous waste disposal.

Examining the Types of Metal Roofing

The two basic types of metal roofing are structural and architectural. Structural metal roofing has load-carrying capabilities and does not require a substructure for support. Structural metal roofing, depending on profile and material thickness, can be applied on very low slopes. The industry minimum standard pitch for structural steel roofing systems is ¼ inch in 12 inches. Other materials, such as aluminum, copper, and various alloys and plated composites, have structural limitations that need to be considered.

Structural, standing-seam roofing is composed of interlocking panels that vary in width from 12 to 24 inches. The panels are designed to form a continuous waterproof membrane across the roof. The configuration of the panels is either a trapezoidal-shaped rib that is 2 to 3 inches high, or a square, vertical, narrow rib that is 2 to 3 inches high. Both of these basic panel configurations (Fig. 11-1) rely on the shape of the panel to provide the structural capability to span structural members.

The standard for structural panel free span is 5 feet. There are other factors to be considered, such as design load requirements and existing structural member locations, particularly in reroofing applica-

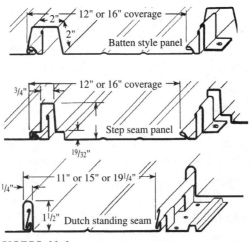

FIGURE 11-1 Basic metal panel configurations.

tions. These factors can alter the free-span capability, but 5 feet is the most common. Having a free-span capability means that no substructure is required to support the metal roofing panels.

The structural, standing-seam panels provide support for construction personnel and maintenance traffic on the roof. In order for these structural, standing-seam panels to be waterproof, the sidelap joints and endlap joints are designed with continuous gasket seals. Most panel manufacturers install mastic in the ribs during the production process. The sidelaps are accomplished by mechanical seaming or snap-together seaming. These seaming methods compress the sealant into the sidelap joint and form a continuous seal along the rib. Endlaps are sealed with mastic or tape sealant placed between two layers of panels and then screwed together to form a compression seal.

Structural, standing-seam panels are attached to the framing members with concealed fasteners and clips (Fig. 11-2). The clips are designed to allow the panels to move back and forth in response to varying roof temperatures.

Architectural panels are nonstructural and must be supported by a substructure. In most cases, the substructural support is plywood decking. Metal decking or fire-retardant treated plywood is used when noncombustibility is required. Architectural panels, due to their design, are water shedders, as compared to the structural standing-seam panels, which are water barriers. The minimum roof slope for architectural panels is generally considered to be 3 inches in 12 inches. Roofing felts are usually installed between the panel and the substructure, to provide additional moisture protection.

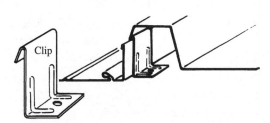

FIGURE 11-2 Typical concealed clip.

Like the other roofing materials described in this book, metal roofing systems are classified into three primary categories.

Steep slope. Also referred to as architectural metal, steep slope is used primarily for visual impact and typically requires a supporting deck with a minimum slope of 3/12. The seams can vary, however, and they usually are hydrokinetic, or designed to shed water.

Low slope. Also referred to as structural, low-slope systems have a minimum slope of $1/4/12$. Typically, these systems can support their own weight without a deck. The watertight, or hydrostatic, seams are designed to withstand water pressure, and the profiles and panels come in a wide variety of shapes and sizes. In structural and architectural metal, thermal movement is accommodated in the field-seamed side joints and clip designs.

Hybrid systems. These commonly are used in retrofit applications. Hybrid systems combine both architectural and structural steel technologies to provide the building owner with long-term roofing performance.

Another major advantage to reroofing with metal is the ability to create slope through the use of a subassembly system (Fig. 11-3). When needed, a sloped subassembly made of light-steel structural elements can be attached to the existing roof surface to create a minimum $1/4$ to 12 pitch and ensure adequate drainage. The attic created by the subassembly affords a convenient area for adding insulation.

The crucial parts of any roof design are the details around projections and the perimeter of most buildings. Metal details, regardless of the application or system used, can be extremely challenging. Aside from normal building movement, the joinery must be able to accommodate the expansion and contraction of the steel membrane. Simply put, metal roofing is less forgiving than most other single-ply, built-up (BUR), modified-bitumen (MBR), or asphalt-shingle technologies. When flashing a projection on a BUR, there can be three or four plies to provide a watertight seal. With metal, there is only one. A contractor must be very precise, from the takeoff right through the final inspection. Because the roofer is dealing with large panels that cannot be easily patched or spliced to correct mistakes, errors can be expensive.

FIGURE 11·3 Creating a slope/subassemby system roof frame.

Selecting Metal Roofing Materials

Metal roofing materials include copper, aluminum, zinc, steel, and stainless steel. The following paragraphs provide a description of the characteristics of each.

Copper

For centuries, copper has been the metal of choice for roofing important structures. It is simple to use, and it offers natural beauty and extreme durability. The familiar green patina on a copper roof is a weathering process that takes many years to complete. Because of the length of time required, chemical treatments have been sought that accelerate the weathering process.

The most common treatments use chloride-salt solutions or sulfate solutions that are either sprayed, brushed, or sponged onto the roof. The success of the chemical treatment often depends on the weather at the time of application. One particular treatment, which uses an ammonium sulfate solution, requires a relative humidity of at least 80 percent while six to eight spray applications are made. A method that uses ammonium chloride requires dry weather for a 72-hour period while the treatment is on the roof. It should be emphasized that the results obtained by any artificial weathering process vary greatly.

Frequently, a convincing-looking patina is achieved by artificial means, only to fade after a few years to one of the intermediate brown shades of the natural weathering process.

Aluminum

A mill-finished aluminum surface rapidly develops a stable oxide coating that protects it from further corrosion. The oxide film is quite thin, on the order of a few millionths of an inch thick, and is clear and colorless. Because of this, the surface appearance does not change much with age. Bare aluminum surfaces are very good reflectors of radiant heat. A bright new aluminum surface reflects up to 98 percent of the radiant heat that strikes it.

Even after weathering for years, the surface still reflects 85 to 95 percent of the radiant heat striking it. This means that a bare aluminum roof reflects most of the sun's radiant energy, thus holding down the temperature under the roof. If the underside of an aluminum roof is left unpainted, it also reflects radiant heat onto the building's occupants for increased warmth in cold weather.

Aluminizing is a process in which a layer of pure aluminum is deposited on steel. The aluminum coating has good weathering characteristics and protects, by sacrificial corrosion, the base steel at breaks in the coating.

Aluminum shingle systems come in several configurations, most of which have either 12- or 10-inch-high exposures. They provide a realistic wooden-shake look because of their deep-drawn graining and their formed-butt thicknesses, which vary by as much as 1¼ inches. Although the panel widths offered by different manufacturers range from 24 to 60 inches, most systems feature upturned and downturned top, bottom, and side flanges that permit the panels to be interlocked four ways. Because of the various aluminum-shingle systems, check the manufacturer's installation instructions carefully.

Stainless Steel

This is a steel alloy that contains at least 11 percent chromium along with other alloying elements such as nickel, manganese, and molybdenum. The higher the chromium content in the alloy, the more resistant it is to corrosion. Some stainless steel alloys contain nearly 30 percent chromium. The type of stainless steel most commonly used for roofing products is Type 18/8, which contains 18 percent chromium and 8 percent nickel as the main alloying elements.

Evaluating Metal Coatings

Galvanized, aluminum-coated, and aluminum-zinc-coated sheet steels have contributed significantly to the growth of the standing-seam roof system. Improvements continue to be made in coating materials, metal treatments, metal substrates, and productivity. This has resulted in a number of quality products that perform essentially trouble-free over the practical life of the roof.

Standing-seam metal roofing is made from cold-rolled, low-carbon sheet steel, typically 0.019 to 0.024 inch thick. A hot-dip metallic coating is added to provide the long-term weather protection needed for atmospheric exposure. Zinc, aluminum, or a combination of both metals usually composes the metallic coating, which is metallurgically bonded to the steel substrate by passing the sheet steel through a molten bath. These coated sheet steels can be roll-formed easily into roofing panel profiles without the coating flaking, peeling, or powdering.

Metallic coatings provide two types of corrosion protection: sacrificial and barrier. With sacrificial protection, the coating oxidizes at a low rate to protect the base steel even at the uncoated, exposed edges. With barrier protection, the coating acts as a wall between the base steel and the environment.

Galvanized sheet steel has a zinc coating that utilizes both protective mechanisms. Sacrificial is the primary protection, and barrier is the secondary. The zinc coating is more reactive and sacrifices itself to prevent the steel substrate from corroding. Zinc coating weights for galvanized sheet are specified in American Society for Testing and Materials (ASTM) A525. Commercial roofing applications are typically G90, or a total of 0.90 ounce of zinc per square foot for both sides of the steel sheet. The coating measures slightly less than 0.001 inch thick per side and usually has a bold, spangled appearance. For painted applications, the spangle normally is subdued.

Aluminum-coated sheet steel relies almost solely on barrier protection. Sacrificial protection is provided only in the presence of chloride ions, such as in a marine environment. Aluminum coating weights are specified in ASTM A463. The minimum aluminum-coating weight for roofing is 0.65 ounce per square foot total for both sides. The coating thickness is 0.0014 inch per side. The durability of aluminum-coated sheets is less dependent on the severity of the environment than that of galvanized sheets. Aluminum-coated sheets have a spangle-free or matte finish.

Aluminum-zinc-coated steel is better known by its trade names: Galvalume and Galfan. Galvalume is approximately 55 percent aluminum and 45 percent zinc by weight. Galfan is 95 percent zinc and 5 percent aluminum. Both materials provide sacrificial and barrier corrosion protection, although the higher the zinc content, the more sacrificial protection provided. As in the case of galvanized steel, the zinc-rich portion of the coating sacrifices itself to protect the steel substrate. The aluminum-rich part of the coating provides a long-lasting barrier between the atmosphere and the base steel. Coating weights for Galvalume are specified in ASTM A792. Galfan, with coating weights similar to those for galvanized steel, is covered by ASTM A875.

Galvalume and Galfan offer a corrosion-resistant sheet that has the durability of aluminum coatings and the sacrificial protection of zinc coatings. The aluminum-zinc metallic surface is less spangled than a galvanized sheet, but is still bright and reflective when unpainted. Galvalume, Galfan, and aluminum-coated sheets can carry a limited 20-year warranty when used for roofing.

Choosing Paints and Laminates

Regardless of the sheet-steel coating, there are instances when the metallic finish appearance is undesirable or extra corrosion resistance is beneficial. In these cases, organic coatings such as paints and laminates are applied over the metallic coating via coil coating, a process of adding an organic finish to a continuous strip of metal.

The most widely used paint systems for standing-seam roofs are polyesters, siliconized polyesters, and fluorocarbons. Polyesters are organic polymers characterized by relatively hard and abrasion-resistant surfaces. Siliconized polyesters are organic polymers that offer exterior durability, principally through chalk resistance and gloss retention. Fluorocarbons are vinyl polymers characterized by formability and durability, heat and chalk resistance, and color retention.

Laminates are applied as a plastic film rather than a liquid, and are three times thicker than paints. They are adhesively bonded to the coil under pressure and heat during the coil-coating process. Plastic laminates prevent chalking, fading, peeling, and other forms of degradation. Acrylic and fluorocarbon laminates most commonly are used in roofing applications.

Both laminate and paint finishes are nearly as formable as the metallic coatings beneath them. They enhance the appearance and increase corrosion resistance. Simply put, the selection of metallic and organic coatings for application over steel substrates is best determined according to specific need. No single coating can be generally stated to be the best. While there are trends in the uses of these products, they are governed primarily by the type of structure and environmental conditions.

Using Metal-Working Tools

Most roofing contractors have an array of more or less specialized metal working tools that range from hand-held shears, soldering irons, steel cutting saws, screw guns, and nibblers to sophisticated break presses, form flashings, seamers, and forming machines. Generally, an abrasive blade is not recommended for cutting panels. The heat it generates can burn away the galvanization and consequently leave the metal open to the development of rust. A nibbler is recommended for cutting panels.

Many special hand tools are available for fabricating a standing-seam metal roof. The tools used today were originally fashioned centuries ago by craftspeople who worked only with hand tools. While the hand tools of old remain functional and have use from time to time, the bulk of the forming and seaming done these days is by power tools.

Roll Formers

Roofers who construct many large metal roofs debate the features of portable roll-forming (PRF) machines and boxed panels manufactured in the factory. This book takes a neutral stand.

With a PRF machine, the roofing contractor is able to go directly to the jobsite to fabricate and custom fit the metal panels. Once made, the panels are sent to the roof to be fitted next to the preceding form. Some PRFs have optional panel profiles that can be changed out in a matter of a few hours. PRF machines are expensive. The metal roofing contractor must conduct a careful cost analysis between manufactured boxed panels and those made on PRF machines.

Power Seamers

Another portable tool that plays an important role in onsite fabrication is the power seamer. These small machines do the big job of forming the individual panels into a monolithic sheet. The seam must be constructed appropriately to ensure a proper fit from panel to panel and to provide a watertight seal.

Power seamers are merely smaller versions of the larger PRF machines. Powered by a heavy-duty electric motor, power seamers do their job by locking onto the two standing seams of side-by-side metal panels. Using a series of driven rollers and idler rollers, the two metal panels are joined to form a lock.

Power seamers can be fitted out to form several types of seams, single or double locks, tee locks, and batten seams. Some models are adaptable to more than one type of seam and can make a double lock in one pass. Depending on the gauge or type of material and the type of seam desired, power seamers do their job at a rate of 30 to 60 feet per minute. Once set, the power seamer works on its own, allowing the mechanic the luxury of attending to other tasks.

The power seamer is a relatively simple device to operate, but it performs an important part of a standing-seam metal roofing job. In addition to greatly affecting the way the finished job looks, it ties the panels together and bonds the panels to the cleats that hold the roof to the deck. With the seams properly completed, the rest of the work has a tendency to take care of itself.

Unloading and Storing Metal Materials

To avoid damage, observe certain precautions when handling the roofing panels before installation. For instance, when unloading the panels, take care not to dent or puncture them. If the material is unloaded with a crane, use a fabric sling and blocking to prevent crushing. Provide adequate support when bundles of long panels are offloaded from a truck. Support usually is required at the third points of a bundle.

If the panels must be stored outdoors, place them off the ground on angled wooden blocking so that water can drain freely. Cover the stack of panels with a breathable tarp that keeps the rain off the panels and allows air to circulate, which lessens condensation buildup.

White rust or staining tends to occur on tightly stacked panels between which moisture is trapped. Do not drag panels over one another or on the ground. Lift off and carry the bundle to avoid scratches that can cause the finish to fail prematurely. A common procedure during installation is to place bundles of panels at intervals on the roof purlins or deck. If the bundles are quite large, however, the roof structure can be overloaded in spots unless extra support is added.

Fastening Panels

Metal roof systems use either exposed or concealed fasteners. Exposed fasteners usually are painted to match the color of the panels, while concealed fasteners are not. Most exposed fastener systems are quite similar in appearance and function, whereas the concealed fastener systems have a wide variety of panel profiles.

Exposed-Fastener Systems

The exposed-fastener method is the easiest and probably least expensive way to secure a metal roof. The method might better be described as a lap-and-fasten system. The panels are lapped at the edges and a screw or nail is used to secure the joint. The fastener is driven through the high point of the lap with a washer to seal the hole.

With exposed-fastener systems, a large number of holes must be created in the roof panels to fasten them to the purlins. Neoprene washers and other steps are taken to seal the fastener holes, but many potential leak sites remain.

Temperature fluctuations also cause dimensional changes in the roof panels, making the panels slide relative to the purlins at points where the fasteners are not tight enough. This process can enlarge the fastener holes so much that the neoprene washers cannot provide a tight seal. Correct tightening of the fasteners is important to allow the neoprene washer to seal properly. An overly tight fastener squeezes the washer out of the joint or possibly tears the washer. In either case, a poor seal is created.

A neoprene plug usually is installed ½ inch up the batten. An aluminum batten closure is slipped in and riveted to the side of the batten. Apply sealer to the inside of the batten before the closure is installed.

If all fasteners are tightened so that the roof panels cannot slide relative to the purlins, then thermal expansion and contraction of the panels is absorbed by a process known as *purlin roll*. The purlins, typically Z-purlins, are of a light enough gauge to flex back and forth slightly as the roof panels expand and contract longitudinally. If the purlins are inadvertently braced in a way that prevents them from rolling, however, repeated expansion and contraction of the panels can eventually snap off the fasteners or cause the fastener holes to leak.

Different methods are used to keep water from creeping under the sidelap joint. Some manufacturers rely on factory- or field-applied sealant to create an impervious barrier. Others form an antisiphon or anticapillary groove in the lower panel of the lap. The groove creates a capillary break that prevents water from being drawn through the seam. In addition, the groove acts as a drain to carry away the water it has trapped.

Concealed-Fastener Systems

As the name implies, the concealed-fastener system leaves no fasteners exposed to the elements on the longitudinal seams. Stamped metal clips are fastened to the substrate and hold the panels in place at the longitudinal seams. There are three types of clips: two-piece, purlin-slip, and panel-slip. There are two main types of concealed fastener systems: two-piece and three-piece seams.

TWO-PIECE CLIPS

These clips are used with long, continuous panels that can undergo major thermal expansion and contraction. They consist of an upper part, which is clamped rigidly into the seam, and a lower part, which is fastened firmly to the substrate. The two pieces are connected by a slip joint that allows longitudinal movement of the panels.

The ideal two-piece clip is self-centering. When it is installed, it is automatically set to allow equal amounts of movement in either direction. Problems can arise when long roof panels are installed in extremely hot or cold weather. Most of the thermal movement from that time on will be in one direction only. In these cases, the sliding portion of the clip may have to be installed off-center to handle the expected thermal movement.

PURLIN-SLIP AND PANEL-SLIP CLIPS

These clips are used for short panel runs for which little thermal movement is expected. The two clips are basically the same. They differ in the way they are fastened. Purlin-slip clips allow for thermal movement in the loose connection between clip and substrate. Panel-slip clips are fastened tightly to the substrate, and thermal movement occurs when the panels slip relative to the clip. This slippage can cause some local damage to any sealant in the seam, but since the amount of thermal movement is small, the damage should be negligible.

Regardless of which type of clip is used, there is some point along the slope at which the roof panels must be fixed to the substrate or else the panels can walk their way down the slope with repeated thermal cycling. Possible fixing points are at the ridge, the eave, or some point near the middle of the slope. If the roof panels are fixed at the ridge or eave, flashing details are simplified at that point because there is no need to accommodate longitudinal movement of the roof panels. This means that the entire range of longitudinal thermal expansion and contraction must be dealt with at the opposite end, however, and movement can be considerable with very long slopes.

A mid-slope fixing point can be desirable on very long slopes because it cuts in half the maximum amount of thermal expansion and contraction that must be accommodated. In this case, both the eave and the ridge flashing details must account for longitudinal movement of the roof panels, but, again, the magnitude of the movement is less than would occur with the panels fixed at the ridge or eave alone.

THREE-PIECE SEAMS

Components of a three-piece seam include the two adjacent panels and the seam cap. The three-piece seam has a variety of profiles, but all are merely variations of three basic approaches: raised, flush, and open seams.

Three-piece seams have an advantage over two-piece seams when it is necessary to replace a damaged panel. A panel secured by a three-piece seam can be taken off by first removing the seam cap at both sides of the panel. The panel then can be lifted off and replaced with little or no damage to adjacent panels. The seam cap may not be reusable, depending on how it was originally applied, but a new seam cap is a minor expense.

If a panel secured by a machine-seamed two-piece seam (Fig. 11-4) needs to be replaced, there is a chance that one or both of the adjacent panels may be damaged in the removal process. On the other hand, some of the snap-together two-piece seams appear to be very easy to sep-

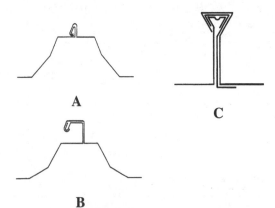

A

C

B

FIGURE 11-4 (*A* and *B*) Two-piece Type I raised-seam profiles; (*C*) two-piece Type II flush-seam profile.

arate, particularly the type shown in Fig. 11-5. Some manufacturers provide a special tool for unlocking the seams on their panels.

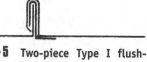

FIGURE 11-5 Two-piece Type I flush-seam profile.

A disadvantage of the three-piece seam is that the seam cap is subject to damage or loss at the jobsite before it is applied to the seam.

Applying Metal Roofing Panels

The application of a metal roof can be considered in three separate stages: substrate preparation, panel membrane application, and terminations. Substrate preparation consists of erecting the purlin framework or solid deck and all other materials that go under the roof membrane. Panel application involves laying and locking the roof panels. Terminations involve the installation of curbs, ridge caps, gutters, and any other finishing elements.

As with all the installation instructions in this book, the procedures and details are general. Always follow, to the letter, the manufacturer's instructions that come with the product.

Preparing the Substrate

Two different roofing situations can occur when preparing the substrate: a new structure that requires a roof or an existing structure that needs to be reroofed. In the second case, the roof membrane that is to be recovered can be a BUR or metal type.

To install a roof on a new structure, two different types of substrates can be used: solid or purlin. The solid substrate can consist of wooden sheathing or metal decking.

Rigid insulation can be placed on top of the sheathing or decking, or blanket insulation can be secured underneath. Certain types of roofing panels require the support of a solid substrate. A solid substrate provides structural bracing for the building and occasionally is used for that reason under roofing panels that do not actually require continuous support.

The purlin substrate can be used only under structural-type panels that have spanning capabilities. The purlins, commonly light-gauge

C- or Z-bar steel members, usually are spaced 5 feet on center and span rafter or roof trusses. Blanket or rigid insulation can be laid over the purlins as desired. Consider the insulation's strength and support requirements when determining purlin spacing. A properly located vapor retardant is important in either type of substrate system to avoid problems with condensation in the insulation or on the underside of the roof panels.

The most common reroofing job is to install a metal roof over a worn-out BUR. The old roof membrane can be left in place unless water has penetrated it and soaked the insulation below, in which case the roof membrane and insulation should be stripped away. The old insulation and roof membrane cannot be relied on to provide a firm, stable base for the new roof. In addition, any contained moisture could cause problems for the new roof system if the roof cavity is not vented properly.

Stripping off the old BUR adds considerably to the cost and labor of installing a new roof and negates one of the attractive features of a retrofit metal roof: the ability to install the metal roof on top of the old roof while causing minimal disturbance to inside activities. If the old roof membrane can be left in place, the loose ballast is usually removed to reduce the dead load.

If the old roof is flat, a wooden or metal framework must be constructed to slope the new roof for drainage. Fasten the purlins to the framework and finish the roof in the normal way. If the old roof already has enough slope, it is not necessary to construct a framework for the purlins. They can be attached directly to the roof surface at the proper spacing. If a thick layer of insulation is used, the purlins might have to be raised up on blocking.

Adding the Panels

Different roof panels form their sidelap seams using a variety of methods, but the main difference in roof panel application methods is in how the panels are laid, from left to right or right to left, or across the slope of the roof. If the roof slope is short enough to be covered by a single panel length, the direction of application is arbitrary and can be left to the designer's and installer's judgment.

If the roof slope is long enough to require two or more roof panels to be endlapped, the lay of the panel usually is dictated by the way in which the ends of the roof panels are prepared for the endlap. Most manufacturers prepare the panel ends that are to be lapped by either trimming away part of the seam on one panel or swaging or die-setting the end of one or both panels. Factory preparation of the panels makes for a close-fitting endlap and, consequently, a better seal. An added benefit is that field installers do not have to worry about trimming the panels.

Once the direction of installation has been determined, the first roof panel or row of panels is laid at, or within, a few inches of the edge of the roof. The important step at this point is to be sure that the first panel or row of panels is laid exactly parallel to the roof. A slightly inaccurate alignment can be corrected as additional panels are laid, but the best method is to avoid misalignment in the first place.

Some manufacturers offer a spacing gauge that can be used to make sure the roof panels are not inadvertently stretched or shortened in width as they are applied. Check panel alignment often as the installation progresses. It is not difficult to skew a relatively flexible metal panel that might be 50 or 60 feet long.

Sealing Terminations

Terminations are components such as ridge caps, curbs, flashings, gutters, and valleys that seal around openings or irregularities in the roof membrane and carry water to the building exterior.

CURBS AND PIPE FLASHING

These terminations seal around penetrations through the roof panels. Curbs are used around larger, usually rectangular penetrations such as skylights, rooftop equipment, and ventilation ducts. Pipe flashings, as the name implies, are used to seal around pipe penetrations, such as a plumbing vent stack. Most pipes, ducts, and equipment items that penetrate the roof respond to thermal expansion and contraction. The curb or pipe flashing must allow for this relative movement while remaining weathertight.

A skylight usually floats with the roof panels, so relative movement is not a problem with a skylight curb. An important feature for a wide curb is a diverter on the curb's upslope side. The diverter directs roof drainage around the curb rather than allowing it to build up against the curb.

Center, as closely as possible, pipe and vent penetrations on the rib to allow maximum movement of the roof relative to the pipe. The rubber boot should provide a tight seal between the pipe and the sleeve. The location of a pipe penetration is not always the roofer's choice, so the pipe may not be centered in the sleeve.

Many installers use a rubber-booted device to seal pipe penetrations. The rubber collar must be field-cut to the proper diameter so that it seals snugly around the pipe. The hole in the panel must be large enough to allow the panel to move relative to the pipe. The pipe boot base should be 2 inches smaller than the distance between the standing seams.

When the pipe penetration is at the ridge cap, the metal flashing is usually clamped and screwed to the ridge cap. The longitudinal thermal movement of the ridge cap is thus prevented at this point (Fig. 11-6). The ridge cap should not buckle from thermal expansion and contraction as long as it is in sections no longer than 10 to 12 feet.

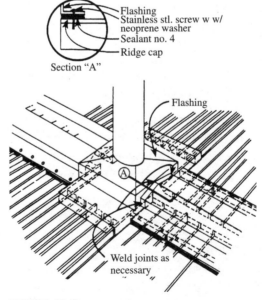

FIGURE 11-6 Pipe flashing at ridge cap.

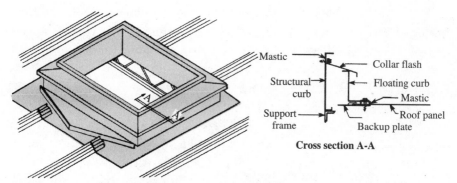

FIGURE 11-7 Curbed opening with diverter that directs water to either side of curb.

There are several ways to use a diverter with a curbed opening. Relative movement is possible between the curb and the roof, since the two are structurally independent of each other (Fig. 11-7). The diverter directs water to either side of the curb. The collar flashing does not overlap the floating curb by much in the vertical dimension. The installed height of the structural curb assembly is crucial to avoiding a gap where rain can enter.

With masonry penetrations, such as the chimney in Fig. 11-8, it would be wise to use reglets. A reglet is a slot cut in the chimney or concrete surface to hold the cap flashing. The reglet runs straight across the front and straight across the back, assuming there is no cricket. Use a diamond blade with a water-spray attachment in a hand-held circular saw,

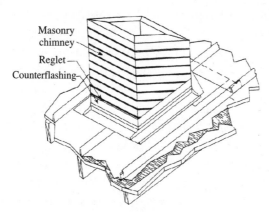

FIGURE 11-8 Use of reglets.

or a portable grinder with a masonry blade. A circular saw with a carbide masonry blade works, but not as fast as a grinder. In many cases, a cold chisel is really all that is needed.

PANEL ENDLAPS

These terminations are used when the roof slope is too long to be covered by one continuous length of roof panel. Some manufacturers specify that endlaps be staggered on adjacent rows of panels to minimize the number of extra plies of metal in the sidelap seam. In addition, one of the panels can be factory-trimmed to reduce the extra plies. Other manufacturers specify that endlaps be aligned on adjacent rows of panels, rather than staggered. In this situation, one or more panel edges are usually trimmed and/or swaged at the factory to provide a smooth fit in the sidelap seam.

On purlin-supported roofs, the endlap can occur either over a purlin, in which case the purlin can provide support, or away from the purlins. When an endlap occurs over a purlin, the fasteners often are driven into the purlin to draw the connection tight. Consequently, the roof panels cannot move to accommodate thermal expansion and contraction at that point. This condition must be kept in mind and the panels should be allowed to float at both the eave and the ridge to avoid buckling. The main concern when a lap joint occurs away from a purlin is whether the lap joint is strong enough to support occasional foot traffic.

RIDGE CAPS

There are several different ridge-cap treatments possible. The typical ridge detail consists of pan closures, anchor strips, and a ridge cap. When designing the ridge, thermal movement must be taken into consideration. Depending on the length of the panels and where they are anchored, the style of the cap also comes into play. Simple ridge ventilation is acquired by using a screen with the step-up style ridge cap (Fig. 11-9A).

Never fasten the cap directly through the roof panel. Fasten the pan closure and the anchor strip to the batten (Fig. 11-9B). The ridge cap then covers the anchor strip and is fastened to the strip between the battens. For thermal allowance, anchor the ridge cap on one side and allow the other side to float. For the float side, hook a return on the cap over a wider anchor strip. Both sides of the ridge cap should be the same size.

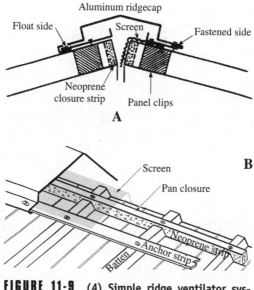

Aluminum ridgecap

Float side

Screen

Fastened side

Neoprene closure strip

Panel clips

A

Screen

B

Pan closure

Neoprene strip

Anchor strip

Batten

FIGURE 11-9 (*A*) Simple ridge ventilator system; (*B*) pan closure and anchor strip.

When a neoprene strip is installed between the battens, the strips can be engaged in a U or J channel, with cutouts over the batten. At this point, fasten them to the roof. Cover the installation with the ridge cap.

For a shorter roof, a low ridge cap can be used (Fig. 11-10). Attach the low ridge cap to the anchored end of the panel. The low ridge cap does not flex with the thermal movement. A float with a wider anchor strip can allow for thermal movement. Using wide low caps without step-ups is not recommended because they have a tendency to wave. Extend the roof panel all the way to the top of the roof when a low ridge cap is used.

Short roof and hip caps can use the one-piece hip cap (Fig. 11-11). Install the hip cap first, and then slip the panel into the channels. Install neoprene weatherstrips or similar closures to prevent water from blowing back under the joint.

Pan Closures at the Ridge: To turn up the pan on the ridge, slash the lock and batten and cut them 2 inches (Fig. 11-12) to make the pan easier to bend. When the pan is turned up the 2 inches, a tuck forms on

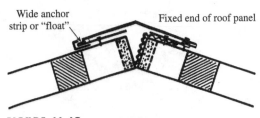

FIGURE 11-10 Low ridge cap.

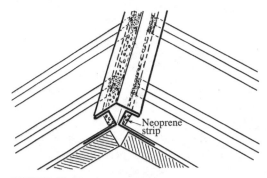

FIGURE 11-11 Roof and hip cap.

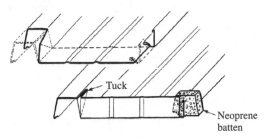

FIGURE 11-12 Pan closures at the ridge.

the batten side, and the lock side is covered with the batten of the following panel. Install a neoprene batten closure under the panel batten and an anchor strip across the length of the roof. Install the ridge cap so that it covers the pan closure and finishes the roof ridge.

Ridge Cap Overlap or Splice Joint: A simple ridge cap can be overlapped. The one-piece ridge cap or the step-up roof cap needs to be butted and spliced. A splice joint can be installed under (Fig. 11-13*A*)

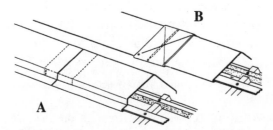

FIGURE 11·13   Ridge cap overlap or splice joint.

or over (Fig. 11-13*B*) the roof cap. Anchor the splice cover to one side of the joint only to allow for thermal movement. Make penetration between the batten and on the outside of the water barrier. The cover can be smooth or an X can be field-formed into the cover for a closer fit.

Shop-Formed Ridge Panel: Roof panels are fixed at the ridge purlins, so panels must be allowed to float at the eaves. The lap joints are not staggered on adjacent panels. Even though the edge of one panel is trimmed back, there is still one extra ply of metal in the seam, which can produce a poor seal.

Field-Cut Ridge Panel: These roof panels are fixed at the ridge purlins, so the panels must be allowed to float at the eaves. The roof panel is trimmed where it meets the ridge section, so the seam cap does not have an extra ply of metal to enclose. Lap joints are not staggered. The vertical portion of the seam contains two extra plies of metal. This design probably has no problems because of the seaming method. It can be difficult, however, to seal the rib cut properly at the ridge.

Field-Bent Ridge Panel: Fix these roof panels at the ridge purlins. Problems can occur if adjacent panels are allowed to move relative to one another near the ridge bend. If the bend is created at the jobsite there also could be some difficulty creating the ridge bend at the proper location on each panel.

Ridge Cap with Profile Closure: Allow the ridge cap to float longitudinally at the ridge. Attach it in lengths of no more than 10 to 12 feet because of its longitudinal thermal expansion and contraction. Provide extra support under the ridge cap to handle foot traffic and other heavy loads. A profile closure is required to create the seal between the roof panels and the ridge cap. The numerous through fasteners required to hold down the ridge cap create opportunities for leaks to form.

HIPS

Hip details are similar to ridge details. The panels leading to the hip are never cut to length. They are cut diagonally and, because of the variation of the angles, must be cut in the field. Allow enough length to cut the required angle. Take special care to match the ribs on both sides of the hip. Real planning must be done when the hip is on both sides. When panels are field-cut for rake edges, valleys, hips, etc., measure each panel before cutting to ensure the right fit and a straight alignment with the flashings.

Due to various hip angles, the closures also are field-cut and assembled. Install a U or J channel between the battens, with neoprene or another weatherstrip, and an anchor strip over the closures (Fig. 11-14). Rivet the anchor strip to the batten. Hook the hip cap over one anchor strip and fasten it to the other anchor strip between battens.

Install the hip cap before the ridge cap, since the thermal movement varies from one panel length to another. Give special attention to the design of the caps to allow for maximum movement. Anchor the panels leading to the hip cap at the eaves.

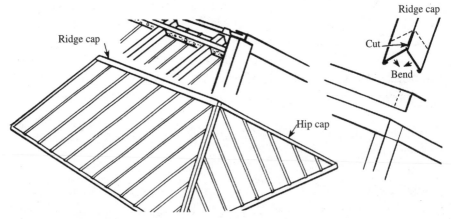

FIGURE 11-14 Closure of the pan on the hip.

VALLEYS

Valley flashings can be the most crucial flashings on a roof. Long panel runs emptying into a lengthy valley, coupled with a low roof slope, require careful design and installation. When designing roofs with valleys, consider using panels with vertical ribs. These panels are much easier to seal at a valley than trapezoidal-ribbed panels. Remember, in valley applications the roof panel and closures are not of standard size because of the variation of the angle. The panels are cut diagonally and must be cut in the field to assure the right fit and angle. Use enough flashing to ensure the complete removal of all rainwater runoff, blowback, or overflow, as well as snow and ice.

Figure 11-15 shows valley details with offset and integral cleats that eliminate exposed fasteners. A valley with integral cleats is one of the best designs because the valley trim is not punctured by fasteners,

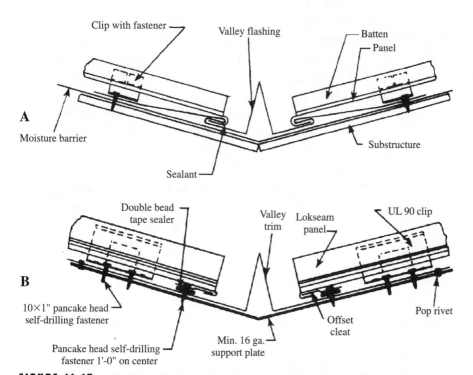

FIGURE 11-15 (*A*) Valley details with offset cleats; (*B*) valley details with integrated cleats.

except at the top of each side. An inverted V in the middle of the valley flashing prevents crossflow and ensures flexing with thermal movement.

This thermal movement varies from one end to the other. While the movement can be ignored on short roofs, on long lengths the movement needs to be considered when the cap or valley and flashing details are designed. To prevent relative movement between panels, give consideration to where the panels are anchored. Anchor valley panels at the ridge. Do not butt the roof panel against the V in the middle of the valley flashing. Leave an open space, or gullie, to allow complete drainage of the panel.

When using valley designs that require fasteners to puncture the valley trim, be sure that the fasteners are set in sealant. On long valleys, allow as much clearance as possible between the panel and the water diverter or the valley trim. This allows the valley to handle more water before it ponds onto the roof panels.

Valleys that terminate before they reach the eave of a roof are especially difficult to install. The most common application of this detail is at a dormer where one or more panel runs must have endlaps at the valley termination point. The lower panel laps under the valley trim, while the upper panel laps over the valley trim and is cut on the bevel to match the valley.

Drive metal and neoprene plugs into the batten to prevent bugs or rodents from entering the panel. Because of the various panel angles, aluminum plugs must be field-made and fitted.

To hide the batten plug assembly, use the two-piece valley system and a cover plate. Install the cover plate over the V and reach across the gullies to the batten. Fasten the cover plate to the batten between the end of the panel and the neoprene plug.

RAKE EDGE OR ENDWALL FLASHING

Use rake trim to seal the side of the roof to the building walls. On roofs designed to float or accommodate the thermal expansion or contraction of the roof panels, design the rake trim so that it floats. For roofs that do not need to float, use a fixed rake trim.

Butt the panel batten up to the wall with the batten intact. Shape the flashing over the batten across the top of the wall and then over to the front and hook it into the fascia cleat. Cut the panel 1½ inches wider than the required width. Bend the 1½-inch piece upwards. Apply two bead strips of sealer and fasten an L channel between the sealer beads. Install the flashing over and onto the L channel.

When a wall is much higher than the roof, use a separate flashing that is independent of the coping. Cut a reglet into the wall about 2 inches above and across the length of the roof. Apply sealer to the cut and slip the flashing into the reglet. Or, use a flashing with a 30-degree return. This acts as a caulking trough. Make sure that these types of applications are sealed thoroughly.

Another way to cope with an unsightly wall is to install a metal panel soffit between the roof and the coping.

PARAPET FLASHING

Generally, if the parapet wall is masonry, the parapet rake flashings are designed to float, even if the roof is fixed. This is due to the difference in rates of thermal movement between masonry and steel. Parapet high-eave flashings can be designed as fixed or floating.

When dealing with masonry parapets, it is better to make a saw cut and insert a counterflashing than to use a surface-mounted counterflashing. Surface-mounted counterflashing can allow leaks due to cracks in the masonry parapet wall or a poor caulk job. In northern climates, snowdrifts can create additional waterproofing problems. Where the possibility of snowdrifts exists, carefully seal the parapet flash, rake, or high eave to the parapet wall. This is in addition to the shingle effect of the counterflashing overlapping the parapet flashing (Fig. 11-16).

Finishing Custom Details

Several variables, such as function and the appearance of the accessories, the length and anchoring of the panel, and environmental conditions, must be considered in the design and selection of these important customized components. Failure to assess these variables can hamper not only the effectiveness of the roof system, but also the aesthetic value of the building.

To help find the optimum combination of aesthetics and function, this section illustrates several alternatives available to the building designer, contractor, or owner. It is recommended that all flashings,

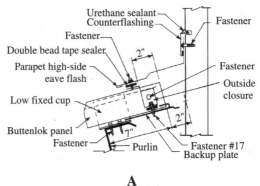

A

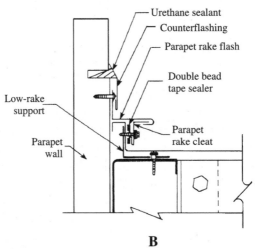

B

FIGURE 11-16 (*A*) Floating high-eave parapet; (*B*) fixed-eave parapet.

copings, etc., be installed as the roof installation progresses to avoid unnecessary traffic on the finished roof.

FASCIA

Different fascia designs require individual applications. Descriptions of the three most common types follow.

Simple Fascia Board: Use this type of installation only if water flow off the roof does not create any problems. There should be adequate overhang and drainage to divert water away from the building.

Batten Strip: In this installation, cut the panel at the batten and lock and bend the pan in the field. Cut and bend a separate batten strip to match the panel and install it over the batten by sealing and riveting it.

Kneecap and Batten Strip: A very striking design is made when the roof rib continues around the fascia and returns with the soffit to the building. Cut and bend the panel at two places. Use a one-piece batten strip to cover both knees. The top pieces always overlap the lower piece to prevent water intrusion.

WALL AND PEAK CAP

This cap is similar to the ridge cap, but converts to a fascia board on the front. If the roof panel is anchored on the top, thermal movement is negligible. When the panel is anchored on the eave, thermal movement needs to be considered when the cap detail is designed. Install the roof panel with a closure and anchor strip prior to installing the cap. Slip the wall panels into the channel of the fascia trim.

For shorter roof lengths or a shed roof, use a one-piece installation. Cut at the batten and lock. Bend at the pan. Install a *kneecap* cover over the batten.

EAVE FASCIA WITH TRIM STRIP

Batten closures are often used along with neoprene plugs to close the roof section. Install the flashing first, and then proceed with the roof panel and vertical panel to align the ribs. Use bottom nose trim and soffit panels (Fig. 11-17). Install the soffit panels onto the trusses, which are made individually and connected with the installation rail, or hat section. This creates a horizontal soffit.

Use a conversion trim on this application. The variation in the angle of the roof makes the conversion trim a custom-formed trim piece. Install neoprene closures in the channels to prevent blowback. Be sure to line up the ribs of the wall panel with the ribs of the roof panel.

DIAGONAL EDGES

Due to variations in roof angles, the panels must be field-measured and field-cut. Make sure enough length is available to cut the required angle. Verify the length and angle of each panel for proper fit and

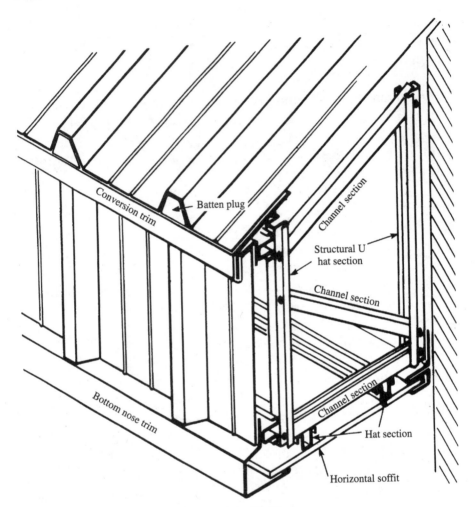

FIGURE 11-17 Eave and fascia detail with trim stop.

alignment with the roof and flashing or trim. Install wide flashing under the roof panel and extend it to form a drip edge or fascia board. Install neoprene plugs in the batten and sealer used under the pan. Because of the angle of the batten, the aluminum closures need to be field-formed to assure an accurate fit.

The trim is installed over the batten and extended past the drip edge to allow for complete drainage off the pan fastened onto the battens. In both applications, water is allowed to fall freely from that part of the roof. If draining to grade causes a problem, consider installing a gutter (see Chap. 14).

BUILDING EXPANSION

To allow for building expansion, the roof needs an expansion cap or crown. Fasten a simple crown to both sides of the expansion joint. Fasten the cap to one side and allow the other side to float. Work the panels outward from the expansion joint. Install a starter strip on both sides of the expansion joint. Allow enough space between the starter strips to hook the panels into place.

An alternative method is to install a flat, or float-anchor, strip on one of the battens and hook the expansion cap into the float. Then fasten it to the other batten. Or, fasten the crown to both battens so that the V in the crown flexes with the thermal movement.

Stop the roof panels at the expansion joint. If the last panel extends past the expansion joint, cut the panel 2 inches longer than the available space. Bend those 2 inches upward for extra strength. If the panel fits flush to the expansion joint, do not cut or bend it. Install a starter strip and continue with the next panel.

For a float anchor, apply a Z-bar onto the cut panel and install a float-anchor strip to the batten of the next panel. Hook the expansion cap into the float and fasten it to the Z-bar.

With the crown application, use an L channel on the cut panel and fasten the crown to the side of the batten and the L channel.

GUTTERS

Gutters are an important accessory for any building. For this reason, the subject of roof drainage is covered in detail in Chap. 14. With metal roofing, it is possible to create two functional gutter systems that are architecturally aesthetic and not easily attainable with other roofing materials.

Hidden Gutters: This gutter system becomes an integral part of the roof. For aesthetic reasons, the gutter can be built into a facade. The downspout or leader can be hidden in a column or the wall.

Recessed Gutters: The recessed gutter system allows the gutter to be installed into the roof, either in the overhang area or within the wall.

This type of installation has to be considered from the start and the structural steel needs to be designed accordingly. Often, a single-ply roof membrane is used as the gutter liner.

Repairing and Caring for Metal Roofs

A metal roof needs occasional care and maintenance to ensure that it lasts as long as possible. Most manufacturers recommend that the roof be washed annually with a strong stream of clear water to remove atmospheric dirt that can make the finish appear dull. This is a solid ancillary business. The roofer is used to heights, the equipment is relatively inexpensive, and the in-and-out nature of the work is excellent. Barring unforeseen damage, a yearly washing is all the maintenance the metal roof requires for many years.

Occasionally, a roof is damaged and one or more panels need to be replaced. This procedure requires the opening of two or more sidelap seams so that the damaged panel(s) can be removed and replaced. In this situation, panels secured by three-piece seams are advantageous because adjacent, undamaged panels are disturbed minimally, if at all. Cut the seam cap for the three-piece seam along its entire length with an air chisel or other equipment and then lift out the damaged panel and lay a new panel in its place. Apply a new seam cap to complete the procedure. This can be a part of the cleaning service.

Panels secured by two-piece seams, particularly those that have been machine seamed on the roof, usually are more difficult to remove without disturbing adjacent undamaged panels. Some manufacturers supply special tools that can be used to pry open the seams. Once the damaged panel is removed, lay the new panel in place and reseal the seams.

Recoating options

Recoating a corroded or badly weathered metal roof can be an economical alternative to total replacement. In general, the recoating procedure requires that any peeling paint and loose, heavy corrosion be removed by either wire brushing or sandblasting. The entire surface of the roof should then be cleaned thoroughly. Use a cleaning solution appropriate for the original finish, and allow the roof to dry completely. Always spot-prime bare metal and tight corrosion. Some manufacturers require that the rest of the roof be primed as well. Then apply the finish coat at the manufacturer's recommended coverage rate.

Although coatings such as urethanes and fluorocarbon polymers often are used in recoating applications, it should be pointed out that these formulations are not the standard factory-applied coatings. Coatings that are intended to be factory-applied normally require oven-baking to cure properly. These coatings would fail if they were applied in the field. The field-applied coatings are specially formulated for air curing. Manufacturers of these urethanes and fluorocarbons claim that lifespans of the field-applied coatings approach those of the factory-applied, oven-baked coatings.

As metal roofs see more common use, the need to reroof structures with worn-out metal roofs becomes more common. Often, the old metal roof becomes the substrate for the new metal roof. This situation raises questions about the compatibility of materials used on the old and new roof systems and the old roof's structural integrity.

As explained in Chap. 7, galvanic corrosion occurs when two dissimilar metals come in contact with each other in the presence of an electrolyte. Moisture that might condense between the old and new roofs could act as the electrolyte and, if the metals are not compatible, cause hidden galvanic corrosion of the support clips or old roof. Moreover, a badly corroded or deteriorated metal roof might need to be removed entirely because it cannot provide a structurally sound surface for the new roof.

If the old roof covering is left in place, check to see how it is attached to the building. The new roof cannot be expected to achieve a high wind-uplift rating if the original roof is not able to achieve the same rating. If necessary, perform tests on the old roof to determine the pullout strength of its fasteners.

Minimizing Oil Canning

Oil canning is defined as a perceived waviness in the flat roof areas. It is an aesthetic problem, the degree of objection being in the eye of the beholder. Normally, structural integrity is not affected. Structural integrity must be reviewed, however, if the distortion is of an extreme nature. Since many uncontrollable factors are involved, no manufacturer can realistically assure the total elimination of oil canning. With careful attention to the production and selection of material, to the panel design, and to the installation practice, however, oil canning can be effectively minimized. Take the following precautions to control oil canning.

Coil. Tension leveling, a process whereby the metal is stretched in coil form, provides a flatter surface that is less prone to oil canning. Also, the heavier the metal gauge, the less likely the product is to wave. The possibility of oil canning can be reduced by ordering tension-leveled material and specifying a minimal amount of allowable camber in the coil.

Design. The addition of stiffening beads breaks up the flat surface and makes oil canning less apparent. Embossing also helps hide surface waviness in the metal. The selection of low-gloss coatings and light colors also tends to minimize the visual effect of oil canning.

Installation. More stringent specifications regarding the alignment of the supporting structure focus attention on this critical aspect. Instructions regarding proper handling, spacing, and fastening should be a part of the manufacturer's delivery packet.

Introducing Plastic Panels

While not metal, this newest roofing material is installed in basically the same way as metal panels, and therefore the method of installation is included in this chapter.

The thermoplastic roof panel is available in wooden shake and slate facsimiles, or in tile configurations. These panels have a better flame-retardant value than the wooden shake. Plastic panels do not rot, curl, grow algae or moss, break, crack or split, or absorb moisture. Roof systems made from thermoplastic resin are not affected by salt spray or by acidic environments. They are virtually indestructible, since they are also rust and termite proof.

The tile and slate configurations do not require structural reinforcement. Temperature fluctuations do not affect the performance of the panel. Its strength, even in subzero temperatures, defies hailstorms. Workers can walk on the roof without fear of denting, cracking, or breaking it.

The light weight of the roof panel makes it an ideal reroof product. Often, tear-off of the existing roof material is not required. This constitutes a tremendous savings of labor and disposal costs. Where there is an uneven roof in place, furring out might be required. Installing the panel over concrete or clay tiles is not recommended. The panels must be installed on, and the fasteners should penetrate into, a solid base for good anchorage. The panels are hooked to each other with overlapping top and sides, and should be installed in a staggered fashion for a more natural look.

Working with the Panels

Recommended fasteners are self-tapping screws, wood screws, or nails, depending on the application and the substructure. Take care to place the fasteners in the nailing slots. When a nail slot is not available, predrill a hole.

The roof panel, which is molded of resin, can be cut with any circular or hand saw. Use a fine-toothed blade with 10 to 18 teeth per inch.

The wooden-shake configuration generally has three courses of shake design per panel, is approximately 8 inches high per module, and has varied shake width. The tile configuration usually has two courses and the scallops that represent the Spanish tile have an 8-inch-wide repeat. The country-slate configuration generally has four courses and each slate is 6 inches high and approximately 8 inches wide. Side overlaps are staggered to achieve a natural look.

Typical Installation Guidelines

Thermoplastic roof panels must be installed from left to right. The minimum recommended pitch is 3/12, and fasteners should be of sufficient length and pullout strength to meet local codes. At least three or four fasteners per panel are recommended. The thermoplastic roofing system is designed to be installed over solid sheeting. The roofing can be installed over low-profile roofs, like asphalt. The recommended underlayment is equal to or better than No. 30 felt. Do not allow water to drain under the roof panels.

PANEL INSTALLATION

To ensure a level installation, snap a chalkline from the eaves of the roof. Install a ¼-inch J channel along the chalkline and fasten every 10 inches. Slide panel tabs into the J channel. Make sure the tabs are engaged into the J channel. Fasten the panel through the nailing slots. Do not overdrive the fasteners. Use fasteners that are long enough to penetrate the structural members or solid underlayment.

Extend the panel ½ inch beyond the edge of the roof at the eaves. Cut the steps from the panel to assure a straight line up the rake edge. The panels can be cut with a fine-toothed saw blade. Install panel seams in a staggered fashion. What is cut from the end of the course can be used to start the next course. This minimizes waste at the beginning and end of each row. Fasten the panel at the end. Cover the end with the fastener with the rake edge. Install the rake edge as work progresses up the roof. The rake edge can be either metal or a specially designed rake edge.

HIPS AND RIDGE

Use a hip or ridge cap to cover the edge. Install the panels to the roof edge and use a metal strip to cover the cap. Or use a specially designed hip or ridge cap.

SIDEWALL FLASHING

Use metal for the sidewall flashing. Make sure there is positive drainage away from the wall and down the roof over the panel.

VALLEY DETAIL

Use a metal valley flashing with generous side flanges. Form an inverted V in the center of the valley sheet. Attach the panels to the inverted V. When installing dormer or mid-roof valleys, make sure that there is a positive drainage of water onto the lower roof section. Overlap the lower panel with the valley sheet, and cut and install the panels to fit the angle.

ROOF PENETRATIONS

A standard vent-pipe boot is available for most penetrations. Follow the manufacturer's installation instructions. Other roof penetrations might need to be flashed out according to field requirements. Use one-ply roof material or other weatherstripping to seal around penetrations. For a chimney, use a camelback, cricket-and-step, and/or reglet flashing, depending on its placement on the roof. All flashings must have positive drainage over the roof panels.

APRON FLASHING

When the panel fits to the wall, install the apron flashing with a reglet or fit it under the wallcover or windowsill. If the panel is too long, cut the extra material, install the panel, and cover it with the apron flashing.

New-Roof Installations

For new roofs, plywood is the recommended underlayment. Apply No. 30 felt paper or another vapor barrier. Install the panels as recommended and previously described.

Reroofing Applications

Solve any existing problems with the old roof before installing the new roof. Pretreat known leaks and remove decayed underlayment or structural members. If the roof is in sound condition and relatively even, install the panel directly over the old roof. The thermoplastic panel is less than 70 pounds per 100 square feet. Make sure the fasteners penetrate the solid underlayment.

Cleaning

Do not clean plastic panels with solvents such as alcohol, ketone, aromatic, etc. These solvents soften the finish. If necessary, wash with a mild detergent and water. Rinse with clear water.

When contracting for any roof work that is new to your company, make certain that all design elements and manufacturer's instructions have been reviewed with care. Talk with the manufacturer about nearby installations and all tricks of the trade.

The Business of Roofing

I t is easy to forget that roofing is first of all a business, and that it is the business of roofing that presents the arena in which the tradesperson can practice his or her craft as a skillful roofer. A roofing company can be a solid, long-term, and profitable endeavor, and this chapter is designed to provide an overview of the day-to-day demands of professional roofing.

Roofing is one of the most important aspects of any building project, and applying new roofs and fixing old ones is one of the more difficult jobs in the entire construction industry. Working high off the ground, vulnerability to the weather during application, continual exposure of the finished roof product to the elements after installation: The roofing trade is full of demanding situations.

A great many people who are involved in the roofing industry begin by working in the field and become professionals only after years of hard work. Because of the continual demands of those many years of strenuous physical labor, many roofers are never trained in business. And because of the never-ending constraints of the day-to-day workload, people often get buried in chores and never find time to quit being mechanics in the field. The important process of overhauling their business to accommodate their own long-term goals may never get addressed.

Thus, when a roofer is trying to fine-tune the business, it can be difficult to adapt to management concepts like learning to work smart rather than hard. The important thing is to look closely at how your current goals in life relate to your existing business and your past work experience.

In the sections that follow, the basic concepts for learning to work smart rather than hard and developing techniques for removing oneself from the stress of day-to-day operations in order to be able to achieve overview are addressed along with the techniques of business. Learning to achieve overview can help a person to focus on the interaction between reaching goals and the daily grind, providing a space in which intelligent decisions can be made. In earlier chapters, we have looked at the craft of roofing in detail. Obviously, the application of roofs is very important, but remember, the work in the field is demanding and dynamic and draws our attention automatically. The hundreds of details that lead up to a project and keep a business running smoothly are not so obvious or dramatic as the work that takes place on the rooftop. However, before the first fastener is driven home, dozens of details—finding the job, estimating, contracting, shopping vendors, managing employees, scheduling, and a myriad of other tasks—that are not dramatic must be dealt with.

It is important to remember that attention to detail before the physical work starts allows the mechanics in the field to apply the roof or perform repairs in a timely, professional manner. All of the behind-the-scenes details are as important as the activity that takes place at the jobsite.

This chapter gives an overview of business. In Chap. 13 we will review some of these concepts as we explore computers. In Chap. 2, "Planning a Job," we took a closer look at what happens after winning a contract, while moving a job from the office to the field. Each professional has a unique method for dealing with the details of the business, so every business is unique. But the following tasks are a part of the territory with any roofing company.

The Art of Management

Management is a lot of work in any business, and roofing is one of the more difficult for a number of reasons: The principal is limited by strict time constraints, construction documents often lack complete details, work product must be integrated with that of other trades, and jobs are highly vulnerable to the elements.

The myriad of peripheral tasks (keeping rein on signed contracts, performance on the jobs, balancing the actual cash flow of the company, hiring, marketing, equipment, aged receivables, etc.) sounds overwhelming—and in the real world it certainly can be.

There are many books on the subject of management; there are courses and seminars of all kinds. But the main ingredient of successful management is the ability to break the business down into parts, and take each—even the most mundane-sounding, such as safety—seriously on an ongoing basis. The highly successful manager filters each of the ingredients out of the big picture, looks at that task carefully, then establishes company protocol by integrating each of them in a simple, efficient manner.

Establishing procedures simplifes and speeds up production, but it is important to remember to stay flexible, to shift approaches to tasks whenever improved techniques can be found, to grow and change with the business's needs on a continual basis.

This brings us to the next rule of successful management: *Assign tasks to the person with the smallest load of responsibility.* However, never think of a person who is low in the chain of command as being a lowly employee. The cleanup person is as important as the principal in a success-oriented company. Often, a person pushing a broom can watch what is going on at a jobsite more clearly than the foreman, the superintendent, the lead worker, or the owner of the company (see Fig. 12-1).

People with light loads of responsibility are naturally standing back with a limited amount of information in their minds. They are able to watch. And powerful managers take advantage of this by having discussions with people who have this unique overview, listening to them, and bringing all of the staff into the loop of success.

One of the biggest mistakes in the construction industry is for the principal to think that she or he can perform all of the tasks better than anyone else, then to hang on to the chores for dear life, refusing to hand them over to other personnel.

This causes many problems, two of which are a breakdown in trust and communication between the boss and other employees, and a

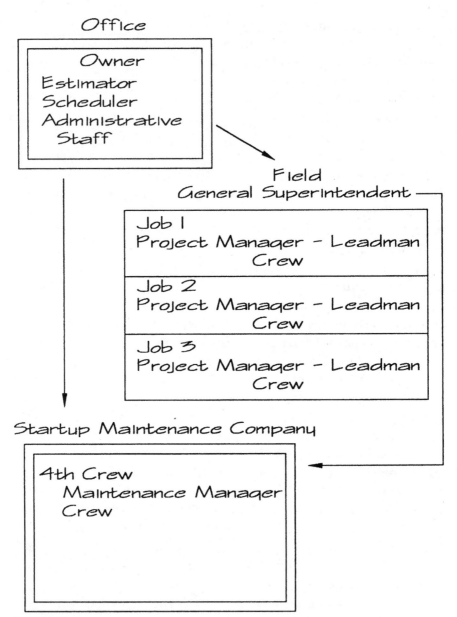

FIGURE 12-1 The art of management.

chaotic mindset for those in charge. Trying too hard causes this muddled view; the boss is always in a hurry and can't pay attention. The owner refuses to let go and continues to perform tasks that others could be doing in their own way, which might even be better than the owner's. If this common management error is not prevented, the principal will continually lose sight of the big picture and of situations that are coming up in the future that could be carefully planned for well ahead of time, and will get stuck putting out little fires all over the place.

The owner of a company should schedule time to get away from it all, to get some distance. This time can be very valuable in helping to clear the employer's head as well as being a time to meet new people, hobnob, and market the company. Naturally, this does not mean that the boss should be off having fun all the time and not paying attention to the details of the business. On the contrary, handing over tasks and taking time off gives her or him an opportunity to listen carefully to that person pushing the broom. Is the broom the correct size? Are the dumpsters being dropped in a labor-saving location? Is recycling and disposal of hazardous waste being treated efficiently?

By letting go of tasks and listening with attention on a regular basis, the roofing contractor can not only fine-tune the business, but also gain and keep the trust and respect of employees. Allowing people to have control and work at their best motivates them to feel pride in a job well done and creates a more enjoyable, efficient, and profitable workplace.

Before Creating a Business Plan

Before we think about the concept of an actual business plan, it is important that you take a look at your life and think through some important factors carefully. Most business plans are simply tools used for getting a loan. They are typically used for startup businesses; however, they can also be used for expanding your business as it now exists. But before you put a business plan together, it is important to think about where you are going with your current business and your life in general. It is good for you to try to really understand why you are in business, why you are borrowing capital that you will have to pay for, and whether you really want to be doing what you are doing.

Like any other part of a person's life, a business works more smoothly if the owner has an idea of where he or she is going and what he or she is doing to help get there. This does not mean that one sets a goal, puts on blinders, and trudges down the path of life like a donkey. No, quite the contrary, setting reasonable goals and being able to shift the approach to achieving them as conditions change will help anyone to attain the overview mentioned above, to relax, and to enjoy all of life more fully.

The First Step Is Knowing What You Want

In order to understand your roofing business fully, there are two questions to zero in on: What do I want? And *how* am I stopping myself from getting what I want? When you begin this exercise, it is important to work on it alone. You can set aside some time in the morning when everything is quiet—get up a couple of hours early. Go fishing, hiking, biking, boating, or walking. Take a trip by yourself. Think about it while you are driving to look at jobs. And don't be in a hurry. Relax.

Let your mind float on the idea, but keep working on it until you can cut through to the truth. Be honest with yourself. You don't have to talk about this with anyone else, but if you really want to get smooth in your professional life, you must know where you are going and have some sense of what you are looking for with the rest of your life.

There are unlimited scenarios, but the point is that it is very important to get to know yourself intimately, to admit to who you really are. Then you can look clearly at what you are doing and make real-world choices about your future. Naturally, you may not like part of what you see. That's okay; don't worry about it. Just keep looking at that real you for a while—don't judge yourself. And don't act rashly. In fact, don't act at all until you have thought about this for some time.

When you first do this exercise, you may think that you actually want to run around and not even have a roofing business. You may want to move to Vegas and become a professional player. But in a few days you will probably see that fantasy as nothing more than a simple flash-in-the-pan thought, an effort to get away from the mundane nature of day-to-day life as a roofer.

If you decide that you do not actually have a real interest in roofing, it may be important for you to start looking for something that you really do enjoy doing and figure out how to attach a cash flow to it. Maybe you really want to train horses, go into computers full time, start up a company on the Net, or work for a big roofer as a superintendent and restore classic automobiles on the weekend. You need to think this through carefully and talk it over with those around you if you finally decide that big changes must take place in order for you to be doing what you want to do with your life. And then, include your loved ones in setting up a plan for changing over to what you really want to be doing. This book is about roofing, so we'll stick with the subject; there are many advisers, books, and seminars available on changing careers.

Start with Long-Term Goals

The important thing is to see that this is *your* life ticking by, and that, if you do want to be a really successful roofing professional, setting some goals can help you get control of your life and your business. What do you want to be doing with yourself in 20 years? This is a very serious question, and if you come to terms with it, you can do some real goal setting. Start with a 20-year look at your goals and work back to the present. Looking at the future can direct you toward what you can be doing in the present to move toward your goal. This will focus your work and give additional meaning to it.

Once you start to see who you are and what you really want, your life will probably begin to change. It is important to remember that your real life is today, not the past and not your goals. Achieving goals doesn't mean putting today aside, it simply gives focus and clarity to today. You are no longer someone who is stumbling along, putting out fires, wishing that things would get better. Things never get better, nor do they get worse; they simply change. More money will simply present you with new dimensions of tasks that must be attended to. Remember that today is very important and set aside time now to do some of those things you want do a lot of when you retire. Also, it is very important to remember that your goals, and the path to implementing them, will probably change—always be prepared to change.

Real World	Things I Really Like to Do	How Can I Make Money at It?
✓	Roofing	Have always enjoyed roofing. Will keep cash flow, work smart, and pursue restoring cars too.
	Bass fishing	Too personal; don't want to make money off it.
✓	Collect and restore 1950s cars	Could start to look for more and buy some and have others do some of the work on them.
✓	Consult about roofs	Could learn more about being an expert witness and put out some cards.

FIGURE 12-2 What I really like to do.

Use Figs. 12-2 and 12-3 to determine your long-term goals. They are available on the CD included with this book for downloading to your computer, so you can change and update your plans as you go along. These goals will become the schematic overview for your roofing business. Figure 12-2 is a sample of how a person might fill in the information. Figure 12-3 is a blank form that you can copy and experiment with. Remember that all of the checklists can be found on the CD.

Setting Short-Term Goals That Guide You toward What You Want

After you have looked carefully at your long-term goals, at what you want in your life, the next question is: How do I start moving in that direction right now? You may come up with goals that are not business-related at all, and you may decide on other businesses. We are covering roofing in this book, but you can take the ideas we are discussing and apply them to the new business.

In relation to roofing, the first thing to look at closely is what the main thrust of your business is going to be at this point in time. During

Real World	Things I Really Like to Do	How Can I Make Money at It?

FIGURE 12-3 What I really like to do.

the beginnings of a roofing business, the owner may be thankful for any work that comes along, just hoping that there will be enough money to pay the bills.

That sense of being at the mercy of events rather than being in control is a very vulnerable position, and part of planning your goals is to move away from catch-as-catch-can—carefully, but as soon as possible. Below are some basic ideas to get you started. For more ideas, see Chap. 14. Be sure to add scenarios of your own if they come to mind.

- Making a transition from a small roofing business to a larger one
- Making a transition to a major, incorporated roofing business and going public
- Maintaining your position as a small roofer, the owner of an excellent company
- Starting new businesses that run simultaneously with the roofing business

Naturally, these goals are unique to each person, but the process of setting goals is what will provide the road map for your success, no matter what path you choose. The main change in attitude that you can make in your business life is from a position of reactive to one of proactive management. You are going somewhere with your life, and you now have a reason to tighten up your business. This approach will be set against a backdrop of long-term goals and their relationship to everyday tasks.

Now that you have a sense of resolve based on long-term plans that really mean something to you, it is time to look at how to put them into action. Take your filled-out copy of Fig. 12-3 and develop the short-term goals that will guide you toward your long-term goal. They must be thought through and implemented.

Figure 12-4 is an example of how a person might fill out the short-term game plan worksheet. You will find a blank copy in Fig. 12-5 and on the CD.

Developing a Business Plan

Now that you have roughed out your long-term goals and begun the short-term strategy for implementing them, we come back to business

Goal	Time Frame	How Can I Make It Work?
Raise profit from existing business by 20%	9 months	Contact existing clients, architects, and general contractors.
Raise profit from existing business by 20%	12 months	Get contracts from the friends of existing clients, plus new architects and general contractors.
Raise cash flow from car restoration business by 75%	1 year	Collect and restore 1950s cars. Start to look for more and buy some and have others do some of the work on them.
Raise total cash flow by 10%	1 year	Become a roofing consultant. Advertise to insurance companies and defense lawyers in the area. Bill reasonably—$90 per hour. Learn more about being an expert witness and putting the business together. Put out some cards.

FIGURE 12-4 Short-term game plan: starting to achieve long-term goals now.

plans. It doesn't matter whether you decide on the consulting business shown on the sample chart, a gutter business, a large roof maintenance business, or something else. You may decide to infuse the startup with some capital.

The business plan is a tool used by professionals for all types of startups. Whether for the transition from a small enterprise to a larger one or for the beginning of a new project, the business plan is a very handy tool. If you are going to a lender, you will most probably be required to have one. If not, it may be a good idea to do one for yourself in order to evaluate the business's potential as a money maker. Don't make it complicated. Keep it simple. Figure 12-6 is an example of a simple plan for a roofing professional. It could be used to grow your business or to add an ancillary business like an inspection service or a guttering company.

Goal	Time Frame	How Can I Make It Work?

FIGURE 12·5 Short-term game plan: starting to achieve long-term goals now.

Subject and Thoughts	Needed Information and Thoughts
The Principal's Current Financial Situation This is basically your financial statement, which can typically be found in your accounting software or with your accountant.	
The Marketing Strategy This is a breakdown of how your service or product will receive notice.	
Goals for This Business The goals here relate to the growth and earnings you expect from this business.	
The Management Group Are you going to be sole proprietor? Who else will join you to manage the business, and what will their positions be?	
Products and Services to Be Offered Give a precise breakdown and where and how they will be procured or facilitated.	
How Much Business Does the Market Offer? This is the most cloudy question. How much business is there, and how much of that can you corner? Talking with past customers while working on the marketing of your maintenance business is a very good way to get feedback. Use local resources like the Chamber of Commerce and the SBA for demographic research.	

FIGURE 12-6 Business plan prep sheet.

Subject and Thoughts	Needed Information and Thoughts
What Are the Financial Projections? How much of the business in the question above will be yours? It may be worthwhile to pay for demographics and some consulting here. This is the bread and butter of your new cash flow.	

FIGURE 12-6 (*Continued*)

There are many books that include the format for a business plan. There is also plenty of software available that will step you right through your own plan. A good beginning assortment of references can be found in App. A; you can update it in your own database. One of the best sources of information pertaining to your particular location is the Small Business Administration—it is an excellent place to find the things that are important to a small business.

If you are doing a plan by yourself, you may find that the headings suggested in books and software are written in financial speak and seem distant, not immediately important to you as a roofing professional. This is your intuition and years of experience speaking to you—trust it. Always be firm with consultants, attorneys, advisers, counsel of any type; they must boil their concepts down into plain

English. If they are not able to do this in a hurry, they are not acceptable for your company.

Many times business plans are created only in order to go to sources of capital for expansion or a startup business. If you are going to move right into an ancillary business, talk with your banker and accountant about whether you need a true business plan at this time or not. It may be that you do not really need to put one together. Or, you may decide that the discipline and the careful look at your current business are worth the effort right now. Provided that taking the time to do this does not affect your cash flow negatively, developing an overview of your business can never hurt.

As explained in the introduction, all of the forms in this book—including these goal setting and business plan coaching forms—are available on the CD that accompanies this book. Appendix A will head you in the direction of groups that can give you advice and direct you toward vendors of software that can step you through creating a business plan.

Forecasting and Leveling

Taking complete control of a business rather than allowing it to control you begins with looking at what you want to do with your life, as we have stated. Taking care of the mechanics of the business so that it runs smoothly requires looking closely at what the business actually does on a day-to-day basis.

In this new century, gathering data has become an absolute requirement. Without it, you simply can't be competitive in the bidding process. If your work is so simple that you can do it on a sheet of paper, you may be able to rely on a pad and pencil for bidding. But that is no longer the norm. Most roofing professionals now use computers. They can make estimating and job cost accounting a great deal simpler and more accurate than in the past.

From your analysis of what you want from your business, you may decide to simplify and keep everything bare bones. However, if you decide to grow your business, you will undoubtedly run into cash crunches. Most roofing firms encounter them, if only during short-lived emergencies—for example, when releases on jobs have been backed up because of the appearance of unforeseen conditions like weather, and weekly payrolls have come due.

On a very simple basis, tight situations like the one above are actually what forecasting and leveling are all about. Forecasting how much money you will have for the next pay period, the next quarter, the next year is very helpful. With the software now available, it is pretty simple to gather the data on your actual figures in the past. To see when you have had ups and downs in your cash flow, you can study your actual track record. For example, you may have had a real valley in August every year, as construction starts were low and there was no fix work.

With the information recorded in and so easily retrieved from your accounting and estimating software, it is fairly simple to forecast when you will have those valleys. Leveling is learning how to put that information to work and keep cash management from getting out of hand.

Forecasting and leveling are basic business concepts that are explained in great detail in many books and even in business courses at your local college. The SBA, templates for spreadsheets like Microsoft Excel, your accountant: There are many places to get help with forecasting your cash and leveling out the peaks and valleys in it.

This is a roofing book, and there is not enough space to go into the complexities of business in detail. However, forecasting may become an important, ongoing part of your goal setting. Figure 12-7 gives you an example of how a roofer might begin to fill out the checklist of information for a beginning look at your forecast and how to level it. An empty list for you to copy is in Fig. 12-8.

Marketing and Sales

"Sales cures all"—there's a lot of truth in the old cliché for any business. But a basic overview can keep the roofer on track, working smart, not hard. If there are several cash flows in place—regular roofing, reroofing, add-on sales, different markets within the roofing business, maintenance and inspection programs—financial stress can be spread across several profit centers. As we all know, marketing is the activity that brings in the potential clients, and sales is getting the client to sign a contract to go with the firm.

✓	Task	Notes
✓	Find software that I can punch figures from accounting into and forecast where we'll be in the next year.	Make sure that we have been inputting correctly; read through printout and see how accurate it looks. Go over it with the accountant and accounting and estimating staff.
✓	Review when we tend to make money and when crunches come.	Find the patterns in our activity.
✓	Find an expert to help us with computers and advanced business tasks.	Check out the expert's credentials.
✓	Talk to accountant about establishing line of credit for leveling.	Should do this before trying to expand my business.

FIGURE 12-7 Checklist for forecasting and leveling.

Marketing

A job well done is the best marketing tool, and word of mouth is by far the most secure and inexpensive form of marketing. As you start to put your new goals into action, take a close look at your business as it stands. If you have a good working relationship with past customers, architects, and general contractors, this is the place to begin overhauling your marketing and sales.

With goal setting under way and forecasting and leveling started, you can begin to see some patterns. How can sales increase as the business stands? How can you handle sales as they increase? What can you do to test the market for a sales increase?

Figures 12-9 and 12-10 can help you understand this important part of your business. Figure 12-9 is an example of how a roofer might start to fill out the marketing checklist, and Fig. 12-10 is a blank form for you to copy for a trial run. There is a working copy on the CD.

✓	Task	Notes

FIGURE 12-8 Checklist for forecasting and leveling.

Goal	Time Frame	How Can I Move It Along?
Expand marketing contacts to raise profit from existing business by 20%	9 months	Contact existing clients, architects, and general contractors. ■ Flyer announcing new plan. ■ Begin sending out a maintenance newsletter. ■ Broadcast newsletter over the Net.
Continue to expand marketing contacts to raise profit from existing business by 20%	12 months	Get contacts from the friends of existing clients, plus new architects and general contractors. ■ Flyer introducing the company. ■ Send out a maintenance newsletter. ■ Put an information site on the Net.
Research markets for cars to raise cash flow from car restoration business by 75%	1 year	Look for places where cars are announced. ■ Announce your cars in the local auto sales paper. ■ Begin a web site for collectors about 1950s cars.
Explore markets for consulting business to raise total cash flow by 10%	1 year	Go to the Bar Association and insurance companies and find out where they locate their roofing consultants and experts. ■ Develop a brochure. ■ Do a mailing. ■ Place a test ad.

FIGURE 12-9   Organizing and expanding marketing.

Goal	Time Frame	How Can I Move It Along?

FIGURE 12-10 Organizing and expanding marketing.

As on any subject that relates to business, there are many books written about marketing. You might find it helpful to browse through them and look for ideas that you can implement. There are many useful ideas that relate to marketing; however, some of them may be more than your firm needs.

Always remember that the basic difference between marketing and sales is that marketing is the activity of presenting your firm to a broad audience and thereby bringing in leads. Sales is closing the deal with those leads, arriving at a solid contract agreement.

For the small to medium-sized roofer, there is no better marketing than word of mouth. Solid estimating followed by a professional contract and a job well done will cause your customers to talk about your firm; they will tell friends and other people about your company and will call you back when the time comes for replacement of their roofs and for any other roofing needs that they have. Word of mouth will not usually happen overnight, but there is no more powerful marketing.

After word of mouth, there are numerous other methods for marketing your roofing business: talks to Realtors or to other organizations about care of roofs; seminars about roofing products; advertising in small publications like local, free newspapers. There are a tremendous number of ideas.

Sales

In roofing, unlike many businesses, sales are almost always contingent on solid estimating practices. As you well know, the bid price is often the sole factor in your firm's getting the contract. This is not necessarily the correct way to go about getting quality roofs put on buildings, but it is the system we have to work with.

Naturally, having a powerful estimating system in place—whether it is a very simple dollars-per-square technique or a sophisticated computerized method—is probably the most important part of sales. We take another look at estimating in Chap. 13, which covers the modern electronic systems.

One factor that can often be decisive in your company's closing a sale rather than another roofer is confidence that your firm will provide a quality roof. Once you develop a solid track record of jobs well bid and executed, with satisfied customers whose roofs have withstood the elements, you should put some time into developing methods for presenting the testimony of these satisfied clients to the new people for whom you are preparing bids. This information will be a part of your sales and marketing presentation package.

In a tight bidding situation, your track record can make or break the deal. No matter how you receive most of your invitations to bid—architects, general contractors, product manufacturers, word of mouth, advertising, or a combination thereof—being able to present a solid history of long-lasting roofs can win the contract in many situations. And this is often the only edge you have over others.

Figure 12-11 presents an example of the way you might fill in the sales worksheet, and Fig. 12-12 is a blank one for you to make copies of.

Peripheral Cash Flows

In Chap. 14, several operations that are closely related to the regular jobs of fixing old and applying new roofs are discussed. A typical roofing firm can jump-start these into parallel businesses with solid cash flows. However, in this chapter, we are taking a more cursory look at the basics of looking for extra income.

It takes a lot of extra energy to set time aside, pull oneself away from the urgent matters that need to be tended to in the day-to-day operation of a business, and work on bringing the tasks of quality management into the loop of current business practices. But if you want to have a really solid or high-growth business, it is essential to put energy into fine-tuning your management skills.

Insurance

The various forms of insurance and bonding are discussed here in simple terms. An understanding of coverage issues, construction defect litigation, claim handling, and liability is an important part of your business. It is important to assign someone in the office who is a stickler for detail and has a knack for explaining what he or she has learned to the task of staying on top of insurance. Work with your vendors and agencies; ask lots of questions, make certain you understand what the policies say, have a broker shop your policies, with the same coverage,

Goal	Time Frame	How Can I Move It Along?
Develop a fast, dependable system for follow-through to close deals generated by marketing—boost existing business by 20%	9 months	Try new methods for improving sales. ■ Offer a maintenance incentive. ■ Hand out a copy of the maintenance newsletter. ■ Try using a dedicated salesperson.
Continue to improve on percentage of closes to presentations by 20%	12 months	Continue new methods for improving sales. ■ Offer a maintenance incentive. ■ Hand out a copy of the maintenance newsletter. ■ Try using a dedicated salesperson.
Keep a list of qualified customers to raise cash flow from car restoration business by 75%	1 year	Follow up on places where the cars have been announced. ■ Massage potential buyers ahead of time by meeting at car events. ■ Learn to close over your web site before the car is restored.
Go to consulting seminars and meetings to learn more syntax for closing sales to increase business by 10%	1 year	Get to know your clients. ■ Send them a brief newsletter about roofs. ■ Do some of your work for free. ■ Keep your hours low.

FIGURE 12-11 Organizing and expanding sales.

Goal	Time Frame	How Can I Move It Along?

FIGURE 12-12 Organizing and expanding sales.

out to various carriers, and make certain that you are figuring fees like federal unemployment tax (FUTA) correctly.

Unemployment Insurance

All states levy an unemployment insurance tax on employers. The tax is based on total payroll for each calendar quarter. The actual amount assessed depends on the history of unemployment claims filed by employees of the company. The tax can range from 1 to 4 percent of payroll dollars.

FUTA

FUTA is the federal government's unemployment tax. This tax averages 0.8 percent of payroll dollars per employee up to a maximum dollar amount per year that is determined by law.

Social Security and Medicare (FICA)

The federal government also requires FICA payments, which amount to 7.65 percent of payroll dollars per employee; for the social security portion, a maximum dollar amount is determined by law each year.

Workers' Compensation Insurance

All states require that employers carry this insurance to cover employees in the event of a job-related injury. The cost of this insurance is a percentage of payroll dollars based on the occupation of the employee. For example, secretarial work has a comparatively low rate because it is typically confined to the controllable environs of an office. The rate for construction jobs, on the other hand, is usually between 5 and 10 percent of payroll dollars.

Liability Insurance

This insurance protects the roofing contractor in the event of an accident. Most general contractors require that a certificate of insurance be presented before a subcontractor begins work. The cost of this insurance is based on total payroll and is dependent on the location of the company, the type of work performed, the history of claims by the company, and the required liability limits.

Make certain that your insurance broker and a coverage lawyer (if needed or if you are just starting out) have gone over what the policies do for you. Discuss this with them in detail. Compare the costs of various carriers.

You must also understand the liability clauses in the general contractor's documents that designate liability.

This list outlines expenses that will be deducted from the gross revenues your firm takes in for any job you do. The costs are clear examples of why the labor burden must be included in every estimate. Since state and local governments pass their own laws and set their own percentages for some of the above-mentioned taxes, and since individual companies set insurance rates, it is necessary to frequently update information and expenses associated with these costs. Failure to update these costs on a regular basis can result in underpriced jobs, reduced profit margins, and perhaps even losses.

Reducing Workers' Compensation Insurance Costs

With a manual system, workers' compensation insurance costs are usually calculated by multiplying the total number of hours worked by the highest premium rate applicable to that type of work. With a computerized system, a roofing contractor can break down the hours worked into risk categories. Insurance premiums are then based on the various categories of work performed, which often substantially reduces insurance costs.

Using Subcontractors

This chapter has touched on the fact that a great many people who are involved in roofing came into the office from the field. And because of the continual demands of the day-to-day workload, people often get lost in small jobs and never find time to quit being mechanics in the field. The important process of overhauling their business to achieve long-term goals may just continue to fall by the wayside.

Learning to work smart rather than hard can be a hefty learning curve. As we explained earlier, the important thing is to look closely at how your ability to work smart relates to your existing business and your past work experience.

Letting go of some of the work and allowing others to take over is a very important step in learning to work smart. The maxim that one should do what one does best and let others do the rest carries a lot of wisdom, as do most old clichés.

Jobbing out work to other contractors is one of the most practical methods of delegating authority. The costs of doing business are related to job costs or business improvement costs and are not an increased, ongoing payroll burden. Just as you are a subcontractor to the general contractor, using subcontractors yourself can be a very wise move.

No matter what tasks they perform—consulting, legal work, toxic waste removal, etc.—it can be very profitable to subcontract work out to others and pay attention to what you do best: roofing. The decisions concerning when to use contract labor are arbitrary, but the best rule of thumb would relate to your new view of your life. If you can do the entire business without any employees, or with relatively few, your job will probably be a good deal easier.

Accounting

Accounting is an interesting part of your business. There is a tendency to make a big, gnarly mystery of it. But in reality it is very much like a road map of your business, except that it tends to change more than maps do.

There was a time when accounting required a good deal of knowledge and an unbelievable amount of attention to detail. A cavalier viewpoint still will not cut it with accounting, but it is much easier to track the money trail of a business than it was 20 years ago.

The main person needed in-house is someone who has knowledge of the accounting software and what it is supposed to do. If this person also has accounting skills beyond managing data, that can be a big help.

However, the main thing that is needed is a good accountant and strong software that is easy to use and can interpret your estimates and offer thorough reports about the financial position of your business.

In Chap. 13 we take a close look at computers. They have forever altered the labor needs for the accounting center of a business.

CHAPTER TWELVE

In the end, try to remember that this is your life that is ticking by, that a business needn't be a somber, funereal arena. It can just as easily be a fun place to spend time and still be efficient and very professional.

As we said in the beginning of this chapter, it is easy to forget that roofing is first of all a business, and that it is the business of roofing that presents the arena in which the tradesperson can practice his or her craft as a skillful roofer. And it should be remembered that the more pride and joy in working together the people in the company feel, the stronger the company will be. Roofing can be very profitable for a long period of time. Everything is contingent on how the owner approaches the business on a day-to-day basis.

Computers

The harnessing of electronics has had a terrific impact on the construction industry. A mere three decades ago, estimates were jotted on paper in the field, carried to the desk, and computed on mechanical calculators or adding machines, or simply extended by hand. If the person doing the math was lucky enough to have a tape coming out of the machine, he or she could go up and down the thin strip of paper with a red pencil, trying to make sure that all of the line items from the takeoff had been included in the calculations.

Obviously, this was a painstaking process with plenty of room for error. It took a long time to put all of these pieces together and do, then recheck, the math. The estimator's concern about mistakes was painful, and margins for error were often put into bids at a high percentage rate to protect the contractor.

The first major breakthrough in electronics was the calculator. Calculators saved roofers time and kept them from being confined to a desk.

Next, the answering machine for telephones appeared. This meant that the small to medium-sized contractor who couldn't afford a phone-answering person didn't miss calls from potential customers—another giant breakthrough for roofers.

Then, around 25 years ago, came the first affordable personal computers, and the roofing industry was completely changed, forever. At first, the PCs—Macs, Osbornes, IBM XTs and ATs—were prohibitively expensive for many firms. But within a short time, clones from the Pacific Rim hit the market, and the roofer could buy one for a few hundred dollars. With an early package of spreadsheet software like Lotus 123 or Visicalc, a simple roof estimate could be done in minutes rather than hours.

The difference in ease and accuracy between estimating with a spreadsheet and using a calculator with a pad of paper was exponential—the estimator could make a template. All of the tasks to be done during the roofing operation could be listed as line items with dollar amounts next to them; the software could remember the cost of applying a square of shingles. The computer extended the numbers, and the laborious task of using a calculator was at an end. The invoices, change orders, and even accounting were stored in files, and the central processing unit could run the numbers faster and more accurately than any person. Human beings are not as good at performing boring tasks over and over again as computers are.

Today roofing contractors must use all the available modern techniques and technologies available to them in order to keep up with the marketplace. The time it takes to learn how to use computers and their software packages is relatively short. For example, a contractor or an employee who has little or no formal accounting experience can learn how to enter the required information. Then a computer can do all of the complex sorting and calculating that would have required years of training in times past.

If the computer system is user-friendly enough, everyone, including roofing contractors and employees, can save time by streamlining paperwork. Accuracy of bidding, winning close bids, massaging aged receivables: The many hours of drudgery eliminated by the correct use of computers can amount to large cash savings on a daily basis.

Employing Computers Rather than Staff

As the brief history above suggests, the days of guesstimating a job's cost and a client's bill are long past. Gone also are the days of keeping records on old invoices and brown paper bags, of tracking labor hours and overtime costs by hand. In the electronic age, these activities are outdated. And obviously, removing considerable chunks of the paper trail can reduce the need to keep increasing payroll.

A one-time investment of several thousand dollars for computers, software, a laser printer, facilities for the Web, and the other items needed for a system can pay for itself quickly in estimating alone—from enhanced consistency and accuracy as well as timeliness. Estimating is only a part of the workload that computers will handle. They represent one outlay of cash with minor upkeep expenses and can slow the ongoing need to add staff in numerous departments: management, accounting, estimating, scheduling, delivery, etc. And, as we all know, adding more staff is not a one-time outlay of cash.

There are many ways in which a construction company can profit from computerization. Estimating, control of payables during jobs, reducing workers' compensation costs, obtaining vendor discounts, staying on top of receivables, eliminating bookkeeping costs—each roofer has his or her own list. But—and we can't emphasize this often enough—one cost savings that runs through the activities of any type of firm is that computers can be used rather than hiring more people.

A wide spectrum of hardware and software is now available at amazingly low costs. The computer and programs required by individual roofing contractors depend largely on the size of the company and how many tasks the computer is expected to perform. Figure 13-1 offers examples of items your firm may want to automate. It is on the enclosed CD; download it into your computer and print out several copies. Figure 13-2 is a blank checklist to use for writing down your own ideas. Have everyone fill it out, then consolidate the responses into a review sheet for decision making about automating your company's paper trail.

Automating the Office

Now that you have some overview of what you want a new computer system or an upgrade of your old system to do, let's take a look at some of the basic overall concepts involved with automating the paper trail.

First of all, think through what happens at your office every day: Possible new jobs come in, typically by phone; reports and requests

✓	Item	Notes
	Keep an updated to-do list	
	Record takeoffs in the field	
	Do takeoffs straight from blue-prints, electronically	
	Calculate estimates while take-offs are being done	
	Prepare a professional-looking bid sheet for the client	
	Control costs during each job phase	
	Reduce workers' compensation costs by tracking work hours by risk category	
	Organize accounts payable for faster collections	
	Speed up and increase the collection of accounts receiv-able	
	Eliminate external service bureau costs	
	Improve clerical productivity	

FIGURE 13-1 Computer wish list—what we would like to be able to do.

from the jobs that are being performed in the field come in; checks come in; takeoffs of potential new jobs that need to be estimated are completed; lawsuits for construction defects claims come in; checks go out to vendors; equipment is purchased; payroll is cut—even for a small business, the list of events that take place each week is immense.

The important thing is to go back and review the wish list. The hardware and software for automating many of the tasks you wish to automate are now readily available in the marketplace. However, there are several obstacles that will arise as you begin to pursue automation seriously. The first problem is where to find the software and hard-

✓	Item	Notes

FIGURE 13-2 Computer wish list—what we would like to be able to do.

ware that will do what you want to do. The next obstacle you will meet is trying to figure out which items are appropriate for your firm. Then there will come several other major hurdles to clear: installation of the hardware and software; learning to use them; implementing their potential as a part of the day-to-day operation of your firm.

There are two major mistakes that firms typically make when they attempt to upgrade computer systems or install new ones: being afraid to invest fully in automation, and failing to establish an overall understanding of what is needed in order to get the tasks on one's wish list taken care of in a real, nuts-and-bolts way.

It is important to remember the idea presented in the previous chapter: It is very easy to forget that roofing is first of all a business, and that it is the business of roofing that presents the arena in which the tradesperson can practice his or her craft as a skillful roofer. When you purchase a truck or a piece of equipment for the field, it is immediately obvious how valuable the item will be. Computers will do many tasks, but all too often the contractor has only a vague idea of what can actually be done and how to use this time-saving equipment.

This is why we have presented the list above, to help focus on what needs to get done. The next step is to find someone who knows a lot about what can be done, what hardware and software should be purchased with the initial investment, where that can be done most reasonably, how it can be paid for, what the installation entails, and how thorough training can most simply be achieved.

The principal will often have difficulty seeing this as a simple part of the business, where the point is to install the computers and keep the business moving with as little interruption as possible. In almost all cases, an outside consultant who is familiar with construction offices should be used as a subcontractor. Just as a general contractor would be foolish to try to use the carpenters to install a roof, rather than hiring a roofing contractor, it is foolhardy for the principal to try to automate the firm—he or she can probably do it, but in very few cases as skillfully as he or she can roof buildings.

The real cost of computers is not the hardware, software, and consultants—keystrokes are what cost money. Keystrokes represent an actual cost of doing business in the form of labor. The more keystrokes involved in getting the daily task of roofing done, the less profitable the computer installation will be.

In other words, if the owner of the company is reading books, calling software companies, and learning code, the actual tasks that create cash flow from roofing are being overlooked. This would be like trying to save money by cutting, bending, and installing sheet metal when your firm has always subcontracted the work out—the crew would be stumbling around the site trying to figure out what to do rather than going ahead with the roof application.

It is imperative that the owner think of the automation process in this same way—that is, like a true business professional. The whole subject of jobbing out work was addressed in Chap. 2—it is so important that we devoted an entire section to it. And it cannot be emphasized strongly enough that a highly knowledgeable computer consultant who can sign a contract for the cost of the installation will pay for the service many times over. Remember, the hardware, software, and consultant are not what is expensive—keystrokes are what cost money.

There are people all over the country who specialize in construction systems. You can find them through roofing organizations, the Yellow Pages, the Internet, software manufacturers, etc. After finding a consultant, sit down with him or her and discuss your wish list. If the consultant has an immediate understanding of what you are talking about and can listen to you carefully, you may have the right person.

Be sure to take the time to get a list of satisfied clients, preferably roofers. Call the clients up and, if they will let you, visit their places of business and review what they are doing with the system. Whether this option is available or not, make certain that the prospective consultant lays out your office for you in a diagram and shows you how all of the components will work together in a language that you can understand.

A Brief Look at Software

A computer system should always be developed to meet the overall needs of the company. The workhorse for accomplishing those tasks will be the software; after it is chosen, appropriate hardware can be configured to the software's requirements.

The most important thing to remember is that keystrokes are what cost money—not the initial investment. Shop wisely, but don't be penny-wise and pound-foolish, and by all means use a consultant with a broad overview to get you up and running.

The ideal system will include software that allows your firm to work as closely to the way it already does as possible, with as few keystrokes as possible. Figure 13-3 is a typical breakdown of what a well-thought-through system will do. This table is available on the CD-ROM that accompanies this book. Open the files on the CD by clicking on the D drive (or whatever drive letter your CD player is assigned). Print the table out for everyone to review and offer feedback. Compare it with your wish list and add anything necessary to adjust it to fit your firm.

Uses of Computers for a Roofing Firm

Job Comes In to Bid

If there is a set of plans, the estimator does a takeoff with a digitizer (a moving wheel like a scale wheel, or some other instrument that automatically takes off measurements from a set of plans), which automatically dumps the measurements of the roof into the estimating package. If there are no plans, the estimator takes the measurements with an electronic measuring device during a site visit and dumps this information directly into a template in a notebook computer; it can then be downloaded back at headquarters or a printout can be done immediately, in the field.

For simple work that does not require meetings, discussion, product pricing, or other complications, the bid can be produced on site, along with a contract document to be signed by the building owner.

Roughing out estimates without extensive office or on-site work in order to land the job can save a good deal of cash on all of the bids the company loses each year. If your rough price lands you the job or puts you in the running, then you can tighten up the estimate before going to contract. This tends to work especially well with customers who

✓	Item	Notes
	Job comes in to bid	Info about jobs should be entered into the system only once. Architects, general contractors, building owners—the software should add them to a database that will pass the information on to all parts of the business (marketing, sales, estimating, accounting, etc.)—always saving keystrokes.
	Estimating begins	Do takeoffs straight from blueprints with a digitizer. Software calculates estimates while takeoffs are being done. Software prepares materials lists and after review sends to vendors.
	Get the bid out to the customer	Software prepares a professional-looking bid sheet for the client and enters the company in the marketing database.
	Bid is accepted	Software allows staff to open a new account linked to all accounts and the entire business. Scheduling is completed. Materials orders are placed.
	Control costs during each job phase	Job is mobilized in the field. Materials deliveries are scheduled.
	Job is tracked by computers	Software does payroll. It reduces workers' compensation costs by tracking work hours by risk category. All payables are done through computers.

FIGURE 13-3   What a well-designed system can do for a company.

✓	Item	Notes
	Organize accounts receivable for faster collections	Payables are handled in a timely fashion. Receivables reports are always available. All tax information is extracted and compiled as receivables and payables come in and go out. Clerical productivity is increased.
	Job is concluded	Demobilization is tracked and profit and loss figures are computed automatically.
	Job-cost accounting is completed	The job becomes a part of the company's overall accounting picture. Reports concerning the real cost of job, profit and loss, etc., are at the owner's fingertips.

FIGURE 13-3 *(Continued)*

can describe the job and site over the phone, architects, property management firms, and others with whom your firm does a large volume of work on an ongoing basis.

This type of estimating is done directly from the blueprints. A digitizer is used for the takeoff from the surface of the plan. The device traces the dimensioned drawing and dumps the areas of the roof into the computer's estimating package. The estimator reviews what is on the estimating sheet and the plans, and then the computer runs a total.

This is an amazing approach for estimators who have never used it before. The digitizer picks up dimensions and counts, and the estimator guides it to the proper line item: demolition, felt, shingles, etc. Then the computer extends everything. The most important thing is that the estimator have an experienced eye for such things as difficult access or disposal of hazardous materials, so that these can be included in the line items.

Moving to Contract and Job-Cost Accounting

The next step will be discussions with the principals who want the job done, and then come contract documents and accounting.

The new generations of software for people in the building industry are very sophisticated, and dedicated software for small systems has been around for well over a decade.

This wrap-around software allows a high-quality printout of the original estimate to be printed or faxed out; if the principals are interested, the estimate can be revised and printed, contract documents produced, and a job-cost account opened for the contract.

Naturally, there are keystrokes involved, but while all parties must still take great care to see that what comes out of the system is accurate, the number of keystrokes and calculations that must be performed by staff can be reduced drastically from that required with hand-done systems, or even earlier spreadsheet/word processor computer systems.

Moving the Job into the Field

After the account is opened in the wrap-around software, the field must be dispatched. Assigning crew and materials; purchasing and/or renting equipment; organizing with the general contractor, her or his other subcontractors, and any subcontractors the roofer must contract with; and scheduling are all time-consuming operations that can be highly streamlined with automation. (See Fig. 13-4.)

Materials Lists Right from the Estimate

This is a typical function of newer software for building industry professionals. The roofer can get a list produced right from the estimate, and the project manager, the job foreman, or even a highly competent gofer can do the shopping—fax it to several vendors and see if they can sharpen their pencils to get the sale.

The Net can be valuable when it comes to looking for interesting add-ons to sell and working on change orders. Appendix A lists products as diverse as photovoltaic shingles (they produce electricity for the home from sunlight that hits the roof) and various types of unique slate shingles that can be bought right off the Net.

On-Site Tasks That Can Be Expedited with Electronics

→ Takeoffs
→ Accurate Materials Orders
→ Rescheduling
→ Change Orders
→ Staffing
→ Selling Extras
→ Produce Documents

Project Manager's Truck
Cell Phone
Laptop w/ Fax and E-mail Capability
Portable Laser Printer
Scanner
Hand-Held Electronic Measuring Device

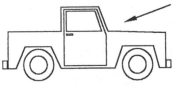

For small jobs, the P.M.'s truck is sufficient for subdivision.

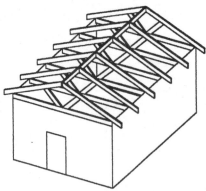

Job shack or space in a building may be needed for larger jobs.

Electronics are now a must at every job.

FIGURE 13-4 Jobsite electronics.

Scheduling and Job Performance

This is another task that the job superintendent can take care of on a laptop. The most important part of scheduling is rescheduling. As we all know, jobs are very often fraught with delays: architectural change orders, weather, poor construction management—the list is endless, often frustrating, and at times devastating to the financial outcome of a contract.

This situation makes the ability to reschedule quickly and accurately an imperative, along with producing any change orders that are required as the schedule changes.

If a foreman at the jobsite is equipped with a laptop, a portable laser printer, and e-mail and fax capability, a great many problems can be ironed out instantly, without labor downtime and/or progress delays.

For example, suppose that the crew has started to place underlayment and the foreman notices that several junctions of vertical walls to pitched and flat areas of roof are not detailed correctly and changes are required. The architect is not in the office, and the general contractor's superintendent is not available.

Without automation, this situation could have serious ramifications: holding up a crew with the meter running; several days of downtime due to inclement weather; missing some extra cash flow from a change order. But at a site that is electronically up to date, the foreman can get through on the architect's cell phone, fax a sketch of the roof junction, get a signed version of the change order back by fax, and move on with the deadline.

After the Job Is Closed

Of course, the final request for payment must be presented to the client and the job must become a part of the accounting system. Once the invoices are out, they become a part of accounts receivable, which must be tracked for aging.

In wrap-around software, reports are available that clearly delineate aged billing. They compute whatever small amount of interest is charged the client for the aged invoice and even print out an addressed invoice that can be inserted into an inexpensive glassine envelope. They can even be faxed or sent by e-mail and funded electronically.

Any of the modern, worthwhile types of job-cost accounting software also offers numerous other reports that can help with management: reports on profit and loss, financial status, average age of billing, names of old customers for marketing campaigns, and a host of other information like accounting trails that is invaluable and very time-consuming to get from the accounting system by hand. Use the list in Fig. 13-5 when you are considering accounting software.

Looking at the Basic Software

Unless you are a very sophisticated computer user or the principal of an extremely large roofing firm that uses big, complex computer systems, there are several pieces of software with which you will probably not need to trouble yourself—the operating system, the shell, communications software, etc.

The operating system is the software (DOS, Unix, Linux, etc.) that coordinates you (your input into the computer through the keyboard, a microphone, or whatever) with the CPU (central processing unit or main computer chip) and all the devices like monitors, printers, etc.

Microsoft's DOS is the most common operating system. Many sophisticated computer people will tell you that it is not the best system; however, it is the most widespread operating software for small systems and therefore the one for which it is simplest to find consultants and office staff familiar with its installation and use.

The shell is the software that interfaces all of the software and gives you swift access to it, including the software that controls outside devices like printers, digital cameras, digitizers, etc. Again, Microsoft has far and away the most widely used shell—Windows. This book is about roofing, not comparing software; therefore, we are not spending a lengthy amount of time on the subject. In order to avoid being distracted from your business as a roofer, sticking with the most widely used shell, which is Windows, may be the most efficient route. When Microsoft becomes obsolete, simply pay your consultant to replace it with a Linux application.

Chances are that MS-DOS and Windows will come installed on your computer and the roofing software will be set up to work with the Microsoft packages. Typically, it will be wisest—unless you already have very strong reasons for doing otherwise—not to worry about the

✓	Feature	Notes
	Profit and loss	Need profit and loss copy for 3 years
	Financial status	Copy of all reports
	Average age of billing	Report that shows
	Old customers, linked to mailing for marketing campaigns	Show us how it works
	Direct link from estimating to accounting	Show us the link
	Accounting trails	Printed for accountant—need example
	Estimating link to digitizers for takeoffs	Show us, and need list of digitizers that work with it
	On-line training and help line	Give us # so we can get list
	Cost of setup on-site	How much—talk with old clients
	Available discounts on hardware	Does it do this from estimating?
	Is scheduling included?	What packs does it talk to?
	Examples of all management reports	
	Other features	

FIGURE 13-5 Questions to ask when reviewing wrap-around software.

operating system and shell, but just go with the flow until you are very sophisticated with computers.

The next type of software is wrap-around generic software packages. WordPerfect Office and Microsoft Office are the two main packages in the market today. Again, since Microsoft has so much presence in the marketplace, it may be best for you to make certain that the computer system you buy comes loaded with Microsoft Office.

These packages contain word processing software, a spreadsheet, and a database as the primary workhorses for generic office tasks. They also include numerous other pieces of software like web connectivity and even marketing presentation packages.

See Fig. 13-6 for a concise look at this information. You will need an operating system, a shell, and a wrap package for general office tasks. Under the following headings, we examine the tasks that are typical for a roofing firm: marketing, bidding, closing the bid, contracts, job management, and accounting.

MARKETING

Typically, the tasks performed when working on marketing are accomplished with word processing and graphics software. Ordinarily, these two functions can be found in the wrap-around software packages like Microsoft Office. There are so many typefaces and drawings available with these packages that the office's word processing person can often handle simple marketing campaigns—newspaper ads, brochures, flyers, etc.

For more sophisticated, large-scale marketing campaigns, it will probably be wiser to shop the artwork out and concentrate on the business of roofing.

Remember that everything about your firm is a part of marketing, not just your ability to estimate and perform. The look of staff members, of equipment, of letterhead and business cards, of bid presentations: We all watch one another subconsciously, and the more confidence we have in the person we are dealing with, the more likely we are to enlist that person's services. Just like a good computer consultant, a good, local marketing person used on a limited basis can save your firm money and be a big help in developing a strong presence in your area of operation.

WORD PROCESSING

Typically, the computers a firm buys today will have a good deal of software onboard already, as mentioned above. If you shop with care, you should be able to purchase a system with Microsoft Office on board.

The word processing software that is included in Office is Word. Word is a quality product and will serve any roofing firm well. The CD with this book is completely compatible with Word.

Field Jobs, Estimating, and New Profit Centers

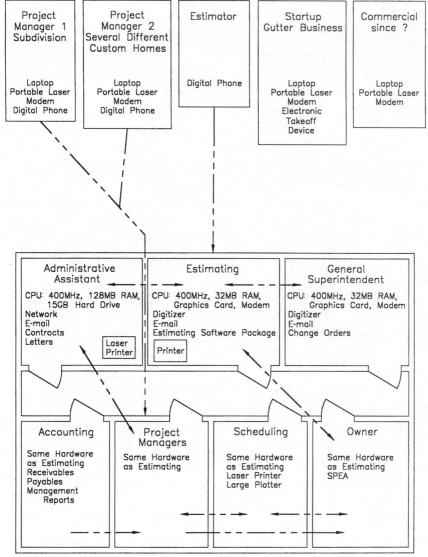

Ideal Current Computer System for Roofing Office

FIGURE 13-6 Computer requirements for a roofing firm.

The main tasks for which you will use word processing are letters, memos, contracts, change orders, and the other documents that are part of the paper trail.

SPREADSHEETS

Spreadsheet software is basically designed for computation of figures. The format is typically a grid of rows and columns, just like a columnar pad of paper. You insert simple codes into the boxes and fill in the numbers, and the software does the math for you.

Again, try to buy computers that have Office onboard. The spreadsheet software that is part of Office is Excel. It is quality software and can be used effectively for the business of roofing.

Spreadsheets are very handy for any math tasks: estimating, tracking real-world estimate-to-actual costs on jobs, leveling cash, forecasting cash, etc. Successful contractors are accurate estimators. Before roofing contractors make any money, they must get a job and complete it. Getting the job usually requires that the contractor estimate the cost and present a bid.

Contractors develop estimating techniques through years of education, experience, and practice. Techniques are often personal and, for competitive reasons, private. However, one truth holds across the board: Roofing success hangs on accurate estimating.

Computer software is imperative in this day and age; it can, without a doubt, enhance the three characteristics of quality estimating: consistency, accuracy, and completeness.

Advanced software systems feature wrap-around work packages. A work package allows roofing contractors to prepare a takeoff by entering the roof's dimensions and any other required items.

To take off a four-ply built-up roof, for example, enter the roof's length, width, and slope. The work package automatically estimates labor, material, and equipment costs for all the felts, asphalt, surfaces, fasteners, and any other items related to the roof. Based on the length of the gravel stop, the work package counts the number of joints required, adds the proper amount of material for the lap, and rounds up to the next 10-foot piece.

From this takeoff, the work package calculates estimated labor and material costs for numerous items, including insulation, fasteners, felts, and surfaces, and then takes into account waste factors. This information is then transferred to a complete bill of materials and a detailed field productivity report.

These intelligent work packages allow the roofing contractor to prepare and generate more accurate bids in less time. The result is, hopefully, more winning bids and more free time during which the contractor can perform other tasks like socializing with potential clients, leveling and forecasting, and rethinking personal and business goals.

Consistency suggests that each estimate follow a similar process. The veteran roofer knows the materials needed, the relationships of various materials to one another (e.g., adhesive to square foot of membrane ratios), the labor required, and other normal costs associated with similar jobs.

Accuracy encompasses mathematical correctness. Most experienced contractors recognize the importance of reviewing all calculations for accuracy. Completeness means accounting for all costs associated with the job. It is here that the experienced contractor might include an unusual cost that an inexperienced contractor might overlook. The contractor's track record of past work is the most valuable accuracy tool. As we have said, that information will come from your past records as you start using new software, and later can be updated continually from the job-cost accounting software.

Software designed to streamline estimating tasks falls into two groups, dedicated estimating software and spreadsheets used for estimating. Prepackaged estimating software usually suggests general relationships between items, but requires the contractor to add to or replace these generic suggestions in order to customize the software.

In the majority of cases, it is wisest to buy dedicated estimating software that works with the job-cost accounting software you choose. Make sure the vendor provides the ability to customize the way it works for your company. Again, it is preferable to use a consultant to customize the estimating software for you. The vendor of the package should be able to do it in house or provide a list of qualified consultants.

Independent spreadsheet programs require the user to have a working knowledge of their most complex functions in order to truly automate estimating and massage data from the accounting software. It also takes a good deal of time to customize spreadsheets so that they perform all the tasks involved with your estimating style simply and quickly. The time invested in learning the software and customizing the estimating sheet is well rewarded, as the result is a professional estimate that reflects the judgments of the user. But this will be time during which you will not be making money roofing—hire a consultant.

If you decide to try estimating with a spreadsheet package, use your manual estimate form as the basis for your computerized spreadsheet. Enter the relationships between the components on your manual sheet into the computer. The relationships are entered as mathematical functions. Consult the instruction manual that came with your software to determine the format to use when entering these mathematical functions. Always read and follow the instructions in the user's manual.

At this point in time, even though you will probably get spreadsheet software for free with your computers, it is generally the best move to buy job-cost accounting software that includes estimating software which is formatted and waiting for you to enter your line items and costs.

Estimating software is a powerful tool. After you have learned to adapt it to your company, you will see its effects all the time. Bids often contain hundreds of variables, any one of which can make the difference between winning and losing a job and the resulting profit or loss. The estimating software can generate accurate estimates quickly and easily. It can calculate the final bid price, list items, apply quantities and unit costs, including overhead and profit margins, and then add the numbers to get a price. The program can even compare your markup margin against those of your competitors.

Testing the System

Once you get your new software up and running, run parallel systems for a while. Whether you are replacing an older computer system or getting automated for the first time, prepare estimates manually or on the old computer and on the new system in order to verify the accuracy of the new work. Once you have made modifications and verified that the new estimating software is functioning to your satisfaction, stop cross-checking your estimates manually.

Like your old computer or manual estimate sheet, the new electronic system requires continual price updates. You will also have to continue to watch for items that are unique to a job that will influence the cost of production; for example, limited access to a structure will increase the dollar amount of a number of line items, such as materials handling. Use Fig. 13-7 as a preliminary list of items that you will need to watch over and change regularly. Pull the checklist up from the CD and add your own line items. Keep it ready for each estimate that is produced by your firm.

Figure 13-8 addresses the concepts that should be clearly understood before you make your estimating software purchases, whether upgrading or buying new. This checklist refers to basic products that are very important when automating the bidding process. Review them with care and add your own items in the empty rows.

Accounting—the Backbone of Management

An integrated, wrap-around financial software package can be programmed to take the appropriate information about percentage of completion punched in by the project manager and, from the original estimate, generate an invoice. This should require only a few simple keystrokes.

The software holds the billing information after the invoice is presented; it remembers that you have invoiced the client. This capability gives contractors the information they need to respond quickly to delinquent accounts and assess late charges. Cash flow can be significantly improved by keeping a tight rein on receivables. The integrated software can produce aging reports that show the accounts receivable balance, as a total, by customer or by overdue period.

The word processing software in Windows or the wrap-around roofing software can be set up to take the information from the accounting modules and produce customized letters to all customers who are 30 days (or any other amount of time selected by the company) or more past due on their payments.

✓	Item	Notes
	Changes in materials vendors' costs	
	Changes in labor costs	
	Seasonal changes to account for temperature fluctuations	
	New roof or reroof job	
	Specific job conditions	

FIGURE 13-7 Checklist of items that must be updated in the estimating database.

TAKEOFFS—FIELD		
✓	**Item**	**Notes**
	Hand-held computers	Comparisons with the unique requirements of our firm have all been made. Pricing from vendors has been reviewed. Projections for purchasing date have been completed and reviewed.
	Measuring devices	Device comparisons with the unique requirements of our firm have all been made. Pricing from vendors has been reviewed. Projections for purchasing date have been completed and reviewed.
	Digitizers	Device comparisons with the unique requirements of our firm have all been made. Pricing from vendors has been reviewed. Projections for purchasing date have been completed and reviewed. Calibrations have been field-tested.
	Estimating software	Comparisons with the unique requirements of our firm have all been made. Pricing from vendors has been reviewed. Projections for purchasing date have been completed and reviewed. Estimator has reviewed software. Price from consultant has been finalized.

FIGURE 13-8 Roofer's computer checklist—takeoffs and estimates.

TAKEOFFS—OFFICE		
✓	**Item**	**Notes**
	Computers	Comparisons with the unique requirements of our firm have all been made. Pricing from vendors has been reviewed. Projections for purchasing date have been completed and reviewed. Integration with software package has been confirmed.
	Estimating software	Comparisons with the unique requirements of our firm have all been made. Pricing from vendors has been reviewed. Projections for purchasing date have been completed and reviewed. Estimator has reviewed software. Price from consultant has been finalized. Integration with digitizer confirmed.
	Download devices from field measures	Device comparisons with the unique requirements of our firm have all been made. Pricing from vendors has been reviewed. Projections for purchasing date have been completed and reviewed.

FIGURE 13-8 *(Continued)*

Contractors who are willing to dig in, pay a consultant, evaluate software, and devote some professional energy to making use of computers can enhance the efficiency of their firm exponentially. For example, your computers can track employee hours and rates of hourly pay, then process payroll and cut paychecks. At the same time, these programs can keep track of bank balances, loan payments, and interest paid or earned. At tax time, reports can be generated that summarize or detail by item all revenues and expenses, categorized by job, by employee, by vendor, by loan number or name, or by any other criterion. This makes income tax preparation easier and less costly. External service bureaus are no longer needed.

As long as the proper figures are entered into the computer system, any reports, analyses, bids, invoices, or other computer-generated documents will be accurate. However, always remember the old saying among computer programmers: garbage in, garbage out (GIGO for short). This is especially true in estimating, forecasting, and leveling, where the information entered is often based on past performance, which must be very carefully tracked in order to be effective.

All in all, when it comes to very complex or long and tedious tasks, computers calculate more accurately than human beings do. But you must wade through the computer hype presented to you by vendors. A computer is just like a roofing knife; it is a very handy tool that speeds things up when it is very sharp. But when it is dull, it frustrates everyone involved.

Taking Advantage of Vendor Discounts

Many accounts payable software programs enable a contractor to take full advantage of available vendor discounts. Input each vendor's due dates and discount, and the computer automatically cuts checks to the individual vendors within the discount time frame. The program calculates the discount and deducts it from the payment. This makes it possible for a project manager to be encouraged to work more and more on profit centers while being professional and diligent in the construction process.

Shopping and Discounts

It should be noted that thorough, disciplined shopping is always a very wise move, both before and after getting a job. Before the job, a rapid quote from vendors with reliable prices should be used in order to get the bid out in a timely manner. After the bid is accepted, further shopping can take place in order to increase the margin of profit on the project.

The estimating software needs to have prices updated on a timely basis if your company's bidding process is to stay highly competitive. Vendors change prices for many reasons: A product line is discontinued, the manufacturer drops or raises prices, the vendor runs a discount to pick up gross sales. Naturally, the more vendors you are receiving prices from, the more effectively you can buy. And, remember that you can shop for value engineering.

The Net also offers a couple of other specific advantages. For example, you can shop anywhere in the country, or in the world for that matter, for a product. If you need a unique red-colored slate from China, the Net is where you will want to start searching. Or, if you just want to search for better pricing on something as common as roofing equipment, you are in a much better position for searching than you were a few years back.

Management Software

For real-world management, the KISS (keep it simple, stupid) principle is always the best policy. At first glance, this expression might seem to be cavalier or to make light of the idea of simplicity. But it is a very straightforward way of making a point that is very important.

The more complex you make your business, the more employees you will most probably need. This extra employee weight can be compounded heavily when middle management is brought into the loop. This is one of the major benefits of well-chosen, properly installed software on which your staff is well trained.

Each job can be treated as a profit center managed by the project manager, who is capable of using all of the company's technology and is in complete and continuous communication with the office, the design anchor's office, and, when appropriate, the building owner. Project managers, estimators, schedulers, and accounting all benefit greatly from software. With the ability to handle shopping, change orders, rescheduling, payroll, job walk records, and progress reports by computer at the jobsite and simply transfer the information to the company's central computer system, a good many jobs can be managed well before extra staff is required.

Reporting on Activities

Accurate reports are one of the many benefits a computer provides. For example, accounts payable software can show, on a monthly basis, which vendor discounts were taken and which were not. Reviewing this report helps in fine-tuning the company's financial picture.

During peak construction periods, it is not unusual for roofers to fall behind on their paperwork responsibilities. The proper software packages can simplify the paperwork so much that difficult chores like extending columns of numbers do not even have to be thought about. Even tasks as ungainly as taking measurements off of a set of plans with an architect's scale are performed automatically.

The important management functions of forecasting and leveling become very straightforward when a company is using high-quality software. Full-tasking job-cost accounting software makes it easy to produce up-to-date reports on accounts payable and accounts receivable, as well as complete financial statements at any time of the year. These reports are imperative when you deal with financial institutions. The computerized reports present information about the contractor's business in the language the banks speak. When detailed, organized, complete financial information is part of the roofing contractor's loan presentation, it reflects well on the contractor's professionalism.

Miscellaneous Software

Make certain to include everything you are interested in doing on your wish list and discuss these wishes in detail with your computer consultant. For example, a daily minder for all of the crews and the principal of the company is very important. Sound this—and all of the tasks that are unique to your company—out with the consultant and the software vendor. Require them to give you a demonstration of these benefits.

Hardware

As was explained earlier, hardware should be chosen in relationship to the software that the firm decides is best suited for its purposes. If the company is so large and complex that a very high-end system is required, then certain items covered in this chapter may change. For example, if very high-end scheduling software is chosen, then Unix workstations could be required and exotic hardware like a high-end electrostatic plotter might be needed. However, the typical roofing firm will probably do quite well with simple scheduling software, which can use a small laser printer quite well. With a good consultant, all of the hardware decisions should be fairly simple—consultants typically know where to get good buys on hardware and on software as well. As with the software, the initial purchase is not the major expense when purchasing computers; it's the keystrokes, which will come after installation. If a person has to sit and wait for a computer that does not handle the job well, the desired savings will not be achieved, and losses from employee downtime and lowered morale will continue to grow.

Developing Cash Flow from Miscellaneous Roof Work

In the last few years, roofing contractors, especially smaller commercial and residential companies, have taken on miscellaneous roof work, such as that described in this chapter. If you are serious about taking charge of your business, setting real-world goals for yourself, and moving on with a day-by-day approach to furthering those goals, small, simple ancillary cash flows can be invaluable. These parallel businesses can be started simply and really enhance your current assets. We have addressed only a handful of the possibilities; if you want more, just open your eyes and start to watch for them. The chances are strong that you will find more possible sources of income than you could ever dream of pursuing.

Warranties and Maintenance Contracts as Additional Revenue Sources

With the new millennium, things are going at a fast and furious pace. In these complex times, roofing contractors must often wear more than one hat. For example, when dealing with construction defects and

443

workers' compensation, the business owner must be both a professional roofer and a bit of a lawyer. And when working with warranties, contractors often find themselves caught between the manufacturer and the customer if there is a roof failure. Manufacturers' product warranties give the customer rights and protection against failure of the product for a given length of time and under certain conditions. Customers often look to the roofing contractor to uphold those rights.

A condition is a provision in the warranty that can, unless met by the owner, lead to the denial of any claims or even the cancellation of the warranty. There is really no standard form or language for these conditions. They are largely a matter of style and policy by the warrantor and are sprinkled throughout all parts of the warranty.

One standard condition is to grant the warrantor, its agents, and employees free access to the roof during regular business hours. Another common condition is that the warranty does not become effective until all bills for the installation of the materials have been paid to the manufacturer, suppliers, and contractor. Other conditions deal with misuse or negligence by the owner or acts of vandalism that damage the roofing membrane. These are sometimes referred to as exclusions to the warranty. Persistent problems along these lines can be grounds for cancellation of a warranty.

With contracts that include the condition that payment be made to the manufacturer, supplier, and general contractor, roofers can easily find themselves in the middle of a legal battle. Assume that the owner pays the contractor, who in turn pays the supplier, who goes out of business before paying the manufacturer. When a roof failure occurs, the owner looks to the contractor to honor the warranty. The contractor then looks to the manufacturer, who denies the warranty due to lack of payment. Guess who gets sued.

Make sure that you fully understand warranty terms and conditions, including the manufacturer's responsibilities and any conditions that might exempt the manufacturer and supplier from responsibility and leave you holding the bag.

Under commercial law, all products carry an implied warranty of merchantability by the manufacturer. This means that the product conforms to the ordinary purposes for which goods of such description are generally used. If the product is presented as a roofing membrane, it must function satisfactorily as a roofing membrane. Any use of this material other than as stipulated can be referred to as implied use. In order to protect themselves from warranty claims resulting from a product failure under circumstances of implied use, manufacturers usually include a provision within warranties that states something to this effect: "The obligation contained in this guarantee is expressly in lieu of any other obligations, guarantees, or warranties, expressed or implied, including any implied warranty of merchantability or fitness for a particular purpose. In no event shall we be held liable for consequential or incidental damages of any kind."

This is also known as a disclaimer. It is legal and can be upheld and enforced in court as long as it is in writing and in clear, conspicuous, or noticeable print.

When you bid a job, make sure that you know what is warrantied and what is not. Do not make any representations to the owner that items or situations are covered by the warranty when they are not. Roofing contractors often make it standard practice to have the owner sign a document that states that the owner has read and understands the warranties that accompany the materials. This precaution, while not eliminating the possibility that the owner will make a claim against the contractor for work that is not warrantied, does give the contractor some protection against such claims.

Another statement common in warranties is a statement of limitations, in which the manufacturer limits liability to an amount not to exceed the original cost of materials. Some statements of limitations go further by providing for inflation. If historical inflation data are any indicator, roofing costs over the next 10 to 20 years can be expected to rise dramatically. By limiting its liability to the original costs, the manufacturer quite intentionally makes no provisions for inflation. With a limitation to the original cost, it is probable that the warranty will cover only a fraction of the replacement cost in the event of a total failure near the end of the warranty period.

Another statement that is generally part of warranties is a limiting statement to the effect that the cost of the repairs shall not exceed in the aggregate, over the life of the warranty, the sum of the original cost of the materials supplied and the labor used to install such materials.

This has the intended effect of accumulating the costs of all repairs over the warranty period and cutting off the spigot for this outflow of cash once the original dollar amount of the deal is reached. Without the in-the-aggregate clarification, the limitation could be interpreted to apply separately to each incident claim.

Again, you must understand that the warranty covers only the manufacturer's materials and the labor to install them. It does not cover insulation, flashings, decking, or other items that might need to be replaced or repaired as a result of a failure. Consequently, a contractor and an owner must be aware of these limitations in order to protect the contractor from spending profits defending unwarranted claims.

At first this situation may look like an additional bother for the roofing professional. But it is preferable to take a proactive stance and turn the situation into a positive by combining warranties with maintenance.

As all roofers know, in most cases, building owners are not careful in maintaining their roofs. This situation can cause drastic problems for all of those concerned. The damages caused by poor maintenance often receive little attention in construction defect litigation. Maintenance can often receive far too little attention in all phases of the construction and life span of a building. This situation, coupled with warranty complications due to legal complexity, can be turned into a positive.

Encouraging Maintenance Programs

Roof maintenance is one of the most neglected areas in the roofing industry. This is unfortunate, since it can be one of the biggest money savers for building owners and one of the most profitable areas for roofing contractors. And many times, when maintenance of roofs is contracted, it is not diligently performed.

Why is it that neither party has seriously dealt with this situation? In most instances, building maintenance departments are understaffed, and the crews that are in place are not highly skilled in the various trades. This situation is so prevalent that in many cases the only time the roof receives attention is when it leaks. On the other hand, the roofing contractor is usually absorbed in reroofing or the installation of new roofs.

There is no reason to assume that the price of roofing materials will not continue to increase dramatically, and there is even concern that the availability of products will become a more and more dramatic issue. If these two factors are combined, roof maintenance becomes very important to the building owner and a very viable business for the roofing professional.

Perhaps the main reason for the void in roof maintenance is a general lack of awareness on the part of building owners, maintenance staff, and roofing contractors. Another deterrent is the roofing bond or guarantee by the roofing material's manufacturer and roofing contractor.

Building owners often believe that these bonds or guarantees are all-inclusive and last forever, or at least 20 years. In the good old days, some did. But today, most guarantees are impressive on the surface, but in the fine print state that the roof must be kept watertight or repaired by the roofing contractor if it leaks.

Awareness of roof maintenance should begin the day the roof is completed and should become part of the overall building maintenance program. The standard two-year guarantee should be explained so that the building owner knows exactly what it does and does not cover.

Roofing contractors need to join other trades in the construction industry and develop guarantees that are explicitly defined and that detail the maintenance that is required by the owner and the consequences if that maintenance is not performed.

Begin roof maintenance with a visual inspection in early spring and late fall on a regular basis to determine if any damage to the roofing membrane or its components has occurred during the season. Roof maintenance is supposed to detect minor problems and arrange for their repair. Minor expenditures can eliminate the need for extensive repairs or premature total roof replacement.

The cost of roof maintenance varies depending on the size, location, and design of the structure. Roofs that are well maintained from their completion through their life span have a lower maintenance cost than those that are permitted to deteriorate for several years.

Applying Maintenance Coatings

Many roofing professionals, having cut their teeth in the hard work and integrity mentality of the field, may find it distasteful to think of selling a client something like a warranty and maintenance package. The thought might be that a job well done should speak for itself. Naturally, there is nothing more important than integrity and an excellent work product. But we live in a very fast-paced world, and the roofing contractor can guide the owner away from the short-term, let-it-sit-there, throw-a-patch-at-it solution by pointing out the long-term savings that can be realized with the application of a complete maintenance coating as a part of the maintenance contract. It is not always an easy sell, given the preference of some owners for the cheapest available option. This kind of straight talk, however, can build trust and eventually lead to increased sales and customer satisfaction.

The straight-talk strategy also requires that the roofing contractor stay current on the technologies affecting business in today's market. The professional must keep abreast of the latest available information if she or he is to respond correctly to customers' needs.

A decade ago conventional systems were practically the only choices on the market. Today that picture has changed radically, and new products are introduced almost constantly. Evidence of this fact can be found in the success of today's innovative synthetic systems, which offer performance comparable to that of traditional products and also extra benefits, such as flexibility, increased durability, and even energy savings. There are a variety of different options in roof maintenance techniques.

An objective listing of different options should include both conventional systems and the nonconventional systems presently available. Conventional choices include resaturants, asphaltic coatings, and aluminized asphalt coatings.

A resaturant is described as a penetrating type of oil. It is recommended for improving the performance of weathered BUR. Available for half a century, resaturants are a quick, inexpensive way to fix an aging roof. Resaturants, which are basically composed of oils, are usually applied over roof membranes where the original oils have leaked out as a result of exposure to the sun. It is believed that these replacement oils revitalize the old roof. The shiny, black appearance of a freshly applied resaturant gives the customer an immediate psychological satisfaction. The utility of resaturants, however, remains a controversial question.

A study on resaturants conducted by the National Bureau of Standards demonstrated that the materials did not penetrate deeply enough to be effective and that the mechanical performance of the resaturated roof did not improve enough to be worthwhile.

Asphaltic coatings are the inexpensive traditional solution. They are composed of asphalt either in a cutback solvent or emulsified in water. Asphalt membranes serve as a protective coating. However, asphalt is extremely susceptible to ultraviolet (UV) radiation and is not long-lasting.

Aluminized asphalt coatings are a more costly solution. They maintain their popularity because of the reflectivity associated with them. These oil-based coatings impart a bright silver appearance. Usually within one year, however, the aluminum pigments oxidize and leave a dull gray color. This causes the coating to lose its reflective capabilities, which decreases the energy-saving benefits. The asphalt in these particular coatings makes them especially prone to UV degradation.

Today, nonconventional, liquid-applied maintenance coatings are gaining acceptance because of their practicality. Aged modified-bitumen roofing (MBR) sheets can be used over traditional BUR systems, as well as over the new single-ply systems, MBR sheets, and polyurethane foam. These nonconventional synthetic coatings place virtually no extra weight on the roof deck or its supporting members.

Even though they are initially more costly than their conventional counterparts, these newer coatings are more effective in the long run, primarily with ethylene propylene diene terpolymer (EPDM) or aged BUR and MBR, because they eliminate the need for radical reroofing. Roofs can be recoated without adding significant weight to the roof load, and the coatings can be formulated in white to impart increased reflectivity to the roof.

Besides contributing to longer service life by reflecting UV rays, the white color keeps the roof surface cooler, which reduces the demand for air conditioning inside the building. The advantages of longer service life and decreased energy costs are the driving forces behind the popularity of reflective roof coatings.

The most distinguishing feature of liquid-applied roof coatings is that they are elastomeric, or flexible. This flexibility enables the coating to withstand the movement of the substrate caused by severe temperature changes. It should be noted that thermal shock is not limited to northern climates. A sudden thunderstorm in Florida can quickly reduce the roof surface temperature by as much as 100°F.

There are three types of nonconventional roof coatings: solvent, water-based, and 100 percent solids. Not all are considered maintenance coatings. Solvent coatings use organic solvents and can be applied over a wide range of temperatures. They dry quickly when humidity is high, but they usually contain large quantities of flammable and often highly toxic organic solvents. Extra care must be taken when transporting, storing, and applying these coatings. In addition, solvent-based coatings are associated with staining problems on certain asphalt roofs.

Water-based coatings use water rather than solvent as the carrier. This eliminates the flammability and potential toxicity hazards linked to solvent-based systems. Cleanup is done inexpensively with soap and water. These coatings can even be applied to a damp substrate without losing significant performance.

The main limitation of water-based coatings is that they can be applied over a somewhat narrower range of weather conditions. They should not be applied when the temperature is below 40°F or when rain is imminent. Since water based coatings contain no organic solvents, they are less likely to create staining problems on aged roofs.

The 100 percent solids coatings have virtually no solvents in them, but they must be applied with special equipment. In addition, they are sensitive to changes in temperatures. Their capacity to form a strong protective membrane depends on the maintenance of the proper reaction temperature and the mixing ratios of the components. If either is wrong, this type of coating cannot dry properly and may even gel prematurely, which could cause a defective membrane.

Within these three classes of liquid-applied coatings, there are four kinds of products available. Each is based on a different chemistry: urethane, Hypalon, copolymer acetate or polyvinyl acetate, and acrylic. These coatings are not asphaltic, and thus they are quite different from bituminous emulsions or resaturants. The urethanes are usually supplied as solvent solutions and can be either one-component or two-component systems. The advantage of two-component coatings is that they offer quicker curing than one-component coatings. But a good deal of skill is required to obtain the proper ratio of the two components. Urethane coatings are also rather expensive compared to other classes of coatings.

Hypalon coatings are based on chlorosulfonated polyethylene. While normally supplied at low solids to permit application with a spray gun, these coatings are comparatively expensive and offer limited durability.

Polyvinyl acetate copolymer-based coatings are less expensive, water-based systems. They can be used when a white reflective surface over an existing roof is desired and long-term tolerance of thermal shock is needed. These coatings are not recommended because they offer comparatively poor durability. In addition, polyvinyl acetate copolymers tend to hydrolyze, which also affects durability.

Acrylic roof coatings can be supplied as either water-based or solvent-based products. They provide an excellent combination of high performance and durability for a relatively modest cost. Acrylic coatings provide superior resistance to UV radiation, as well as good resistance to dirt pickup and water ponding.

An acrylic coating forms a seamless, breathable membrane. It can be applied readily to roofs with irregular shapes. For best results, acrylic coatings should not be applied when there is high humidity or if rain is imminent.

Almost all roof maintenance coatings are best suited for one type of project or another. Some are recommended only for specific situations. It falls on the shoulders of roofing contractors to educate their customers on the best overall option. The benefits of being able to point out the performance and cost features of the many available products cannot be stressed enough. And, though it may take some changes in the roofing professional's approach to everyday business, selling maintenance and warranties can create excellent customer relations and lock in some strong profits.

Selling Warranty and Maintenance Contracts

You have a limited duty to warrant the products you provide to the public as a roofing professional; that is a given. However, you also have the opportunity to provide a warranty service for your clients that offers them protection well beyond the obligatory, statutory requirements. There are firms in place that sell warranties for many services—for example, home buyer protection warranties.

You will have to think this through and determine what your firm can offer and what is needed by your particular clientele. The idea is that you must provide a limited warranty; however, your clients can purchase a more extensive warranty if they prefer. For example, the warranty might cover failure of plumbing vents from the top surface of the roof sheathing up, breakage of tile (with limitation of visits), and all other elements of the roof.

This may seem a bit risky at first, but the package should be available only to those customers who choose to sign up for your company's maintenance service contract. This service product will also have to be designed specifically to fit your company's mix of types of roofs and clients, and also to fit your staff or the staff you intend to hire.

The basic idea, however, is that your firm will visit the building twice a year and perform a visual inspection from the ground and from the roof as well. The inspector will provide the client with a signed inspection sheet and an estimate for service and repairs. The service and repairs may include treating galvanized flashing for rust, replacing damaged gutter, and a general cleaning of the gutter.

This is a proactive approach that benefits all parties. Let's face it, the average building owner is simply not aware of how much punishment a roof takes. Water, sun, wind, foot traffic, trees: Most people are not at all aware of just how tough a roof has to be, or that it must be cared for in order to extend its lifetime.

Use photos of damage to get the client's attention; nobody wants to throw money away, and how many building owners ever get on top of their buildings? In the case of commercial buildings, the owner should be made aware of how difficult it is for professional building maintenance firms to hire people who are both knowledgeable and diligent when it comes to roofs.

The basic proactive approach is that your customers may purchase an extended warranty from your company, but only if they sign up for the maintenance service. This closes two sales and the sale of any future work the roof may need before it requires replacement.

The arrangement is excellent; you have developed an ancillary cash flow, and you can make a real-world commitment to the promises in warranties because you know that the roof will be installed properly and attended to properly as it ages. The client will get a quality installation that will be monitored by professionals throughout the lifetime of the roof.

Reroofing Can Be a Healthy Cash Cow

In contrast to new construction bids, in the reroofing market the lowest price is not always the key to success. Quality, service, and long-term performance are the factors that attract reroofing business. And to a firm that is revamping its business practices, this can be a real blessing.

Many building owners rely on referrals when searching for a roofing contractor. As a result, it is essential for the roofer to establish strong customer relations in his or her locale. Naturally, this need fits hand in glove with all of the concepts discussed in Chap. 12. Reroofing is always needed. It is like toilet paper, sheetrock, light bulbs, etc.; roofs are always wearing out and needing to be replaced. This is an excellent business medium for developing an endless cash cow. The two imperatives are that you develop an excellent team for the work product and sales of roof products and extra work and that you are consistent in asking existing clients to spread the word.

Another item that is important to remember is that not all roofing systems are well suited to reroofing projects. Simply because a contractor has experience with a few systems does not mean that those products can be installed on every job. When the smart roofer runs into invitations to bid on work that is not suited to the company, he or she is ready to walk away, while guiding the prospective client to a steadfast contractor in the area who will do a good job.

The proactive benefits in this decision are that the intelligent roofer does not intend to work hard any longer. He or she knows which jobs make the company money and does not take on work that can cause negative cash flow.

Walking away from work is very important. It is a strong stance. The prospect will notice this integrity, and you can ask him or her for the names of friends who have roofs on which you know you will turn profits. You can also work out a marketing trade with contractors in your vicinity who install roof systems that you do not want to incorporate into your business.

The principles for reroofing are essentially no different from those for new construction. However, there are additional factors that must be taken into consideration, such as what caused the previous roof to deteriorate and the extent of any damage that resulted from that deterioration.

If either the insulation or the membrane is wet, they must be removed. Perform a thorough inspection of the roof to determine the requirements for rebuilding the structure. But, do not automatically assume that a reroofing job requires a complete tear-off. Just because the shingles are curled at the edges, or because most of the granules are gone and the shingles are brittle, does not mean that the roof has failed—but be sure and discuss any weaknesses of the roof with the owner, precisely. Explain what life the system has left and sign a maintenance contract; this will probably help secure the reroofing for your firm.

Check the sheathing and insulation from underneath the roof for signs of moisture penetration. Inspect the roof to find out how many roofs are currently on the structure. If the roof has failed and only one or two roofs are currently in place, a new roof might be placed directly over the existing roof. This assessment depends on the structure and the type of existing roofing material.

Winning Reroofing Jobs

Take into account the function of the building and the building's need for roof insulation and ventilation. Incorporate these findings in your reroofing bid when appropriate. They are essential components of a good roofing system, and they contribute to the bottom line.

Evaluate the need to create a sloped roof if none exists. Check all masonry, stucco, copings, etc., for needed repairs that might be pertinent to the total waterproofing of the building.

When a tear-off is required, address the job in as professional a manner as possible. Take all necessary precautions to ensure that no collateral damage is done to the property through the careless removal of the current roofing system.

Plan out the most unobtrusive route for getting heavy equipment onto and off the property and into an effective working position. Be

✓	Item
	Which roofing products are best suited to individual reroofing jobs
	Which products offer the best profit margins
	The customer's interests and desires concerning the property
	The type and location of the property
	The surrounding environment
	How the customer can contribute to your marketing needs
	Sell extras: gutters, ground drains, etc.
	Sell an extended warranty for mutual protection
	Sell a maintenance plan to the client

FIGURE 14-1 Factors to consider when reroofing.

aware of the occupants of the building and their needs, as well as the roofer's requirements, including access to utilities and debris removal.

Make certain that you carry the policies of the preroofing conference into all parts of your business. The reroofing meeting may not be as long or as intense as the meetings for new work—many of the design elements will be dictated by the existing roof and structure—but the meeting should cover all questions and details and is an excellent opportunity to sell extras.

Roof Drain Systems

More and more roofing contractors are installing roof drain systems. Drain systems are the single most vulnerable area for leakage on the roof, and, while the contractor might have had nothing to do with their installation, they are high on his or her callback list.

The architect or specifier usually designs the drainage system by taking into account such factors as the area to be drained; the size of gutters, downspouts, and outlets; the slope of the roof; the type of building, and appearance. The design capacity for a roof system depends on the quantity of water that needs to be handled. The quantity of water, in turn, depends on the roof area, slope, and rainfall intensity.

When evaluating the roof area, it must be remembered that rain does not necessarily fall vertically and that maximum conditions exist only when rain falls perpendicular to a surface. Since the roof area increases as the pitch increases, it is not advisable to use the plan area of a pitched roof to calculate drainage.

Though it is imperative that drains be designed for a capacity that will fully handle the "100-year rain," it is important to keep calculations accurate. Experience has taught us that using the true area of a pitched roof often leads to oversized gutters, downspouts, and drains. To determine the design area for a pitched roof, consult Table 14-1.

When sizing gutters, the following factors must be considered: spacing and size of the outlet openings, roof slope, gutter style, and expansion-joint location.

Since no gutters are effective for their full depth and width, the gutter must be designed so that water from a steep roof does not, by its

TABLE 14-1 Design of Gutter Systems

Areas for pitched roofs	
Pitch	**Factor***
Level to 3 in./ft	1.00
4 to 5 in./ft	1.05
6 to 8 in./ft	1.10
9 to 11 in./ft	1.20
12 in./ft	1.30

*To determine the design area, multiply the plan area by this factor.

own velocity, spill over the front and rear edges. This can result in numerous types of damage to the building: scouring of siding, rot at rafters, and water intrusion below the gutters. Consequential damage is very important; it can result in complex legal actions.

The location of gutter expansion joints is important because water cannot flow past an expansion joint. If it backs up at joints and overflows, the same damages are apt to occur as with overflow caused by improper size.

The size of half-round gutters is directly related to the downspout size. If the downspout spacing is 20 feet or less, the gutter size can be the same as the downspout size, but not less than 4 inches. If the downspout spacing is between 20 and 50 feet, add 1 inch to the downspout size for the gutter. Although draining more than 50 feet of gutter into one downspout is not usually recommended, whenever it is necessary, the gutter should be 2 inches larger than the downspout.

The required sizes of other leaders can be determined by finding the semicircular or rectangular area that most closely fits the irregular cross section. The depth of a gutter should not be less than half or more than three-fourths of its width. Hence, gutter sizes are usually expressed in width only.

Half-round gutters are economical and highly efficient. They are commonly used at eaves and troughs, and less often as built-in gutters.

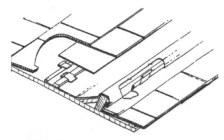

Other common types of gutter are the pole gutter (Fig. 14-2), molded gutter (Fig. 14-3), and built-in gutter (Fig. 14-4). Typical flat-roof drains are shown in Fig. 14-5.

FIGURE 14-2 Pole gutter.

Sizing Outlets

Outlets or leaders (Fig. 14-6) should be elliptical or rectangular in plan, with the longer dimension in the direction of gutter flow. The longer dimension should be the same as the gutter width and the shorter dimension about two-thirds of the gutter width (Table 14-2).

Modern, factory-made gutters are made of wood, metal, or plastic. At one time, wooden gutters were the favorite for residential drainage systems. Their popularity has decreased rapidly in the last few years, however, because of the weight, cumbersomeness, and cost of these systems, as well as the fact that wooden gutters require frequent appli-

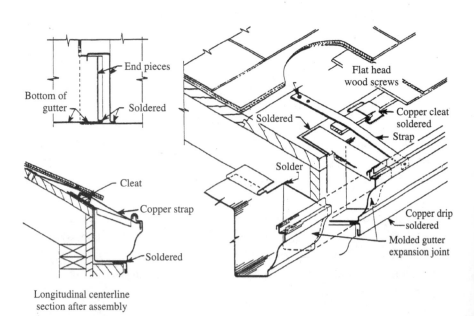

FIGURE 14-3 Molded gutter.

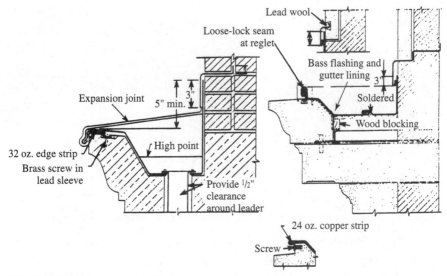

FIGURE 14-4 Built-in gutter.

cations of wood preservative to prevent rot. As a result, metal and plastic gutters are most frequently used for residential drainage replacement jobs.

Metal gutters are available in four varieties: galvanized steel, enameled galvanized, aluminum, and enameled aluminum. Figure 14-7 shows the various metal shapes. Galvanized steel gutters,

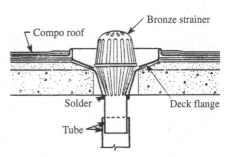

FIGURE 14-5 Typical flat roof drains.

the least expensive option, must be primed with a rust inhibitor and then painted. Care must be taken to treat all exposed areas to prevent corrosion.

Aluminum gutters do not rust when left unfinished, since they have their own natural protective coating. However, certain chemical reactions do cause the deterioration of aluminum, and the installer should clarify all areas of vulnerability with the manufacturer and explain these thoroughly to the owner. This may be an excellent opportunity to sell a maintenance contract.

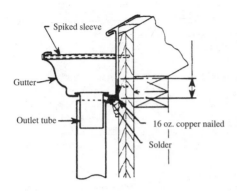

FIGURE 14-6 Outlet or leader.

Typically, the long-life, baked-on enamel finishes on products permit color matching without the hassles of painting. Keep in mind, however, that steel gutters rust if their enameled surface is scratched and the metal exposed. Also, remember never to mix aluminum and steel gutters, since contact between the two metals starts electrolytic action and corrosion (see Chap. 7).

TABLE 14-2 Dimensions of Standard Leaders

Type	Area (sq. in.)	Leader sizes (in.)
Plain round	7.07	3
	12.57	4
	19.63	5
	28.27	6
Corrugated round	5.94	3
	11.04	4
	17.72	5
	25.97	6
Polygon octagonal	6.36	3
	11.30	4
	17.65	5
	25.40	6
Square corrugated	3.80	$1^3/_4 \times 2^1/_4$ (2)
	7.73	$2^3/_8 \times 3^1/_4$ (3)
	11.70	$2^3/_4 \times 4^1/_4$ (4)
	18.75	$3^3/_4 \times 5$ (5)

TABLE 14-2 Dimensions of Standard Leaders (*Continued*)

Type	Area (sq. in.)	Leader sizes (in.)
Plain rectangular	3.94	$1^3/4 \times 2^1/4$
	6.00	2×3
	8.00	2×4
	12.00	3×4
	20.00	$3^3/4 \times 4^3/4$
	24.00	4×6
SPS pipe	7.38	3
	12.72	4
	20.00	5
	28.88	6
Cast-iron pipe	7.07	3
	12.57	4
	19.64	5
	28.27	6

Plastic gutters do not corrode, and the colors of the new products tend to be long-lasting. The major problem with plastic gutters is their high coefficient of thermal expansion, which means that if they are not properly installed, they can buckle in hot weather.

To determine the amount of material needed, measure the length of the gutters around the house. Most gutters are available in 10-foot lengths. Figure how many multiples of 10 feet are required. Some manufacturers produce longer lengths of gutters, but these are more difficult to handle. Once you figure out the

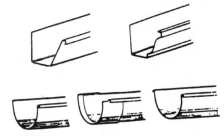

FIGURE 14-7 Various metal gutter shapes.

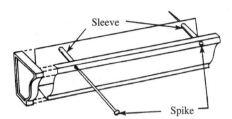

FIGURE 14-8 Installing metal gutters with large spikes that pass through the sides and into the fascia board.

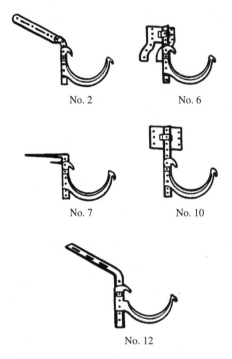

FIGURE 14-9 Sickle-shaped hangers.

length of gutters and downspouts needed, plus the various accessories such as elbows, corners, and hangers, order the parts necessary for the installation.

Installing Metal Gutters

Most metal gutters in residential drainage systems are installed with large spikes that pass through the sides and into the fascia board (Fig. 14-8). A metal tube or sleeve, called a *ferrule,* mounts around the spike and maintains the spacing of the gutter sides. Locate these spikes 24 to 30 inches apart, depending on snow conditions.

Two other hanging methods can be used for metal gutters. Sickle-shaped hangers can be fastened to the fascia boards about every 30 inches. The gutters are laid on top of the hanger (Fig. 14-9). This method eliminates the need to drill holes in the gutter and thus is easier than spiking the gutters in place. The sickle-shaped hangers are more expensive than spikes.

The second way to hang metal gutters is to use strap hangers. These hangers have flanges that are nailed in place under the roofing material (Fig. 14-10). This type of hanger is best used on new work, since there is always a chance of damaging the existing roofing when it is pried up to install the hanger.

Elbows usually connect outlet tubes and collector spouts. Squeeze the end of the downspout and insert it into the large end of the elbow. Down-

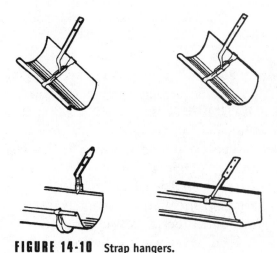

FIGURE 14-10 Strap hangers.

spouts are fastened to the wall with downspout straps. Be sure the straps are long enough to anchor the downspout securely to the wall. If the gap between the wall and the downspout is too wide, shim the downspout strap 1 inch away from the wall with a product that is weather-resistant or waterproof. Use two straps on each 10-foot length of downspout and three straps on two joined 10-foot sections.

To enable downspouts to perform their maximum service, install them so that they carry water as far away from the building's foundation line as possible. This is very important—foundation problems are common triggers of very complex and expensive litigation. There are several ways to achieve this, but the most popular are to use a concrete splash block directly under the downspout, which runs the water toward the driveway or a similar draining surface, or to connect the downspout directly to an underground line that leads to the storm sewer system.

Installing Plastic Gutters

To install the gutter hangers, line them up with the chalkline and nail them at 18-inch centers. Be sure the section of each hanger that comes in contact with the fascia is perpendicular to the chalkline. Use only noncorrosive screws or nails.

Locate hangers so that when the gutter is snapped in place, it catches the rain as it runs off the roof. For installations where there is no fascia board, use special gutter hanger rods. Nail these rods to the roof edge and bend them to the desired angle.

Use collector boxes or expansion joints on long runs of gutter in order to avoid problems with expansion and contraction of the run. Make allowances for movement by properly aligning hangers and placing them at least 18 inches from corners adjacent to long runs.

When the gutter ends are not confined to prevent excessive lateral shifting, secure the gutter to prevent creeping. This can be accomplished by drilling a nailhole as high as possible in the midpoint of the back wall of the gutter section. Then nail the gutter to the fascia, using shims, if necessary, to maintain the proper alignment. Where the ends of the gutter are confined between walls or rake boards, maintain a clear space at each end to prevent water from standing between the gutter end and the wall covering.

On buildings with inside or outside corners, always start a gutter system installation at a corner. Attach the gutter end securely to one corner before placing the section into the hangers. Once the section is clamped firmly into the hangers, attach the other gutter end. On a small, hip-roof building, plan for the required expansion-contraction allowance using properly spaced hangers only. This is done by placing, on successive corners, one hanger 6 inches from a corner and the next hanger 18 inches from and around the corner. If this practice is followed on each corner, allowance is made for the typical thermal motion.

As a rule of thumb, you will use rivets and polyvinyl chloride (PVC) cement to fasten sections together. Installation procedures for elbows and downspouts are similar to those used for metal sections. Always follow the manufacturer's assembly direction with care.

Use only aluminum screws or nails. Other kinds can corrode and stain the vinyl. Coat each screw or nailhead with PVC cement to make it match the rest of the drainage system.

Replacing Gutters and Downspouts

When replacing gutters, remove the run all along one side of the house; otherwise you may distort the old gutters when you remove them, and they are very useful as a guide for fitting the replacement.

Once the old gutter is on the ground, line up the appropriate fixtures and the required number of new gutter lengths parallel to it.

Mark the new pieces and cut them to the appropriate size. Use a carpenter's square to mark the cutting lines. Use a hacksaw or power saw with a metal-cutting blade for metal gutters. File the edges and apply a coat of rust inhibitor to the cut surfaces.

Cut plastic gutters with a sharp, fine-tooth handsaw. To prevent the gutter wall from flexing while it is being cut, slide a wooden filler block into the gutter and position it as close to the cutting point as possible. After cutting, file the rough edges.

Before installing the new gutters, check the fascia board for rot. If it is defective, replace it. Be sure to paint both sides and the edges of the replacement board.

Locate the downspout end of the gutter at the lowest point on the fascia board. To achieve this, snap a chalkline on the fascia board to show the desired slope for the gutter.

Solving Ponding Problems

Ponding water on a roof is the number one enemy of the serviceable life of any low-slope roof system. Ponding water is the direct result of improperly located roof drains and drain supports placed directly on joists. After the building settles, drains rise well above the roof's drainline.

In the 1950s, totally flat roofs were typically of a boilerplate design, usually with unsatisfactory results. Unless *camber,* or deflection, was built into the structural system to allow for anticipated dead and live loads, the deflection between bearing points resulted in low points on the roof.

Typically, roof drains were located along beams or bearing walls, which end up being the high points in the roof system. This left little, if any, room for construction error. Everything, from the footings on up, had to be constructed to very close tolerances in order to end up with equal elevations across the building.

The roofing industry gradually learned that ponding water acted to reduce the serviceable life of a built-up roof (BUR). Much more water entered the building if leaks did develop, and getting the roof dry enough to patch a leak was also a problem.

Most major material manufacturers in the United States eventually came to recommend that slope be applied to all new roof designs using

BUR membranes or manufactured single-ply membranes. There are many, many buildings, however, that still have no slope and excessive water ponding. The most opportune time to correct this condition is when reroofing becomes necessary.

Ideally, eliminating all ponded water would be best for the long-term performance of the roof system. There are times, however, when this is not practical because of the complexity of the solution or because of limitations imposed by the existing conditions. It is important to keep the design as simple as possible. A complex design might be the most effective, but costs rise proportionately, and the roofer might have difficulty executing the work. In these cases, the cost to eliminate the ponds exceeds the advantages of doing so. Many experts believe that a roof pond that dries within 48 hours after a rainfall has little effect on the long-term performance of a roof membrane.

Once the existing conditions and the causes of the ponding problem have been determined, approaches to the reroofing design can be developed. Often, the addition of roof drains or scuppers is not feasible. Adding insulation of varying thicknesses, not necessarily tapered insulation, can be a viable solution in some cases.

In response to the industry's movement toward sloped roof systems for complete drainage, manufacturers are marketing tapered insulation to give slope to relatively large sections of a roof area. Tapered insulation board is available in several different types, including polystyrene, perlite-aggregate, isocyanurate, cellular glass, and fiberglass. The slope available with each type of tapered insulation varies.

The selected slope should be sufficient to achieve positive drainage, yet not so great that installing the insulation is difficult and costly. It may not be necessary to use tapered insulation to slope the entire roof area. There might be an existing deck slope, no matter how slight, that can be enhanced. However, it is best to bring in an architect to make decisions for the principal, removing design responsibility from the roofing professional's shoulders and placing it where it belongs.

Let us look at a simple example of a 50-year-old building with an essentially flat concrete deck. The deck has gradually deformed and sagged between supports through the slow process referred to as *creep*. To compound the problem, new heating, ventilation, and air conditioning (HVAC) equipment has been suspended from the deck. The net result is a slight slope in the roof section from the supports to the center.

The roof is drained at one roof edge through scuppers and over one of the bearing walls. The result is substantial ponding in the central portion of the roof. The reroofing design might include tapered insulation from the scuppers to the opposite roof edge to eliminate ponding. Since half of the existing roof already slopes toward the scuppers, however, why not take advantage of it? Placing flat insulation over this half of the roof gives a net positive slope toward the scuppers.

Tapered insulation with enough slope to counteract the negative deck slope on the other half of the roof provides a positive slope across the entire roof area at the membrane level. This approach minimizes the insulation thickness so that roofers can better deal with height limitations at existing flashings, such as through-wall flashings and at large mechanical units. Likewise, blocking requirements at the details might be less. The final product provides the desired drainage at less cost than a fully tapered system.

Another typically inexpensive method to reduce ponded water is to increase the thickness of insulation in steps, or tiers, from the drainage points to the perimeter area. These tiers are typically made in increments with tapered edge strips that provide a smooth transition from one tier to the next.

The same basic concept as in the first example is applicable here. Take advantage of the existing deck slope on half of the roof. In this case, the insulation is stepped down in four increments rather than having a continuous taper (Fig. 14-11). The steps are located more closely in the area of greatest ponding, or negative slope, to minimize the existing highs and lows.

The disadvantage to this approach is that it does not completely eliminate the ponding in all cases. It can reduce the depth of ponding to a tolerable level, however, by displacing the water in gradual increments. As with the tapered insulation example, height restrictions at the details are more workable with this method.

Cost is the primary advantage to this approach. Depending on the type of roof system and the complexity of the insulation layout design, the installed insulation cost for the tiered system can be 10 to 40 percent less than the tapered system. The cost savings can go a long way to offset any loss in service life of the roof system caused by minor ponding with the tiered system.

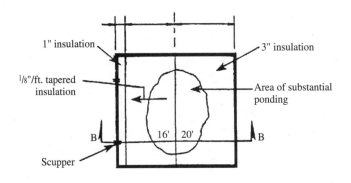

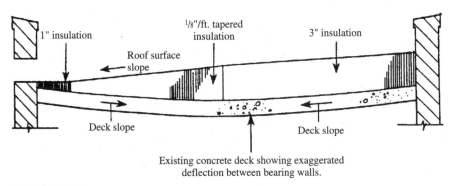

Existing concrete deck showing exaggerated
deflection between bearing walls.

FIGURE 14-11 Insulation stepped down in four increments.

Once the surface water gets to the drain, it must be able to flow off the roof. The extra materials used to flash in the drain or scupper often create a buildup around the drain, which results in ponding water. This effect can be minimized by placing the drain in the center of a depressed area that is enhanced by tapered insulation. The tapered area should be slightly larger than the largest flashing sheet.

In climates where freezing temperatures often occur, ice damming at scupper outlets and ice damage to the downspout are often problems. The open-faced downspout was developed to minimize the potential for plugging while continuing to provide the wall protection and waterflow control afforded by a totally closed downspout. Using a dark color for the downspout helps solar heating keep the downspout as ice-free as possible.

Certain types of buildings, such as schools, can have other special considerations. It is amazing how a new downspout can provide children with access to the roof. A W-shaped, open-faced downspout securely anchored to the wall is a design that does not leave much for children to grab. Eliminate all projections or sharp sheet-metal edges that can cause injury. The W-shaped design has an advantage over the more conventional open-faced downspout when factory color-coated sheet metal is used, as the entire exposed face shows the same run of color.

The inside of the more conventional open-faced downspout shows the reverse side of the metal and produces a "racing stripe" on the building unless the inside is painted. This "child-proof" downspout provides a reasonable solution to several common downspout problems.

Many roof contractors have found the W-shaped downspout easy to fabricate and install, and are satisfied that it provides essentially the same function as the more conventional open-faced downspout. Likewise, building owners are satisfied with the reduced potential for injury and traffic on the roofs.

A better approach to edge drainage in northern climates is to install an inside-outside drain. This is usually more expensive than an exterior scupper and downspout, but its ability to remove snowmelt is much better, and the risk of damage from accumulated ice falling off the wall is eliminated. It is also relatively vandal-proof.

This type of roof-drain system is simply a roof drain installed at the low point of the roof surface. The drain leader runs down the inside of the exterior wall and discharges through the wall 1 or 2 feet above grade (Fig. 14-12). The interior location keeps the drain and leader open during cold weather. Ice buildup at the discharge is minimized because the drainage water has warmed slightly. Damage from any ice development is minimal because the ice forms at grade level and not at roof level.

The best locations for roof drains depend on several conditions: the highest water level and good accessibility from the ceiling or along the predominantly low side of the roof. Since a roof-membrane installation starts at the drain, it is ideal to line up new and old drains whenever possible. This way, the plies in the roof system do not buck the flow of water. From a cost standpoint, the best time to install a drain is while the old roof is being replaced. If the roof is in fairly good condition but suffers from large areas of standing water, definitely add new drains.

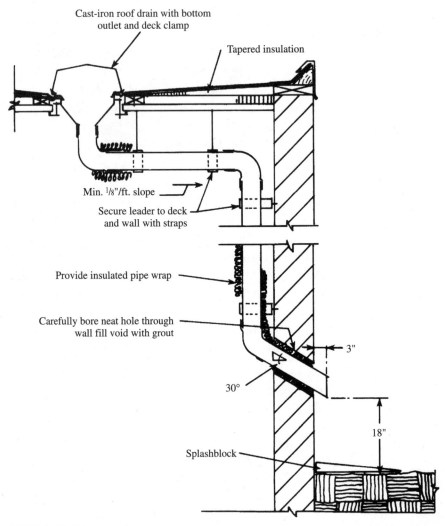

Cast-iron roof drain with bottom
outlet and deck clamp

Tapered insulation

Min. 1/8"/ft. slope

Secure leader to deck
and wall with straps

Provide insulated pipe wrap

Carefully bore neat hole through
wall fill void with grout

3"

30°

18"

Splashblock

FIGURE 14-12 Inside-outside roof drain system.

Getting rid of ponding water reduces leaks, makes the roof last longer, and pays for itself in the long run.

Select the drain that best fits the building conditions and the existing roofing system. Drains can be made of cast iron, plastic, or copper. Consult local roofing supply houses for the many alternatives. To ensure that the proper components are obtained, use the checklist in Fig. 14-13.

✓	Item
	Appropriately sized drain body
	Roof membrane clamping device
	Underdeck clamping assembly
	Expansion connection coupling
	Drain grate
	Roof-membrane flashing skirt

FIGURE 14-13 Gutter component checklist.

On new, large construction jobs, the plumbing contractor is often responsible for the installation of the internal drainage system. In addition to providing and installing the roof drain, the plumbing contractor should ensure that the major components, such as a clamping device and grate, are made available to the roofer. The plumbing contractor's line of responsibility is normally completed once the drain is securely fastened to the roof substrate.

Enter the roofing contractor, who must properly disassemble the parts, cut insulation, install flashing, and proceed with the installation of the roofing materials. The roofer then must fasten the clamping device securely over the in-place membrane. Future problems arise for the roofing contractor if the architect or specifier has not carefully selected the proper roof drain or the plumbing contractor has not installed it to specifications.

This situation is slowly changing through grassroots efforts by some drain manufacturers and others in the roofing industry. Because the roofer is responsible for the drains in the end, it makes sense for him or her to be given the task in the contract and be paid for their installation. Manufacturers who now offer their own drains to accompany their roofing systems have recognized this and are the catalysts, at least in offering the roof drains to the roofing contractor. Perhaps one day roof drains will be entirely the responsibility of roofers.

Retrofitting Drain Systems

For retrofit projects, the prejob walk-through should include a look at the existing drains to determine whether a corrosion problem has occurred with the bolts. During the walk-through, ask yourself some questions. Can you, in fact, count on the added time and labor involved with retapping, or should a complete drain replacement be addressed? Is the existing drain compatible with the specified roofing system? Address these questions beforehand so that you have more control over both the job and the associated job costs and potential profits. Discuss your concerns with the project manager or building owner before they become an issue. This enables you to better serve the client.

Change Orders as a Profit Center

Change orders are another potential problem area for roofers during any construction project. In theory, design drawings and specifications are executed by the design team for the building owner. These plans and specifications are the blueprint for what is to be built and each party submits a cost for its component, and the building project is started.

However, anyone who has any building experience knows that the original plans and specifications are almost always changed during a construction project because of additions to the structure, alterations in the design of the building by the designer and owner, or onsite requirements for changes in the design because of unforeseen construction problems.

In a perfect world, the construction of the building would be orchestrated by an onsite project manager who was extremely diligent and had good weather and good performance by all other parties, so that there are no surprises for the roofing contractor when he or she arrives.

An example of a surprise that the roofer might find is an alteration in framing, such as changing the slope direction, which causes problems with drain locations. The plumber and the general contractor's project manager have not seen this, but the roofer's project manager spots it on a prework site visit. The roofer's project manager returns to the roofer's office and prepares a change order, rather than getting bombastic with the general contractor or the design team. The principals accept the change order. The roofer has brought the job back on track, prevented future infiltration of water and possible construction defect litigation, and made several hundred extra dollars in a few moments time.

In most cases, the communication is adequate in new construction jobs, and relatively few problems occur. Yet, with all of this coordination and communication, roofers are often the last to know about many things. Since the drain is the plumber's responsibility, the roofer's input is rarely sought. Even if everyone is consulted and all perform their expected duties, after the plumber completes his or her tasks on the roof and the roofer reassembles the drain to the existing roof system, who inspects the final drain assembly?

Because the plumber's responsibility is almost nonexistent on the roof, he or she is naturally reluctant to go back to the job and check the roofer's work. Very often, plumbers do not have any real expertise in roof drains, so they probably could not find any problems even if they undertook an inspection at completion. Thus, the matter is largely ignored until there is a leak. Precautionary measures can be taken if proper preparation is observed beforehand.

The important thing is to make this a part of your ongoing bidding process. Review bid specifications carefully, and then submit change orders to the principal and the design team to apply for addendums aimed at alteration or replacement of faulty design criteria. A page of change orders can add a good deal of money to the bottom line of any estimate, and this is often money that is earned at terrific profit margins.

Pay close attention to developing the sale of change orders as an ordinary part of doing business, and always get a valid signature to back up the request.

The Roofing Consultant

Within the last decade or so, construction defect (CD) litigation has become a huge business on the periphery of the AEC (architecture, engineering, contracting) industry. Many of you are well aware that this is an immense web of complex civil actions and the cash flow is huge.

If your firm has been sued directly or in a cross complaint, you are well aware of what a long, drawn-out process CD litigation can be. The earlier chapters in this book took a proactive approach to roofing, and appropriate maintenance contracts will help to keep your firm from being ensnared in this complex web. But the ideas in the book must be implemented in a professional manner, and continual diligence must be applied to make them a seamless part of your company.

Because of their extreme vulnerability to the elements, roofs have been a large part of the tidal wave of CD litigation. Typically in these lawsuits, the plaintiff has noticed something wrong with the building, generally water intrusion. The owner hires a plaintiff's lawyer. The plaintiff's lawyer hires an expert team, of which one member is typically a roofing expert. The plaintiffs sue the developer or general contractor, who hires lawyers, who hire more experts. The developer sues the subcontractors in cross complaint, and their lawyers hire more experts and file notice with the insurance carriers, who may hire more lawyers and experts.

This is a very brief rundown of this situation. Naturally, the circumstances can vary greatly, but it is obvious that CD lawsuits require lots of hours from roofing experts. And the discovery on the roofs is one of the parts of the field work that is most heavily laden with billable hours.

The time billing for this work is typically set at a fair rate for the roofing expert, and there can be a good deal of it available. CD work is only one part of being a roofing consultant; there are numerous other services that your company can provide. Inspections for Realtors and corporations, feedback for architectural firms, expert witness at trial—there are many ways to offer your knowledge.

The cash flow is a part of your general game plan, and at the same time you will reap excellent side benefits from this work—your understanding of what your roofing firm should not do will increase dramatically, and you will have the opportunity to associate with numerous people to whom you can market your other services.

How Consulting Works

Typically you will be hired to investigate the condition of an existing roof. And generally speaking, there are two ways in which this is done: visual inspections from the ground and on the roof, and thorough investigations during which parts of the roof area are opened up. These are often referred to as nondestructive or nonintrusive and destructive or intrusive site inspections or investigations. As with everything related to the AEC industry, different people use different terminology.

The usual scenario for discovery on site is that the plaintiff experts walk the site and produce a rough list of possible construction defects. This is usually called the preliminary defect list. Destructive work is not typically a part of this process.

The defense team may then walk the site for a round of nondestructive testing, and then the two parties will meet to discuss their findings. At that time, they often agree on what parts of the buildings need to be cut open in order to determine what has actually caused the listed defects, which cannot be learned from nondestructive observation.

The next step in the process will be to schedule the first round of destructive testing. During this time, parts of the roof may be opened up in order to determine what has caused the leaks and other problems that have been observed. In certain cases, instruments are used to avoid the costly process of cutting buildings open, then patching them back up, that is involved in destructive work.

Performing Nondestructive Testing

Millions of dollars are spent each year to make roof repairs because of leaks. This creates an ever-increasing demand for more accurate moisture damage assessment. Roof leaks that result in wet insulation, yet cannot be seen as a drip, cause real problems for roofers. A successful moisture analysis can significantly reduce overall repair costs while developing and securing mutually beneficial relationships between roofing professionals and their customers.

Nondestructive testing (NDT) is designed to locate moisture in roofing systems, including interply and substrate moisture, without causing physical damage to the roof's surface or substrate. Wet insulation in roofing systems is often compared to a cancer. NDT acts as an x-ray or biopsy of the roof to determine the location of the cancer. If left untreated, this cancer causes the roof membrane, fastener system, and structural deck to deteriorate. Figure 14-14 shows some elements of nondestructive testing. Descriptions of the most popular NDT methods follow.

INFRARED TESTING

Infrared instruments, both hand-held and aircraft-mounted, are used to detect trapped moisture by identifying areas of heat loss through wet insulation. One of the advantages of infrared testing is that large areas of the roof can be scanned relatively quickly. Using an infrared scanning camera, the operator takes a photograph, called a thermograph, that indicates energy loss due to wet insulation. However, it is important to take into consideration the following disadvantages of infrared testing:

- In most cases, tests must be conducted at night.

- Exacting weather requirements dictate when tests can be run and achieve accurate results.

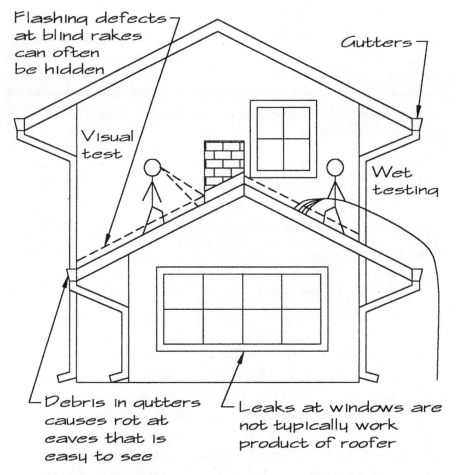

Flashing defects at blind rakes can often be hidden

Gutters

Visual test

Wet testing

Debris in gutters causes rot at eaves that is easy to see

Leaks at windows are not typically work product of roofer

Roof problems can sometimes be seen without destructive work. If there are doubts about cause of water intrusion, destructive work can be planned during nondestructive work.

FIGURE 14-14 Nondestructive testing.

- The cost of the equipment is significant.
- Readings can be affected by hot air escaping on a roof, heat sources suspended below the roof, or differences in insulation thicknesses and BUR, wet surfaces, and ponded water.

OPERATING NUCLEAR EQUIPMENT

Moisture is made up of hydrogen and oxygen. Nuclear moisture detection operates on the principle of neutron moderation, which means that fast neutrons emitted by a radioactive source are slowed by hydrogen in the roof. The slowed neutrons are detected, and the count displayed is proportional to the hydrogen content. Nuclear testing also has its disadvantages:

- Because asphalt and bituminous roofing compounds contain hydrogen, variations in material density can be misinterpreted as moisture.

- Nuclear surveys are very slow.

- Nuclear testing is regulated by the Nuclear Regulatory Commission (NRC), which requires operators to be licensed and strictly controls the use, transportation, storage, and disposal of these instruments.

TESTING WITH THE IMPEDANCE METHOD

The impedance method combines the principles of resistance and capacitance. These instruments operate by emitting low-frequency electronic signals from rubber electrodes fitted at the base of the instrument. When there is no moisture present in the substrate, these signals are insulated from each other, and no reading is recorded. When moisture is present, the electrical conductance is dramatically increased and the circuit is completed, giving instant readings. The greater the moisture content, the higher the reading. Advantages include:

- Versatility and ease of operation

- Instant and continuous readings

- Ability to trace the path of leaks and pinpoint the place of entry

- No effect from temperature differentials caused by wind, vents, or variable roof layer and asphalt thickness

- Safe and easy to transport and use

A disadvantage is that these meters do not operate on roofs with a metal cap sheet, through ponded water, or on inverted roofing systems without the removal of ballast and insulation.

Providing Preventive Maintenance

Consulting in CD litigation is not the only area where nonintrusive work can be a valuable service. The life span and energy-saving performance of a roofing system are greatly influenced by the presence or absence of a roof-maintenance program. For this reason, nondestructive testing is becoming increasingly popular among roofing professionals. These cost-effective diagnostic methods provide vital information that can expand business opportunities and ensure long-term profit potential.

Many roofers agree that more than one method of study is necessary if roofs are to be properly surveyed for leaks. Each system complements the others and can be used for verification of their findings. Anything that does not show up using one method usually shows up when another method is used.

However, there are times when water testing and intrusive work must be resorted to in order to fully understand a roof problem. Destructive testing is fairly straightforward for a veteran roofer (see Fig. 14-15). The problem area is studied before the intrusive work is done, and the experts instruct the mechanics doing the work on what areas to open, and in what order. The experts observe this with care, photographing, taking notes, and sometimes running water in order to understand what the exact problem is, how it was caused, and how it can most effectively be fixed in a cost-efficient manner.

There is no substitute for know-how and roofing knowledge. No machine can replace the roofing contractor's experience and knowledge about the many different roofing systems. For example, some roofs require insulation that is not butted firmly. A roofer must know that when inspecting the roof so that false readings do not send the team on a wild goose chase. There is a host of knowledge that only years of hands-on experience can bring to the table, and that is why you should be well paid for using this base of information to help building owners work through their difficulties with the various concerned parties.

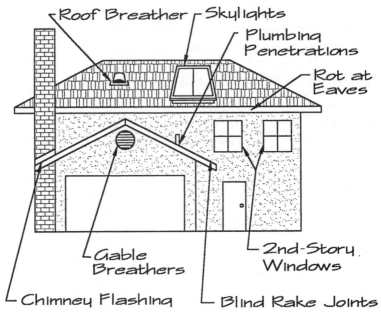

Roof Breather
Skylights
Plumbing Penetrations
Rot at Eaves
Gable Breathers
2nd-Story Windows
Chimney Flashing
Blind Rake Joints

→ Skylights are typically well-flashed, but removal of tile at curb flash is sometimes needed.

→ Breathers in walls are not typically the roofers' work product.

→ Roof breathers will leak if lapping is not well executed.

→ Eaves and open rakes must be studied closely to allocate resultant damage - - home owner and other trades can contribute to cause of damage.

→ Stains below 2nd-story windows are not typically resultant of roofers' work product. Careful wet testing with destructive exploration is often required.

FIGURE 14-15   Destructive testing.

Promoting Roof Accessories

There are several roof accessories that roofing contractors can install as part of their normal work operations to help their profit picture. These accessories help prevent damage to a metal roof when a curb is used to distribute the weight of a rooftop unit, or prevent damage from foot traffic when the installation of a roof walkway between equipment located on the roof is specified (Fig. 14-16). Roof walkways, especially when combined with handrails, also provide surer footing for future visitors to the rooftop, which can be a valuable safety feature for the building owner.

Roof curbs enable roofing contractors to effectively flash a roof penetration. Some curbs that are installed on a metal roof panel can have a water diverter and matching ribs welded into the curb base.

Other roof accessories that enhance safety and protect a building owner's investment are access hatches, smoke hatches, and gravity ventilators. Access hatches allow workers to get onto the roof safely to install or repair rooftop equipment.

Several states have upgraded their fire code requirements, and architects are specifying that smoke hatches be installed on an increasing number of projects. Smoke hatches typically have a release mechanism

FIGURE 14-16 Roof accessories/walkways.

that is attached to a fusible link. When the temperature inside the building reaches a predetermined level, the fusible link breaks. This activates the release mechanism, and the hatch opens to release the smoke. Releasing the smoke helps firefighters bring a fire under control more quickly, more safely, and with less water damage to the building and its contents.

Gravity ventilators not only provide for the safety and comfort of building occupants by regulating the temperature and air quality, they also help protect the building contents from moisture damage. When properly utilized in conjunction with adequate air-intake devices, gravity ventilators can provide an inexpensive way to ventilate a building.

Descriptions of other roof accessories that a roof contractor can install follow.

Keeping Snow from Sliding

Snow-retention devices can be invaluable when snow must be prevented from sliding off roofs. Snow guards come in many types, but they are all too often omitted from new construction projects in order to save money and dismissed on existing structures on the grounds that the event that prompted their consideration in the first place might not happen again.

On textured roofs or structures with limited liability exposure, these justifications might be reasonable, but on metal and slate roofs, this kind of thinking can be dangerous as well as costly. While snow initially accumulates on even the steepest of metal roofs, it is not going to remain there. Unless it is restrained and allowed to dissipate as snowmelt or the amount of snow that can slide free at any one time is restricted, the accumulated snow can cause considerable injury or damage.

Even low-slope metal roofs experience snow movement. Although research indicates that sliding is unlikely until the roof slope is 14 degrees or more, snow massed on shallow-pitched roofs can behave like a glacier and creep slowly downslope, even on a very gradual slope. The first line of defense should be good planning during the design of a structure. This can reduce the potential for harm to persons, property, and even the building itself when snow or icing is anticipated.

A host of criteria determine the placement of snow guards on a metal roof. Climate obviously is the first of these because it affects the amount, frequency, and physical characteristics of the snowfall, i.e., wet, dry, or mixed with sleet or ice. Roof height, shape, type, pitch, lengths of runs, pan widths, locations of other physical features such as gables, changes in roof elevation and pitch, vents, stacks, chimneys, other penetrations, the directions of prevailing winds, and even the presence of snowmaking machinery at ski resorts all have an impact on snow-retention measures.

The purpose for which snow or ice is to be retained must also be evaluated. Are building occupants, patrons, or passersby to be protected? What about foundation plantings, vehicles, adjacent structures, or equipment? Is there a need to control snow buildup on roofs below or to protect gutters or stacks? Of equal importance to many owners and every architect is the visual impact that these devices have on the structure. Given all of these variables, one can see that device placement certainly requires some thought. The checklist in Fig. 14-17 is on the companion CD-ROM and offers a beginning set of guidelines that you can adapt for your locale and company.

The traditional fastening method, used for almost 100 years, requires that a pair of prongs be driven through the seam. This method often results in deteriorated metal, sheathing, and framing. While this type of snow guard is still used on slate roofs, the three most popular designs for metal roofs are fences and surface-mounted and seam-mounted devices.

FENCES

Fences are field-assembled systems designed to hold snow at the eaves. They consist of rods and various other components. The height of the bottom rod above the roof surface is such that these systems are not particularly effective in retaining ice or light accumulations of snow. Care must be exercised during design. Unless sufficient rows are employed, fences tend to concentrate snow loads rather than distribute them over the roof.

SURFACE-MOUNTED DEVICES

Surface-mounted devices are secured to the pan of the metal roofing system with solder, fasteners, and adhesive sealants. While effective in

✓	Item
	Review the structural capacity of the roof to verify that it is adequate to support the loads that might develop.
	Place snow guards at or within the interior face of insulated exterior walls to reduce the potential for ice dams at the eaves.
	Space snow guards with the goal of holding snow in place, rather than trying to catch it at the eaves. On long runs or steep pitches, this means spacing the devices at suitable intervals up the slope. It is easier to retain snow or ice on a roof than it is to resist the dynamic loads generated once it breaks free.

FIGURE 14-17 Snow load guidelines.

retaining both ice and snow, many of the devices are not well engineered and lack material strength. Those of quality are particularly useful when there are no suitable seams to which a seam-mounted device can be fastened.

Unfortunately, the use of adhesives is limited by weather conditions. In hot weather, temporary supports or tape may need to be employed until the sealant sets up. In cold weather, the adhesive may not cure properly. Generally, a minimum temperature of 50°F for a period of 30 days is required for a professional application. In any case, the surfaces must be clean and dry.

SEAM-MOUNTED DEVICES

Seam-mounted, one-piece snow guards are secured to the standing seams. This one-piece device is clamped to the metal roof seam with stainless steel set screws and installed with a ratchet and hex socket. This makes the device simple to install and easy to retrofit on existing roofs.

CONTROLLING BIRDS

On some roofs, bird control is serious business, and gimmicks, toys, or other gadgets seldom work. If your client has this problem, you might suggest the installation of a stainless, humane bird barrier such as the type shown in Fig. 14-18.

As you begin to notice services that you can easily provide for your clients that will enhance your cashflow, more of them will fall into place. Take one step at a time, but do not procrastinate. Remember your goals and pursue them diligently—this will add a whole new sense of interest and vitality to your business.

The Future of Roofing

The traditional roofing practices described in this book are by no means on the way out; whether one-ply, urethane, fiberglass, or conventional commercial BUR systems, each has had its day. And those that remain have done so because they evolved as roofing systems that meet the needs of commercial and residential roofing contractors. In this ever-changing world of construction and architecture, however,

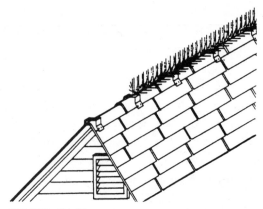

FIGURE 14-18 Stainless humane bird barrier.

new concepts, systems, materials, products, and ideas are definitely on the way. They are bound to be introduced.

An excellent example of a brilliant and far-reaching new concept is fabric roofing. In the end, being able to protect the tops of our built structures from the elements with lightweight sheeting is a powerful concept—a step forward in the evolution of one-ply systems.

Permanent fabric roof structures are past the experimental stage. They are undergoing the technological evolution that takes place in every segment of this industry. They are not the best solution to every design program, but they are an excellent option when the design requires long, clear spans.

For example, the Denver International Airport is a unique architectural statement. The distinctive silhouette above the terminal roof, which shows 17 sets of pure white peaks, evokes a sense of the Rocky Mountains (Fig. 14-19).

When a clear span exceeds 100 to 150 feet, fabric structures generally cost less than traditional structures. Proponents of fabric structures cite the fact that the weight and cost of the supporting structure, per unit of plan area, do not increase with the roof's clear span, as in conventional construction. Furthermore, the cost advantage occurs because the fabric envelope performs the function of several conventional building components, including roofing materials, roof deck, insulation, structure, wall cladding, waterproofing, acoustical absorber, interior finish, and daytime light source.

FIGURE 14-19 Denver International Airport evokes a sense of the Rocky Mountains with its fabric roof.

It should be noted that fabric structures usually require specialized engineering, which can make them a more economical option for larger structures than for smaller structures. Special consideration also should be given to the use of fabric structures in cold climates, despite the fact that additional insulating materials may need to be used to expand their climatic range.

Fabric structures must be held in tension to be structurally stable. There are two methods to accomplish this goal: Internal air pressure can be supplied, resulting in an air-supported structure, or the fabric can be mechanically pretensioned from a mast or frame to create a tension structure.

Curves are a distinctive feature of fabric architecture because, in both air-supported and tension structures, they are necessary to translate applied loads into tension forces and to ensure the stability of the structure. Tension structures use a double curvature that is *anticlastic,* or in opposite directions. An air-supported structure also has a double curvature, but the curves are *synclastic,* or in the same direction.

Air-supported structures are engineered with an increased interior pressure of 0.3 percent to inflate and posttension the fabric roof. A network of cables across the roof relieves stress in the canopy fabric by transferring wind lift and inflation loads to the ground or walls. A compression ring on the ground or walls resists the anchor loads of the cables. Doors are airtight, and small fans operate continuously to maintain interior air pressure.

Practically speaking, there are no limits to the area an air-supported structure can enclose. The cost per square foot of conventional construction methods increases dramatically as clear-span distance increases. However, with air-supported structures, the cost per square foot might actually decrease.

Fabric in a tension structure is pretensioned into anticlastic shapes. Structural stability results from the double curvature. Applied loads from any direction are resisted by a modified stress field in the fabric canopy. Tension structures are supported by masts, arches, cables, or edge beams. They create a dynamic visual appearance because they allow endless variations of shape and form. Unlike air-supported structures, tension structures can be open to the outside because internal air pressure is not required for stability.

Fabric roofing is not the only material that the roofing contractor will consider in the future. With computers as design and engineering tools, products can be invented and tested more quickly than at any time in history. New products and roofing structures are now in the design and experimental stage, and they will be the cutting edge in the field soon enough. Photovoltaic roofs, which have the ability to produce electricity from the sun, are already available and will become important products as the population doubles and the cost of energy keeps going up.

To keep abreast of the changes in the roofing industry, the roofing contractor should read such publications as *Professional Roofing, Western Roofing, Insulation and Siding,* and *Roofer Magazine.* And, as part of becoming more professional in your business, you might also benefit from joining organizations such as the National Roofing Contractors Association.

This is a very dynamic time in the history of the roofing industry. There are many directions for growth available to the roofing professional—a healthy mix of old tried and true products and services and work in the forefront technologies like fabrics and photovoltaics may be the best route for your firm. The new products and services might just inspire everyone at the firm, as learning keeps us all vital and interested. The door to success is wide open, but the decisions are all yours to make.

Using the CD

The compact disk is found in the sleeve on the inside of the back cover of this book. Simply insert the disk in your CD drive and boot it up by selecting the CD drive icon with your mouse. On most systems this will be drive D. If the drive has trouble reading the disk, call your consultant. Older drives may not work with newer CDs.

If you are new to computers, you will need to be working in a word processing software package. The software must be Microsoft Word 97 or later, or word processing software that is fully capable of importing MS Word 97 files and translating them automatically to work *fully* in your software. If you have any doubts about these instructions, ask your consultant to do this with you the first time in order to make it a "working smart" endeavor.

Once you have determined that you can boot your CD drive and that your software *imports* and *automatically translates* Microsoft Word files, you are set. The magic of computers is at your fingertips. You can open any of the worksheet files found in the book. They become word processing files automatically. You are then free to print them out and hand them out to your staff for issuing at job sites and throughout the streamlining of your firm.

The APPENDIX file opens in exactly the same way and is designed for very simple activity. When you open the drive by clicking on the icon for the CD, you will find a list of files. Each file has the W symbol for Microsoft Word. You will need to click on the APPENDIX file.

The file will display this appendix as it is printed out here. After you have opened the file, save it to your ordinary word processing directory. If this is confusing to you, have your consultant visit and do it with you in order to adapt it to your system.

Go into any of the listings and explore it at your leisure. You can then save the e-mail address in your e-mail software. For example, if America Online is your Internet provider, you will save the address in your AOL address book. Again, if you are a new user, get your consultant to help you wade through this process at first in order to work smart rather than hard—never waste your time playing with computers when you are a roofing professional. That would be like trying to manufacture shingles for one of your jobs—it would affect your cash flow drastically.

Welcome to the Net. You will never regret the money you have spent, and you will never turn back.

Resource Directory

The resource directory is a true twenty-first-century tool for the roofing professional. If you are not online currently, we suggest that you go ahead and do it immediately. The worldwide capabilities are staggering; they will inspire you to move in the direction that the handbook is designed to guide you.

Go ahead and buy a fast, modern computer and use the Net to guide you—the savings in time and its usefulness for research, estimating, and endless other activities will pay for this first investment.

Once you are online, open the CD and read what the vendors say about themselves. All you have to do is click on the sites and you will be right there. For example, look below for roofhelp.com; it's listed under marketing. It is a good starting place to explore the idea of using the Web to your advantage. The resources are simply a beginning; you will undoubtedly add many personal selections to your database of net contacts.

Asphalt Roofing

Roofing: CertainTeed: Related Web Sites
The CertainTeed Roofing Collection consists of 18 different styles of asphalt shingles, including a wide range of multilayered architectural shingles, as well as standard styles in more than 150 colors.
http://www.certainteed.com/pro/roofing/ctroof/websites.html

Elk Premium Roofing: A leader in laminated shingle technology, Elk originated
Elk is The Premium Choice in roofing for homeowners, contractors, architects, and builders across North America....
http://www.elkcorp.com/

Associations and Agencies

Adhesive and Sealant Council
ASC serves as a leading marketplace for the exchange of information about products and services and interfacing among manufacturers, suppliers and others in the industry.

Asphalt Institute
U.S.-based association of international petroleum asphalt producers, manufacturers, and affiliated businesses.
http://www.asphaltinstitute.org/aboutai/about.htm

Asphalt Roofing Manufacturers Association
ARMA's membership includes approximately 90 percent of the nation's manufacturers of bituminous-based roofing.
http://www.asphaltroofing.org/contact.html

Associated Roofing Contractors of the Bay Area Counties
Construction industry trade association of roofing, reroofing, and waterproofing contractors providing information on roofing and waterproofing and contractor referrals.
http://www.arcbac.org

ASTM
Standard testing labs for products.
http://www.astm.org/CONTACT/

The Cedar Shake and Shingle Bureau (CSSB)
The Cedar Shake and Shingle Bureau (CSSB) is a trade association representing member manufacturers of cedar shakes and shingles.
http://www.cedarbureau.org/index.htmhttp://www.cedarbureau.org/index.htm

Chicago Roofing Contractors Association
Roofing contractors (by areas of expertise), suppliers, manufacturers, and consultants in the Chicago area.
http://www.crca.org/index.html

Construction Estimating Institute, Training, Education, Building, Conc...
Construction Estimating Institute—Education and Training for the Construction Industry.
http://www.estimating.org/

Construction Technology Centre Atlantic
This sixty- (60-) hour course is designed to help estimators to maintain a competitive edge by developing the skills required to complete a construction cost estimate through the use of a...
http://ctca.unb.ca/CTCA/document/estimate.html

EPA
Contact the EPA Notice of Use Questions?—EPA's Information Resources Center can offer assistance in finding EPA information that doesn't seem to be on the Web, information about EPA programs, and general environmental information.
http://www.epa.gov/epahome/comments.html

About the EPA Public Access Server
About the EPA Public Access Server. System Status Update. Privacy and Security Notice Retrieval Statistics What's HOT on the EPA server About...
http://www.epa.gov/epahome/about.html

EPA, Clean Air, and Small Business
Where to find information on small business insurance, from your About.com guide. New EPA clean air standards will have a large impact on small business, from your About.com guide.
http://sbinformation.about.com/library/weekly/aa072897.htm?iam=m

Search the U.S. EPA Internet Site
Search the EPA Internet.
http://search.epa.gov/

EPA's Office of Solid Waste
U.S. EPA Office of Solid Waste Home Page
http://www.epa.gov/epaoswer/osw/

Midwest Roofing Contractors Association
The MRCA web site requires the use of a frames-capable web browser. We recommend Microsoft Internet Explorer 3.
http://www.mrca.org/

North/East Roofing Contractors Assn., Inc. Massachusetts Associations
Associations North/East Roofing Contractors Assn., Inc. Quincy, Massachusetts. Tel: 617-472-5590 Fax: 617-479-1478. Click the Links Above. Geographical Areas Serviced: Connecticut, Massachusetts, Rhode...
http://ww1.thebluebook.com/ne/htm/0011672100000.shtml

NRCA, The National Roofing Contractors Association's Web Site
Welcome to the National Roofing Contractors Association's (NRCA's) official Web site. NRCA is an association of roofing, roof deck, and waterproofing contractors; industry-related associate members; and international members worldwide.
http://www.roofonline.org/

OSHA
Extensive information from the Occupational Safety and Health Administration.
http://www.osha-slc.gov/SLTC/asbestos/index.html

Roofing & Sheet Metal Contractors Association of Georgia
RSMCA serves the needs of the roofing industry by providing information on products, installation techniques, government regulations, building and energy codes, and other topics of importance.
http://www.rsmca.org

Roofing Contractors Assoc. of WA
The Roofing Contractors Association of Washington exists for the purpose of advancing the roofing industry in an ethical and professional manner throughout the state of Washington. It is made up of

hundreds of contractors....
http://www.rcaw.com/

Roofing Intelligence Member Associations
Associations affiliated with Roofing Intelligence.
http://www.roofingintelligence.com/assoc.htm

Small Business Loans, SBA Loans, Government Loans, Real Estate
Loans
We provide loans to small business through SBA's 7a program.
http://sba.governmentloans.com/

Small Business Help for the Year 2000
Information and links from the U.S. Small Business Administration.
http://www.sba.gov/y2k

SBA: Ombudsman
This office of the SBA is where any small business person can
engage the ombudsman program, which is aimed to help such persons cope with the enormous bureaucratic mazes and complexities
of doing business.
http://www.sba.gov/regfair/

SBA: Introduction
Small businesses are the backbone of the American economy. They
create two of every three new jobs, produce 39% of the gross
national product, and invent more than half the nation's technological innovation.
http://www.sbaonline.sba.gov/intro.html

SBA homepage
UConn School of Business Administration...
http://www.sba.uconn.edu/

SBA.NET.WEB
SBA.NET.WEB® Internet World Wide Web Consulting A division
of SBA * Consulting® Email: Info@sbanetweb.com Special pricing
just for the users of Who's Who in New York City High-End Residential Construction www.nyresidential.com SBA.NET.WEB® provides: Web...
http://www.sbanetweb.com/residential/

Stone Roofing Association
Giving information on the geology, history, and use of sandstone and limestone roofing in the UK of interest to architects, conservation professionals, and roofers.
http://www.brookes.ac.uk/geology/stoneroof/

Western States Roofing Contractors Association: Serving the Roof...
Western States Roofing Contractors Association [207.171.243.35] promotes roofing industry professionalism by providing leadership to benefit members.
http://www.wsrca.com/

Clay Tile Roofing

International Roofing Products, Inc.
Our products include group 117 glazed clay roofing tile and slate roofing tile.
http://www.hybgroup.com/irpweb/htm/

Renaissance Roofing, Inc. 1-800-699-5695
Visit the link for details.
http://www.claytileroof.com/

Sandtoft Roof Tiles—clay—concrete—slates—cpd—roof tiles
Sandtoft Roof Tiles, Britain's largest supplier of roof tiles. Sandtoft has over 7000 roof tiles and fittings in their range of clay tiles, concrete...
http://www.sandtoft.co.uk/index.html

Computers

SOFTWARE

AccountPro 98 Roofing Estimating Software
The AccountPro 98 Roofing Estimating software module is a powerful estimating software package which will help to standardize your estimating, eliminate the tedious task of extending and summarizing totals, and, in addition, analyze cost and labor units.
http://www.accountpro.com/mod178.htm

AppliCad Australia—home of Roof Wizard, Roof Magician, and Cadair
The home of Roof Wizard, Roof Magician, and Cadair estimating software.

http://www.applicad.com.au/breaking.htm

ARCAT—Section 07550—Modified Bituminous Membrane Roofing
www.ARCAT.com Home/Search Site Click Here! Specifications
CAD Details Building Products New Products Building Product
News Companies Associations ADA Listings Hard/ Software CES
Providers Architectural
http://www.arcat.com/divs/sec/sec07550.cfm

ASR Home Page—Tapered & other roofing software
AUTOMATED SYSTEMS RESEARCH (A Division of Mainland
Research Inc.). Automated Systems Research (ASR) develops and
markets quality software to the construction industry. Our current
focus is software that automates the estimating, design, and draft-
ing of...
http://www.asrsoft.com/

Baarns Publishing: Excel Product—Cost Estimating Form—Microsoft
Office Supe...
Cost Estimating Form. Cost Estimating Made Easy. An updated
template for estimating job costs, based on unit material costs
and/or labor hours. You...
http://archive.baarns.com/excel/products/costest.asp
Bruco Enterprises, Inc.
Offering roof evaluation and management software, as well as consulting services.
http://www.flash.net/~bruco

Computers Roofing
Roofing. Bottom. Quick-est, a builder's estimating package. Stun-
ning training package from CITB on cd-rom. Software for scaffold-
ers. Cartoon...
http://www.build.co.uk/cm_rmy97.htm

Computing Advice from Xpertsite.com
Ask an Expert at XpertSite.com, where users can get advice and
answers from real people for free. We're not a search engine and
we're not a bulletin board service. With over 20,000 experts, we're
the place to go when you have questions.
http://www.xpertsite.com/ShowCategory.asp?category_id=483

Data Trak Maintenance and Roofing Software
Data-Trak's computerized maintenance management systems
(CMMS) enable you to control your maintenance process. Each is
customizable for your needs and available at a price you can afford.
http://www.maintenance-software.com

Download Material Estimating Spreadsheets
Construction estimating software. Lots of Animated Graphics for
use on your web site. Need banners, then stop here and ask us
about our banner swap.
http://www.geocities.com/SiliconValley/Heights/8519/down-
load.html

Estimating Software for Restoration and Cleaning—Point and Click...
The leading edge in property loss estimation software solutions
built specifically for restoration contractors and property adjusters.
Point-and-click simplicity and 32-bit power in a fully customizable
system.
http://www.qirra.com/Simplicity.ht

Excel Spreadsheets in Estimating
Excel Spreadsheets in Estimating. [Follow Ups] [Post Followup]
[Message Board] Posted by `akreeves@hotmail.com' on August 20,
1998 at 12:53:00...
http://www.baanfans.com/baanfans/forum/project/messages/19.html

FastEST, Inc./Microsoft Excel. Support
Microsoft Excel. Support. The FastPIPE system allows you to
download the labor and material takeoff to a Microsoft Excel
spreadsheet for your final bid summary. Our standard spreadsheet
provides support for equipment, fixtures, labor rates, rentals.
http://www.fastest-inc.com/excel.htm

FAST Planner for IT—Press Release
IT project plans and cost estimates are now easier with FAST Planner for IT. This add-on software for MS Excel and MS Project produces Gantt charts, cost estimates, and reports for management.
http://www.rms.net/fast_planner_news.htm

FORUM MESSAGE: Re: roofing estimating software—posted at JLC's
Computer Solutions Forum

Re: roofing estimating software An Interactive Message Posted at JLC's Computer Solutions Forum [Follow Ups] [Post A Follow Up] [JLC's Computer Solutions Forum] Posted by: Michael Leistiko/Builder on 5/30/99: In Reply to: roofing estimating software.
http://www.jlconline.com/forums/computers/messages/513.html

Guide to computer security from 4anything.com
Protect yourself from viruses and hackers with Guardian, McAfee, Norton, IBM Antivirus, RSA Data Security, Secure Computing, and the Electronic Frontier Foundation.
http://www.4computersecurity.com/?%3B019039a

Guide to computer software from 4anything.com
Programs for business, education, and fun from Microsoft, Micro Warehouse, TUCOWS, Egghead, CompUSA, Accolade, Corel, Ambrosia, Best Buy, Cyberian Outpost.
http://www.4computersoftware.com/?%3B019039a

Guide to data storage from 4anything.com
Don't be sorry—protect your data with Safeguard Interactive, Total Recall, eVault, @Backup, Iomega, and JPC Zip Drives. Information is a terrible thing to waste!
http://www.4backup.com/?%3B019039a

Links to all construction cost estimating software downloads on the We...
More than 60 programs in 11 estimating categories.
http://bidshop.org/

Microsoft Word contractor and distributor proposal and quotation software for the contractor in hvac, roofing, electrical, plumbing, painting, or
Construction general contracting 1Q for the Contractor [Previous Page][Contractor Partnering] [Contractor Links] 1Q for the ContractorEquipment/Systems MfrDocument Assembly Special 1Quot™ for Contractor/Distributor Equipment/System Manufacturer. See a...
http://amplifyllc.com/onequot/contractor/index.html

Products
The home of Roof Wizard, Roof Magician, and Cadair estimating software.
http://www.applicad.com.au/products.htm

Project Cost Estimates
TEMPEST COMPANY ESTIMATING CAPABILITIES (Key Highlights) SITEWORK. Site, Building, & Selective Demolition Asbestos Removal Site Preparation, Dewatering & Earthwork Piles & Caissons Railroad...
http://www.tempestcompany.com/est1.htm

Roofing Contractors Software by FunnelVision. Designed for Roofing Contractors
Software systems for roofing contractors to manage profile The Best Software for Roofing Contractors.Funnelvision is designed to allow Roofing Contractors to track every aspect of your daily informati...
http://www.funnelvision.com/WEB_297.HTM

Roofing Software: Tapered, roof estimating, and quotation
Specialty software for the roofing and construction industries. Quotation tracking and customer database, tapered insulation estimating and drawing, and roofing estimating.
http://www.tsnorthland.com/

Roofing Materials Suppliers, Building Products—Roofing Materials
Roofing Materials. The complete source of roofing material suppliers. ARCHITECTS & BUILDERS! Get the software you need at discount prices! CLICK HERE PLYCEM—"NEW" noncombustible, struct...
http://www.affordablewebdesign.com/roofing.htm

Roofing System
Computer hardware, systems, and software computer system software multiplexer pcs disk support safety software sprayer cmms software test kit maintenance software software communication package cmms computer monitor computer enclosure software...
http://www.impomag.com/B_comp/TEMPLATE.HTM

Swiftest
Welcome to Swiftest, the home of the estimating software for roofing contractors in the UK. We provide leading-edge technology for the roofing contractor or estimator to create quick, accurate quotes so that you will have the competitive edge in an expanding market.
http://www.swiftest.com/

UDA Spreadsheet Templates for Estimating Construction Costs
Simple and accurate way to estimate construction costs using
Microsoft Excel or most any spreadsheet application.
http://www.uniteddesign.com/excel_spread.html

HARDWARE

Computers@Home
Setting Up a Home Office Setting Up a Home Office Laser Printers
Multi Function Devices Getting Your Office Online What's New in
Office Suites Productivity Software Managing Information on Your
PC Rev...
http://computersathome.com/sep97/5ii0061101.html

Dell Computers: Easy-to-use site!
Buy computers direct from Dell on this easy-to-use site. One click
gets you to your category. Many other great products and services
on one site. Click here and try us!
http://www.MetalShopper.com/page9.html

Guides for all computing needs from 4anything.com
Find the hardware, software, and shareware you need to fill your
computing needs. From monitors and modems to e-mail services
and web design, all of your computing needs can be filled right here.
http://www.4computing.com/?%3B019039a

Guide to renting a computer from 4anything.com
Try it out before you buy with Bit-By-Bit, Micro Computers, or
CRCA Corp. Plus, product information from IBM, Gateway, Apple,
Dell, Hewlett-Packard, Compaq, NEC.
http://www.4computerrental.com/?%3B019039a

ROOFING SYSTEM
Computer hardware, systems, and software computer system software multiplexer pcs disk support safety software sprayer cmms
software test kit maintenance software software communication
package cmms computer monitor computer enclosure software...
http://www.impomag.com/B_comp/TEMPLATE.HTM

Trecc International Industrial Computers and Network Services
Live Web Cam What's New Industrial Computers Budget Comput-

ers Ace Ver 3.5 System Check Us Out Shop Our Online Store The Trecc Forum World Link System Domain Check Trecc Search Site Stats Visitor Number Recognition Express Embajador Adventure Tours Balboa...
http://www.trecc.com/trecc/index.htm

Dewatering Equipment

Dredge pumps from Piranha Pumps—dredges, submersible pumps, dewatering pumps
Dredge pumps from Piranha Pumps: floating dredges, dredges, floating dredge pumps, submersible pumps, dewatering pumps, sump pumps, floating dredge,...
http://www.piranhapumps.com/

Griffin Dewatering Corp. Texas Contractors Equipment—Pumps-Dewaterin...
Blue Book Ad for Griffin Dewatering Corp. Under Contractors Equipment—Pumps—Dewatering.
http://www.thebluebook.com/fs/htm/0037207011000.shtml

National Manufacturers of Pumps—Air-Rotary Valve
Page contains list of national manufacturers of Pumps—Air-Rotary Valve.
http://www.constructionnet.net/manu search/results/55221.shtml

Wanner Hydra-Cell Pumps
Wanner Hydra-Cell Pumps, Distributed by Texas Pump & Equipment 8489 Jacksboro Hwy., Wichita Falls, TX 76310 1-800-234-1384, FAX 940-322-6661 Contact us at : hydracell@texaspump.com Wanner Hydra-Cell Models Texas Pump also sells other brands of pumps.
http://www.texaspump.com/

Environmentally Responsible Roof Subjects

Big Green Machine, The
Manufactures machines that recycle fiberglass insulation and roofing products.
http://www.empnet.com/BigGreen

EM waves (was ferrocement roofing /Gore-Tex)
[Subject Prev][Subject Next][Subject Index] Re: EM waves (was ferrocement roofing /Gore-Tex) To: Abby Nance <Abby_Nance,@maccomw.org> Subject: Re: EM waves (was ferrocement roofing /Gore-Tex) Fr
http://solstice.crest.org/efficiency/strawbale-list-archive/9702/

Winpower Generators
Winpower Generators for Y2K problems, any electrical power problems. PTO or Power Take Off Generators work from your tractor or other PTO. Our portable power generators set the standard: Winpower Inc.
http://www.winpowerinc.com/

General Information

About RCI-Mercury Roofing Documentation Library
About RCI-Mercury RCI-Mercury provides you with instant, inexpensive access to thousands of documents. It provides fast access to years of roofing knowledge, compiled and organized into a single, searchable library. We update RCI-Mercury every day, to...
http://www.rci-mercury.com/about.htm

E.L. Hilts Home Page
A sincere desire to offer the roofing industry the best full service available has been the foundation of E.L. Hilts & Company's success. Our experience and background in the roofing industry await you.
http://www.elhilts.com/

Home Page
America's Favorite Place on the Internet for Roof Help! Interested in some great products to help with your roofing project? Click on the banner below.
http://www.roofhelp.com/

Roofing, Siding, & Insulation Materials Wholesale—The DIRECTory
Looking for information on Roofing, Siding, & Insulation Materials Wholesale? Try this!
http://www.okdirect.com/Biz/5/5033.html

Generators

Americas Generators
Wholesalers/retailers of new and used electrical generators for home, business, or mobile uses.
http://www.americasgenerators.com

Best Power Generators
Supplies tractor-driven and portable engine-driven electric generators.
http://www.best-power-generators.com/

Bowers Power Systems builds generators—gas and diesel generators for...
Bowers builds generator sets using Lister-Petter, Lombardini, VW Industrial, and Continental Diesel and Gaseous (propane or natural gas) engines.
http://www.bowerspower.com/

Circle H Generators, Inc.
Quiet, mobile generator rentals for the Los Angeles entertainment industry. Our generators are quiet, "crystal sync," and fully portable behind your lighting or grip truck or production pick-up.
http://www.circlehgen.com/

Fischer Panda!
Extremely small, light, and quiet Diesel Gensets. Welcome to Fischer Panda. Here we have information about our products and hope you find it useful.
http://www.fischerpanda.com/

Generac Power Systems generator powers up emergency generators for electrical energy.
A leading manufacturer of standby generators and prime power generator sets for industrial, telecommunication, commercial, small business, mobile, and residential applications.
http://www.generac.com/

Generators
Monster links page for sites about the Alternative Survival Information.

http://www.cairns.net.au/sharefin/Markets/Alt2.htm

Generators Unlimited
Offers rentals of quiet generators to the entertainment and outdoor events industry.
http://generatorsunlimited.com/

House of Generators
Sells propane-, diesel-, and gasoline-fueled generators for home, business, and resort backup.
http://www.powerpony.com/

Master Generators
Portable generators powered by Honda gasoline, tri-fuel, propane, natural gas, and Yanmar diesel-powered air cooled electric generators.
http://www.mastergenerators.com/

Portable Generators from USA Light & Electric
Highest-quality, best-performance portable generators. Lowest prices with on hand stock of 800 units from...
http://www.yamahagenerators.com/

Power Generators
Power generators; we sell what the world desires—power! We buy and sell internationally the very best in power generation *Waukesha *Caterpillar *Detroit *Cummins *Allison *Kato *Ruston *Solar.
http://globalgenerator.com/index.html

PTO Generators—Power Takeoff Generators/Power Take off Generators From SSB
We offer PTO generators / power takeoff generators / power take off generators delivering superior power at lower cost...
http://www.ssbtractor.com/PTO_generators.html

Weber Generators—preferred by the motion picture industry
Leader in producing quiet, efficient, reliable generators. Specializing in generators up to 1500 kW. Renowned in the motion picture industry for dependability and quiet power. Burbank, California.
http://www.webergenerator.com/web13/

Winpower Generators
Winpower Generators for Y2K problems, any electrical power prob-

lems. PTO or Power Take Off Generators work from your tractor or other PTO. Our portable power generators set the standard: Winpower Inc.
http://www.winpowerinc.com/

Young Generators, Inc.
Supplier of home generators, industrial/agricultural generators, cogeneration, transfer switches, microturbines. Elliott Magnetek and Wacker distributor. Rental of silent mobile power to the entertai...
http://www.younggen.com

Gutters and Accessories

Ryan Seamless Gutters
Also offers windows, siding, tools, and other building materials.
http://www.ryangutter.com/

Weather Guard Building Products
For over 25 years, Weather Guard Building Products has been manufacturing top-quality metal gutters and roofing accessories. Our commitment to quality has earned us a reputation for excellence in products and service.
http://www.wgbp.com/

Insulation

Polycoat Systems, Inc.—Roofing, Insulation, Coatings, Waterproofing, Equipment, Accessories
Visit the link for details.
http://www.polycoat.com/framset2.html

Sprayed polyurethane foam roofs last long, do not leak, and save money...
Sprayed polyurethane foam roofs and roofing techniques. Sprayed polyurethane foam insulation. Advantages of sprayed polyurethane foam roofs over built up roofs (BUR). Warranties and experience of contractors. Capabilities, experience, and customer references of Mainland Industrial Coatings.
http://www.spf-roofs.com/

Team Industries, Inc.
Manufacturer of expanded polystyrene foam (EPS), rcontrol building panels and diamond snap foundation forms, roofing materials, along with OEM and packaging.
http://www.teamindustries.com

Marketing

RSI—Roofing, Siding, Insulation, etc. Advanstar Communications Roofing Market
Reaching the Roofing Markets Advanstar's Roofing Group has played a key role in the roofing, siding, and insulation market since the mid-1940s, and our people are considered by most to be the industry experts. Add industry leading magazines, high...
http://www.advanstar.com/markets/roofing.cfm

AEC Resource Guide: A Directory of Architecture, Engineering...
Companies, resources, and organizations serving the Architecture, Engineering, and Construction (AEC) industry in the Pacific Northwest.
http://www.nwbuildnet.com/nwbn/resourceguide.html

Architecture Shopper
Looking for an architectural product or service? Good spot to get some information out about your company.
http://architecture.about.com/library/blq%26a.htm?iam=ma

B & M Steep Roofing Profile
BIX—AEC / Construction Industry Directory of Architects, Engineers, Building Suppliers, Contractors, Developers, Products, Software, Equipment Tools B & M Steep Roofing Category: Services Addres...
http://www.building.org/texis/db/bix_search/+fnWePY85mrmww eR2zw/p

BuildingOnline: 949-496-6648 Search Engine Version 3.0
The building online directory search page visitors since January 1, 1997: Appliances Architects Associations Baths Builders Manufacturers Contractors Flooring Hardware
http://www.buildingonline.com/blsearch.shtml

Build.com Home Page
  Build.Com ...your personal web site directory for building and
  home improvement products and information. Home Improvement
  Web Site Directory. Be sure to check out HomeTalk, Build.com's
  building and home improvement bulletin board/discussion forum.
  http://www.build.com/

DIRECTORY OF BUILDING AND ROOFING BUSINESS OPPORTUNI-
  TIES—A free online report—From 177 Business, Mail Order and
  Personal Reports Software
  DIRECTORY OF BUILDING AND ROOFING BUSINESS OPPORTU-
  NITIES—A free online report—From 177 Business, Mail Order and
  Personal Reports Software. Click here for the Best Mail Order Mail-
  ing Lists available. Write to these companies for more information
  on...
  http://www.infoarea.com/vitamins/small-business-informa-
  tion/58.ht

Holland Roofing, Inc.—Burlington, Ky Profile
  BIX—AEC / Construction Industry Directory of Architects, Engineers,
  Building Suppliers, Contractors, Developers, Products, Software,
  Equipment Tools Holland Roofing, Inc.- Burlington, Ky Click...
  http://www.building.org/texis/db/bix/+0wwrmwxerd1wrmwx-
  Ceuxww/prof

Home Page
  America's Favorite Place on the Internet for Roof Help! Interested
  in some great products to help with your roofing project? Click on
  the banner below.
  http://www.roofhelp.com/

RoofHelp
  Information for homeowners.
  http://www.roofhelp.com

roofing.com
  roofing.com, serving the roofing and construction industry through
  the Internet. [submit your site] [drop us a line] [advertise here] Consultants...
  http://www.roofing.com/

Roofing Consultants Online
An Internet resource giving you helpful tips to identify problems
with your slate and tile roofs and other types of roofing problems.
http://www.iroofer.com/

Roofing Resources, Guides, and Do It Yourself Tips and Information at
the Remodel Online Home ...
Roofing Directories at the Remodel Online Home Improvement and
Do it Yourself Center
http://www.remodelonline.com/directories/construction/roof-
ing/gen

Roofing WebRing Homepage
Welcome to the Roofing WebRing Home Page! This ring, started in
January of 1999, was developed to link and showcase roofing contractors, suppliers and manufacturers on the web on a virtual
online tour. By using the ring, you can travel from one roofing...
http://www.latch.com/webring.html

RSI—Roofing, Siding, Insulation, etc. Advanstar Communications
Roofing Market
Reaching the Roofing Markets Advanstar's Roofing Group has
played a key role in the roofing, siding, and insulation market since
the mid-1940s, and our people are considered by most to be the
industry experts. Add industry leading magazines, high...
http://www.advanstar.com/markets/roofing.cfm

S & K Roofing Profile
BIX—AEC/Construction Industry Directory of Architects, Engineers,
Building Suppliers, Contractors, Developers, Products, Software,
Equipment Tools S & K Roofing Click Here to Find Out More (Ad...
http://www.building.org/texis/db/bix/+0wwrmwxerA1wrmwx-
exhMw/prof

Magazines

Construction Magazines Online
Architecture.
http://www.geocities.com/Paris/Louvre/1776/magazines_con-
struction

Iroofer.com, roofing services, slate and tile roofs, slate and tile roofing
Home slate roofs tile roofs asphalt roofs contact Telephone: (630)
257-9394 e-mail:iroofer@aol.com To learn about roofing and how to
spot potential problems, click on your style roof: Slate Roof Tile
Roof Spanish Tile Roof Tile Roofs Tile roofing comes in...
http://www.iroofer.com/tile.htm

Professional Roofing Magazine, National Roofing Contractors Association
An association of roofing, roof deck, and waterproofing contractors;
industry-related associate members; and international members
worldwide.
http://www.professionalroofing.net/past/may99/

Roofing Contractor Online
National magazine that focuses solely on the roofing contractor.
http://www.roofingcontractor.com/

Roofing
Roofing magazine. Roofing was founded by the National Federation
of Roofing Contractors. TM To visit the NFRC site, click the above
logo. The NFRC is based at 24 Weymouth Street, London W1N 4LX
Phone +44 (0) 171 436 0387 Fax +44 (0) 171 637 5215 May 1998.
http://www.build.co.uk/roofing.htm

Roofing Contractor Online
Roofing Contractor magazine is edited to help the roofing contractor
succeed. We are dedicated to achieving this mission through concise,
aggressive coverage of the commercial and residential industries.
http://www.roofingcontractor.com/

Steel Roofing Web Site—Newswire
Home and Building Owners: Where to Buy Galvalume® Steel Roofing. To Our Guests: We'd like to learn more about the current and
future roofing needs of home and building owners like you. Please
take just a few moments to answer the following brief questions...
http://www.steelroofing.com/where.htm

Western Roofing

Insulation and Siding

Roofer Magazine

Metal Roofing

Advanced Metal Roof Systems—Main
Advanced Metal Roof Systems, the best source for metal roofing.
http://www.amrs.w1.com/

AEP-SPAN Metal Roofing Solutions (Excite)
The leading manufacturer of architectural metal products utilizing state-of-the-art technology to set industry standards for roof systems, equipment screens, fascia, and soffit systems.
http://www.aep-span.com/

Copper Page
A service of the worldwide copper and brass industries.
http://www.copper.org/

Custom-Bilt Metals: Metal roofing, equipment, gutters, tools
Supplier of metal roofing, standing seam, steel roofing, aluminum roofing, copper roofing, roofing tools, roofing equipment, gutter systems, flashings, copings, tools, and training.
http://www.custombiltmetals.com/

Lianro Metal Roofs, Inc. specializes in commercial and reside...
Lianro Metal Roofs specializes in commercial and residential steel roofing. Steel roofing, especially stone-coated steel roofing by Decra, looks more like tile, shakes, or slate roofs. Steel and stone roofing makes a beautiful alternative roofing system for your new home, reroofing needs, home improvement plan, or your remodeling project.
http://www.lianro.com/

Metal Roofing Tools and Equipment (Infoseek)
ROOFING ACCESSORIES We carry a full line of metal working tools and accessories for each product line we carry, including clips, screws, closure strips, and Dektite flashings for all of your ...
http://www.custombiltmetals.com/tools.htm

Pioneer Roofing Systems
Pioneer Roofing Systems...Call us @ 1-800-646-3601 for prompt quotes and service. We specialize in commercial roofing and metal.
http://www.pioneerroofing.com/

Recycling

Authentic Roof

Authentic Roof™ is the world's first true environmentally responsible building/roofing material for the construction industry. Made entirely from 100% Recycled Material and is also completely recyclable. Specifically...
http://www.authentic-roof.com/

Asphalt and Tar Recycling—Roofing Tar Chips

Click Here to Goto the Table of Contents Recycler's World [Main Menu I Asphalt and Tar Recycling I On-Line Market Prices] GARM: Recycler's World provides Global Access to Recycling Markets Grades &...
http://www.recycle.net/recycle/spec/gr080215.html

Asphalt and Tar Recycling—Roofing Shingle Scrap

Recycler's World [Main Menu I Asphalt and Tar Recycling I On-Line Market Prices] Grades & Specs.: Roofing Shingle Scrap. Roofing Shingle Scrap....
http://www.recycle.net/recycle/spec/gr080210.html

Big Green Machine, The (Yahoo)

Manufactures machines that recycle fiberglass insulation and roofing products.
http://www.empnet.com/BigGreen

Specification Chemicals, Inc

Q: What is Recycled Rubber Roofing? A: Recycled Rubber Roofing is a liquid emulsion formulated with the use of ground rubber from recycled tires. The cured product produces a monolithic elastomeric membrane. This...
http://www.spec-chem.com/rrr/

Roofing Fabric

Single-Ply Roofing, Membrane Roofing, Coated Industrial Fabric, and Fa...

Bondcote produces coated industrial fabrics and roofing systems for your toughest challenges.
http://www.bondcote.com/

Roofing Systems

GenFlex Roofing Systems
  GenFlex Training Programs. Find the location and date of the Gen-
  Flex Authorized Contractors Training Programs in your area. New
  Products &...
  http://www.genflex.com/

Interlocking Roofing Tiles
  GCI Eagle Interlocking Roofing Tiles. GCI is the *first* clay roofing
  tile plant in Asia which utilizes the latest Franco-German machinery and technology, inclusive of the patented, computerized
  Hydrocasing tunnel kiln firing system and horizontal tiles firing.
  http://www.goldenclay.com/rooftiles.html

Johns Manville Commercial & Industrial Roofing Systems
  Manufactures and markets insulation products for buildings and
  equipment, commercial and industrial roofing systems, high-effi-
  ciency filtration media and fiber and nonwoven mats.
  http://www.jm.com/roofing/index.html

Pioneer Roofing Systems
  Pioneer Roofing Systems...Call us @ 1-800-646-3601 for prompt
  quotes and service. We specialize in commercial roofing and metal.
  http://www.pioneerroofing.com/

Polycoat Systems, Inc.—Roofing, Insulation, Coatings, Waterproofing,
  Equipment, Accessories (FindWhat)
  As the nationwide marketer of the DOW CORNING
  Silicone/Polyurethane Foam Roof System, Polycoat Systems, Inc., is
  a full-service supplier specializing in the sale of moisture and thermal protection products for the building envelope, including...
  http://www.polycoat.com/

Product: roofing system
  Roofing system literature highlights properties of a Dow Corning®
  roofing system for both new construction and renovation. The protective system resists sunlight, heat, cold, and numerous chemicals
  to form a leak-proof seal. Polycoat Systems, Inc.
  http://www.impomag.com/A_build/0698A010.HTM

Single-Ply Roofing, Membrane Roofing, Coated Industrial Fabric, and Fa...
Bondcote produces coated industrial fabrics and roofing systems for your toughest challenges.
http://www.bondcote.com/

Roofing Supplies, Miscellaneous

Heely-Brown Roofing Equipment & Supplies
Heely-Brown Leister Hot Air Equipment Powered Hand Tools Tool Boxes Hoisting Specialty Fasteners E-Mail Heely-Brown About Heely-Brown.
http://heely-brown.com/hbroof1.html

Reeves Roofing Equipment Main Page
P.O. Box 720 Helotes, Texas 78023-0720, 800-383-3063. Reeves Roofing Equipment Co., Inc., was started in June of 1965 by Joe S. Reeves Sr. "Jody" began working for Blackwell Burner Co., in San Antonio, Texas, in 1937 at the age of 13. His brother Preston...
http://www.reevesequipment.com/main.htm

Robseal Roofing—Case Study
PITCHED ROOF FAILURES SOLVED WITH FLAT ROOFING TECH-NOLOGY. Since the publication of the BRE Construction Quality Forum report into industrial roof failures highlighted that there are more than twice as many failures of pitched roofs than with flat roofing
http://www.robseal.com/case1.htm

Roof Center, The
Distributor of residential & commercial roofing, siding, windows, doors, accessories, and tools.
http://www.roofcenter.com

Roofers Mart
Directory of commercial and residential roofing products and equipment for the professional contractor.
http://www.roofersmart.com/

roofing.com
roofing.com, serving the roofing and construction industry through the Internet. [submit your site] [drop us a line] [advertise here] Consultants...
http://www.roofing.com/

roofing material
Housemart.com—The very first online home improvement store; 35,000 products; lowest prices available anywhere in the world. roofing material...
http://www.roofingmaterial.webscour.net/

Roofing Materials and Hot Rubber
Roofing Roofing Materials | Contractors and waterproofing If you are searching for any of the following topics: roofing roofing contractor kettles waterproofing look no further. You'll find it at...
http://kettles.web-scape.com/roofing.html

Roofing Siding & Insulation Materials Wholesale—The DIRECTory
Looking for information on Roofing, Siding, & Insulation Materials Wholesale? Try this!
http://www.okdirect.com/Biz/5/5033.html

Roofing Wholesale Co., Inc.
Irish flax, cotton fabric, glass fabric, peel and seal, peal and seal, peel & seal, peal & seal, spray primer, single ply, gravel spreader, roofing tools, roofing equipment, roofing material, high boy, copper nails, nails, stainless steel nails, stainless nails, monel nails, lift conveyor, mastic sprayer, glass chopper gun, roof master, red dragon, dragon wagon, felt layer, midget mopper...
http://www.rwc.org/index.html

Roofing wholesale co., inc.
Distribute roofing, roofing equipment, tools, and supplies.
http://www.biw.co.uk/BIW/register/2510.htm

Roofing Siding & Insulation Materials Wholesale—The DIRECTory
Looking for information on Roofing, Siding, & Insulation Materials Wholesale? Try this!
http://www.okdirect.com/Biz/5/5033.html

Shop CRS: Your On-Line Roofing Distributor: Specializing in Tools, Drains, Equipment, Hardware, Accessories
Everything for roofs and roofing contractors, commercial and residential. Items include roofing materials, tools, and equipment.The site includes links to manufacturers of roofing supplies and equipment.
http://www.shopcrs.com

Stoneway Roofing Supply
Supplier of roofing products, including asphalt composition, shakes, metal roofing, tools, and accessories.
http://www.stonewayroofing.com

Structural Materials has been providing quality roofing materials to the Southern...
] Click here for truck for sale Structural Materials has been providing quality roofing materials to the Southern California...
http://www.roofs.net/smc/index.htm

Safety

Ace Fire Protection
Residential, commercial, and industrial fire protection products, including fire extinguishers, restaurant fire suppression, and first aid sales and service. Ansul, Pyro Chem, and Afassco distributor.
http://www.acefireprotection.com

Able First Aid
Industrial and consumer kits available.
http://www.ableaid.com

Adventure Medical Kits—Innovative leaders in Emergency First Aid
Adventure Medical kits are innovative emergency first aid kits for practicing wilderness medicine. Our lines of first aid kits are designed by Dr. Eric Weiss for outdoor recreation.
http://www.adventuremedicalkits.com/home/home.htm

American Red Cross—October Health and Safety Tips
Red Cross Halloween Safety Tips for Kids and Adults. With witches, goblins, and superheroes descending on neighborhoods across America, the American Red Cross offers parents some safety tips to help prepare their children for a safe and enjoyable trick-or-

treat holiday.
http://www.redcross.org/tips/october/octtips.html

Ekman Safety: Fall protection equipment for the highest safety
Welcome! You are visitor: since 26-11-1998. This is Ekman Safety, leading manufacturer of personal fall protection equipment, we proudly like to invite everyone who have a need to protect themselves from falling from a height to study all our products. We...
http://www.ekmansafety.com/index.html

Equipment safety
You'll find all the supplies you need to get the job done with resources like Caterpillar, Metal Forms, Sylvania Lighting, and many others on 4buildingsupplies.com.
http://www.4buildingsupplies.com/?%3B019039a

Fire Safety and Fire Extinguishers
Things you need to know about using fire extinguishers in a chemistry laboratory...
http://www.chem.uky.edu/resources/firefighting.html

First Aid Direct—First aid kits for home and industry
First Aid Direct—First aid kits and supplies for commerce, industry, sports, and home use.
http://www.firstaid-direct.co.uk/

First Aid Kits
A good selection of top quality first aid kits. Secure on-line ordering available.
http://www.healthproductswh.com/hpw/firstaidkits.html

First Aid Kits—Johnson & Johnson First Aid Kits
First Aid Kits—Johnson & Johnson First Aid Kits...
http://www.easyshopn.com/first_aid_kits.htm

First aid kits
Discover why we were voted one of the top first aid kits sites on the web. CLICK HERE NOW.
http://www.firstaidkits.lookandgetit.com/

GovSearch: Government Agencies, Departments, & Resource Searc...
Use these fast and subject-specific government search engines to find information about the government, IRS documents, US Law

Code and OSHA regulations and rulings.
http://www.nwbuildnet.com/nwbn/govbot.html

Halon 1211 Fire Extinguishers
Reliable Fire Equipment Company provides options for dealing with ozone depleting halon 1211 fire extinguishers....
http://www.reliablefire.com/portablesfolder/halon_1211_i.html

Korkers—Safety Sandals For Wood Shakes, Shingles, and Underlayment
WET, FROSTY, OILY, STEEP MOSSY SHAKES, SHINGLES, UNDERLAYMENT * Increase Productivity. * Increase Employee Safety. * Increase Employee Mobility. * Increase Available Work Time. * Decrease Employee Stress and Fatigue. * Decrease Accidents and Injuries
http://www.dumarbusiness.com/korkers.htm

North East Fire and Safety Equipment Co.
Full-service company for portable fire extinguishers, fire hoses, emergency lighting, fire alarm systems, sprinkler systems, and fire suppression systems.
http://www.northeastfireinc.com

OSHA Safety Training
OSHA-approved safety and industrial training, DOT and EPA approved. Address in Dallas, Texas.
http://www.burtonandassociates.net

Portable Fire Extinguishers
Important points of using a portable fire extinguisher.
http://www.vbg.org/FIRE/Fire-ext.htm

Portable Generators: How to Use Them Safely—Date: 12/28/99
CPSC issues these tips for safe use of portable electrical generators, from your About.com Guide.
http://usgovinfo.about.com/library/news/aa122899a.htm?iam=ma&term

pdxfire]
Web site by TAS Mania Graphics Fire Extinguisher Sales & Service with 12 years of experience! Bookmark for easy reference: www.PDXFire.com. Our mission is to provide superior fire extinguisher service at an affordable price throughout the greater Port-

land and Salem metro areas.
http://www.pdxfire.com/

Pyrotec Fire Protection
Provides safety signs, fire training courses, servicing of fire extinguishers, and fire protection equipment.
http://www.city2000.com/md/pyrotec.html

Quips—Safety signs required for roof anchorages for portable suspension equipment
OSHA Interpretive Quips Safety signs required for roof anchorages for portable suspension equipment. Title: Safety signs required for roof anchorages for portable suspension equipment Standard Number: 1910.145(c)(3) Interpretation Source Description.
http://198.252.9.95/oshaweb/IQ_data/IN3D4B~4.HTM

Safety Tips: Portable Fire Extinguishers
Safety Tips: Portable Fire Extinguishers. This is a brief overview of the important points of using a portable fire extinguisher. Fire can be devastating, but when used properly, a fire extinguisher can save lives and property.
http://www.vbg.org/FIRE/fire-ext.htm

Wheeled Fire Extinguishers
Reliable Fire Equipment Company provides on-site sale and service of hand-portable fire extinguishers in the greater Chicago Metropolitan area....
http://www.reliablefire.com/portablesfolder/wheeledunits.html

Willis Re-Roofing Safety
Willis Roofing is very proud that our low accident rate has earned us special status with the SIIS, and lower rates, too! Workers In Harness On All Roofs. WILLIS ROOFING SAFETY STATEMENT: The personal health and...
http://www.willisroofing.com/safety.htm

Slate

American Slate Company
American slate, slate roofing, slate tiles...
http://www.americanslate.com/

Aspigal Roofing Slate
ASPIGAL—Roofing slate producer and supplier to the whole of Europe.
http://www.aspigal.com/ie/index.asp

Black Diamond Slate Company
Black and green slate roofs for your home.
http://blackdiamondslate.com

Buckingham-Virginia Slate Roofing
ROOFING CHARACTERISTICS OF BUCKINGHAM® SLATE. This famous roofing slate is quarried from natural stone of exceptional enduring qualities and has proven its permanent nature by continuous use on roofs for over 150 years without signs of fading or decay.
http://www.bvslate.com/roof.htm

Echeguren Slate, Inc.
Importer and distributor of roofing and flooring slate.
http://www.echeguren.com

EJ Roberts Roofing—Slate
EJ Roberts Roofing—Slate Home Page | Slate | Tile | Lead | Asphalt | Felt | Guttering | News Londons Smithfield Market Safeway Palmers Green http://www.e-j-roberts-roofing.co.uk Web Pages by Trah Computer Services Limited.
http://www.e-j-roberts-roofing.co.uk/slate.htm

Evergreen Slate Company
Providing roofing slate.
http://www.evergreenslate.com

Greenstone Slate
Greenstone Slate Company is a 45-year-old family-owned and operated manufacturer of quality Grade S-1 architectural roofing slate. We offer all the Vermont colors, along with a new unfading black to enhance our ability to provide you with the finest roofing material available today.
http://www.greenstoneslate.com/

International Roofing Products, Inc.
Our products include group 117 glazed clay roofing tile and slate roofing tile.

http://www.hybgroup.com/irpweb/htm/

Renaissance Roofing, Inc., 1-800-699-5695
Visit the link for details.
http://www.claytileroof.com/

The Slate Book—How to Design, Specify, Install, and Repair a Slate
Roof
The Slate Book, how to design, specify, install, and repair a slate
roof.
http://www.oldworlddistributors.com/pix_slatebook.html

Slatecraft—Handcrafted switchplates and outlet covers
Switchplates and outlet covers handcrafted from century-old antique
barn slate....
http://www.slatecraft.com/

SRB
The first book to be written on the subject of slate roofs since 1926.
Thirty years of slate roofing, five years of meticulous research, and
thousands of miles travelled (and slate quarries), produced the
Slate Roof Bible....
http://www.jenkinspublishing.com/slate.html

Slate Roofing
Llechwedd Slate Quarries Greaves Portmadoc Slate Greaves has
been mining and quarrying natural slate at Llechwedd Slate Mines
for over 150 years. Each slate is individually produced by a skilled
craftsman with the experience and expertise of a previous...
http://llechwedd.co.uk/nfslate.htm

Stoneside Slate Roofing Tile Application
Photo of Stoneside Slate roofing tiles.
http://www.stoneside.com/samp11.htm

Universal Slate International Inc.
Supplying slate for flooring, roofing, cladding, signs, and landscap-
ing.
http://www.universalslate.com

Virginia Slate
Offers slate products, including roofing, flagstone, and crushed rock.

http://www.virginiaslate.com

Vermont Slate Sales, Inc., Homepage
Vermont Slate Sales, Inc. Vermont roofing and flooring slate is still quarried mainly by hand and split by hand as it has been for hundreds of years.
http://www.sover.net/~slate/index.html

Snow Guards

Real-Tool, Materials, Finishes
Year Round Application RT and AP Snow Guards Require no Adhesive or Solder Can Be Installed in Any Season Materials and Finishes All Real-Tool® Snow Guards are available in Certified ALMAG 35...
http://bergerbros.com/real-tool/Materials.html

Removing Snow
Removing Snow & Ice From Roofs. The Better Business Bureau has received a high volume of calls this winter regarding water damage caused by ice dams and snow on area roofs. There may be several reasons...
http://www.buffalo.bbb.org/alerts/snowremoval.html

Tools

HAND TOOLS

E.L. Hilts Home Page
A sincere desire to offer the roofing industry the best full service available has been the foundation of E.L. Hilts & Company's success. Our experience and background in the roofing industry await you.
http://www.elhilts.com/

1999 Tool & Equipment Guide
E.L. Hilts & Co. Hickory, North Carolina. Hand tools: brushes, slate cutters, snips knives/blades, shears, trowels, etc. Power support tools: generators, gage box. Heat Welding: hand held,...
http://www.roofingcontractor.com/tool-equipment_guide.htm

AJC Tools and Equipment
Manufactures roofing tools and equipment.
http://www.ajctools.com

H & H Tool Co.

Makers of hand tools, including custom, for the sheet metal and roofing trades.

http://www.hamlettools.com

HVAC, Pumps, Roofing, Air Quality, CAD/CAM, Pumps, Packaging, Protective Coatings, Waste...

PRODUCTS & SERVICES PLANT ENGINEERING Building & Grounds Maintenance Communication Equipment & Systems Compressors & Air Dryers Computerized...

http://www.proshows.com/attendee/exhp&s.htm

Leister Roofing Hand Tools

Leister hot air tools, World-wide Leader in Hot Air Technology! ISO 9001 Certified CLICK AND GO! Leister Roofing Hand Tools—Examples—Shrinking Drying Plastic Welding Civil Engineering—Product...

http://www.hotairtools.com/Roofing.htm

Metal Roofing Tools and Equipment

ROOFING ACCESSORIES. We carry a full line of metal working tools and accessories for each product line we carry, including clips, screws, closure strips, and Dektite flashings for all of your...

http://www.custombiltmetals.com/tools.htm

Roofing Tools

RM Virtual Store.........001-ASG27M. Adjustable gauge, 19 oz. Measures Metric. 20.69. 002-SG100. Stationary gauge, 18oz. 20.69. 001-ASG27S....

http://roofersmart.com/rmstore/ajc.htm

Roofstripper Roofing Tool

Roofstripping tool.

http://www.sowashco.com/roofstripper/index.shtml

Stoney Cedar—Roofing Tools

Tools used in the Roofing Trade.

http://www.renweb.net/stoneycedar/tools01.htm

Roofing Tools and Roof Products by THOR

Roofing Tools, Roof Shingle Layout Tapes, Roofers Double Chalker,

Magichalk, Permachalk by THOR.
http://www.thorsystems.com/

Roofing Tools—Roofer's Edge Shingle Placement Tool by Rocket Equipment
Rocket Equipment, roofing tools and supplies, features the Roofer's Edge Shingle Placement Tool for shingles, shakes, slate, and tile by professional roofers, builders, contractors, and homeowners.
http://www.rocketequipment.com/

Roofing with Senco Fasteners and Tools
Roofing with SENCO Fasteners, Tools. This is one in a series of articles written by our sales representatives (Steve Berry) on SENCO tools and fasteners they recommend for...
http://www.senco.com/tech_tips/nailsvsstaplesroofing.html

Rusko
Manufactures a roofing tool designed to tear shingles, pry nails, and remove old tile.
http://www.rusko.com/

TENT, TARP AND ROOFING EQUIPMENT
Send us an e-mail.
http://www.wegener-na.com/tent.htm

Werner Ladder Co.
Manufacturers of fiberglass, aluminum, and wood climbing equipment, extension and attic ladders, scaffolding, step stools, roofing and painting contractor tools.
http://www.WernerLadder.com

HEAVY ROOFING TOOLS

Acro Building Systems, Inc.
Supplier of construction roofing industry tools and equipment.
http://www.acrobuildingsystems.com/

Cleasby Roofing Products
Cleasby Roofing Products—Manufacturing the finest quality roofing equipment and products for the professional roofing contractor.
http://www.cleasby.com/

Commercial roofing equipment
Commercial roofing equipment commercial roofing equipment
Click Here! Click Here for More Info Interested in commercial roofing equipment? You've come to the right place. We're the ultimate
resource f...
http://garlockequipc.hlkj.com/commercial_roofing_equipmentlbd/

Commercial Roofing Specialties, Inc.
Everything for roofs and roofing contractors, commercial and residential. Items include roofing materials, tools, and equipment. The site
includes links to manufacturers of roofing supplies and equipment.
http://www.crsroofingsupply.com/

Garlock Equipment Company—Roofing Equipment
Garlock Equipment Company.
http://www.garlockequip.com/

Heely-Brown Roofing Equipment & Supplies
Visit the link for details.
http://www.heely-brown.com/hbroof1.html

Hoisting and On-Deck Roofing Equipment
From the light-duty single ladder hoist to the Hydraulic 2000# R&G
Swing Hoist, Heely-Brown has a hoist for your roofing needs.
Heely-Brown is a stocking distributor of Cleasby Kettles and On
Deck.
http://heely-brown.com/hoist.html

New Tech Machinery Corp—Roofing equipment
Page contains data on New Tech Machinery Corp, located in Denver, CO. They offer the following services: New Tech Machinery
Corp Roofing equipment...
http://search.commerceinc.com/profile/id3108027136.html

==PANTHER ® NO FRAMES==
Panther Products—Roofing Equipment Online Catalog.
http://www.toto.net/panther/nonframe.html

Phoenix Sales, Inc.—Roofing Equipment and Materials—800-766-
9104
Equipment I Roof Systems I Products at a Glance I Areas and Services I Contact Us Phoenix Sales, Inc., founded in 1990 by Scott

Hall and Paul Schauer, provide quality roofing products to the contractor while helping owners and specifiers solve roofing
http://www.phoenixsalesinc.com/about.htm

Reeves Roofing Equipment Inc. Distributors
P.O. Box 720, Helotes, Texas 78023-0720, 800-383-3063 Here are a list of Reeves Roofing Equipment distributors in your area | ALABAMA | ARKANSAS | CALIFORNIA | FLORIDA | | GEORGIA | INDIANA | LOUISIANA | MICHIGAN | | MISSISSIPPI | MISSOURI | NEW MEXICO | NORTH...
http://www.reevesequipment.com/distributors.htm

roofing equipment
BT site equipment FT materiel de couverture NT tar boilers [Previous Item] [Next Item] [Search] [Help]
http://www.nrc.ca/irc/thesaurus/roofing_equipment.html

roofing equipment
Roofing equipment roofing equipment. Dedicated to bringing you the best on roofing equipment. Whether you currently know a lot about or are looking for information on roofing equipment, see what we have.
http://garlockequip.hlkj.com/roofing_equipmentlba/

Roofing Equipment
Extra long life carbide blades for roofing equipment, such as roof cutters and high speed saws. D'AX is manufactured by Diamond Systems Inc. of Ontario, Canada.
http://www.dsidiamond.com/roofing.htm

Roofing Equipment
Roofing Tools and Accessories, The Leister Variant. A hot-air automatic welding machine for heat seaming of all single-ply and modified bitumen roofing materials. The Leister Universal 6K The Leister...
http://columbineint.com/prod_4.htm

Roofing Equipment
Model LH200 Material Hoist Model LH2000 Material Hoist Model LH400 Material Hoist Material Hoist Accessories Roof Brackets 5901 Wheaton Drive Atlanta, GA 30336. Toll Free 800-241-1806,

Office 404-344-0000, Fax 404-349-0401 info@tiedown.com.
http://www.tiedown.com/roofing.htm

Roofing Equipment & Supplies
Chicago Construction Network, Construction Guide containing 970
construction related categories....
http://www.chicago-construction.com/search/r-htms/roofsup.htm

Roofing Machinery for Residential and Commercial Applications
Visit the link for details.
http://www.rollformers.com/pages/frameset.htm

Roofstripper Roofing Tool
Roofstripping tool.
http://www.sowashco.com/roofstripper/index.shtml

Rentals and Leasing

Leasing options
Explore the alternatives to owning equipment with Keystone Leasing, Anthony Crane, Fleet Services, Computer Rents, American Capital Leasing and others.
http://www.4equipmentleasing.com/?%3B019039a

Roofing Equipment
WIRTZ RENTALS CO. 773-523-1901 Chicago Location 708-594-9292 Summit Location ROOFING EQUIPMENT Roofing Equipment Daily Weekly Monthly Asphalt Kettle, 115 gal cap/ kero 100.00 n/a n/a Roofers Hoist wel
http://www.wirtzrentals.com/roofing.htm

Roofing Equipment
Roofing Equipment. Daily Rates. 42' Conveyor $130. 52' Conveyor $175. 77' Conveyor $250. Roofing Saw $65. Spud Machine $65. 130 gallon Propane Kettle...
http://www.fastrentals.com/roofing.html

Wood Shakes and Shingles

LifePine Wood Shakes and Shingles
LifePine Wood Shakes and Shingles. Tamark Manufacturing, LLC, manufactures these fine wood shakes and shingles from yellow

pine, then pressure treats these quality products to ensure many years of use. For detailed information on the product, as...
http://www.olesmoky.com/p_lifepn.htm

Steadfast Cedar Shakes and Shingles
For roofs, siding, and paneling.
http://www.steadfast.com/

Wesco Cedar, Inc.—Cedar Shakes & Shingles—FTX & Osmose CCA
EVALUATION REPORT ER-5404. Copyright ®1997 ICBO Evaluation Service, Inc. Issued October 1, 1997. Filling Category: ROOF COVERING AND ROOF DECK...
http://www.wescocedar.com/evaluations.html

Metric Conversion Guide for the Roofer

Property	To convert from	Symbol
Application rate	U.S. gallon per square	gal (U.S.)/100 ft²
	U.K. gallon per square	gal (U.K.)/100 ft²
Area	square inch	in.²
	square foot	ft²
	square	100 ft²
Breaking strength	pound force per inch width	lbf/in.
Coverage	square foot per U.S. gallon	ft²/gal
	square foot per U.K. gallon	ft²/gal
Density, or mass per unit volume	pound per cubic foot	lb/ft³
Energy or work	kilowatt-hour	kWh
	British thermal unit	Btu
Flow, or volume per unit time	U.S. gallon per minute	gpm
	U.K. gallon per minute	gpm
Force	pound force	lbf
	kilogram force	kgf
Heat flow	thermal conductance, C	Btu/h·ft²·°F
	thermal conductivity, k	Btu·in./h·ft²·°F
Incline	inch per foot	in./ft
Length, width, thickness	mil	0.001 in.
	inch (up to ~48 in.)	in.
	foot (~4 ft and above)	ft

*Exact conversion factor.

to	Symbol	Multiply by	Remarks
liter per square meter	L/m²	0.4075	= 0.4075 mm thick
liter per square meter	L/m²	0.4893	= 0.4893 mm thick
square millimeter	mm²	645.2	1,000,000 mm² = 1 m²
square meter	m²	0.09290	
square meter	m²	9.290	
kilonewton per meter width	kN/m	0.175	
square meter per liter	m²/L	0.02454	
square meter per liter	m²/L	0.02044	
kilogram per cubic meter	kg/m³	16.02	water = 1000 kg/m³
megajoule	MJ	3.600*	J = W·s = N·m
joule	J	1055	
cubic centimeter per second	cm³/s	63.09	or 0.0631 L/s
cubic centimeter per second	cm³/s	75.77	or 0.0758 L/s
newton	N	4.4448	N = kg·m/s²
newton	N	9.807	
watt per square meter kelvin	W/m²·K	5.678	
watt per meter kelvin	W/m·K	0.1442	
percent	%	8.333	3 in./ft = 25%
micrometer	μm	25.40*	1000 μm = 1 mm
millimeter	mm	25.40*	1000 mm = 1 m
meter	m	0.3048*	

Property	To convert from	Symbol
Mass (weight)	ounce	oz
	pound	lb
	short ton	2000 lb
Mass per unit area	pound per square foot	lb/ft^2
	pound per square foot	lb/ft^2
	pound per square foot	lb/1000 ft^2
	ounce per square yard	oz/yd^2
Permeability at 23°C	perm inch	grain·in./ft^2·h·in. Hg
Permeance at 23°C	perm	grain/ft^2·h·in. Hg
Power	horsepower	hp
Pressure or stress	pound force per square inch	lbf/in.^{2} or psi
	pound force per square foot	lbf/ft^2 or psf
Temperature	degree Fahrenheit	°F
	degree Celsius	°C
Thread count (fabric)	threads per inch width	threads/in.
Velocity (speed)	foot per minute	ft/min or fpm
	mile per hour	m/h or mph
Volume	U.S. gallon	gal (U.S.)
	U.K. gallon	gal (U.K.)
	cubic foot	ft^3
	cubic yard	yd^3

*Exact conversion factor.

to	Symbol	Multiply by	Remarks
gram	g	28.35	1000 g = 1 kg
kilogram	kg	0.4536	1000 kg = 1 Mg
megagram	Mg	0.9072	
kilogram per square meter	kg/m^2	4.882	
gram per square meter	g/m^2	4882	
gram per square meter	g/m^2	48.82	
gram per square meter	g/m^2	33.91	
nanogram/pascal second square meter	$ng/Pa{\cdot}s{\cdot}m^2$	1.459	ng = 10^{12} kg
nanogram/pascal second meter	$ng/Pa{\cdot}s{\cdot}m$	57.45	1 grain = 64 mg
watt	W	746	W = N·m/s = J/s
kilopascal	kPa	6.895	Pa = N/m^2
pascal	Pa	47.88	
degree Celsius	°C	$(t_F - 32)/1.8*$	32°F = 0°C
kelvin	K	$t_c + 273.15*$	273.15 K = 0°C
threads per centimeter width	threads/cm	0.394	
meter per second	m/s	0.005080*	
kilometer per hour	km/h	1.609	
cubic meter or 3.785 L	m^3	0.003785	
cubic meter or 4.546 L	m^3	0.004546	
cubic meter	m^3	0.02832	
cubic meter	m^3	0.7646	

INDEX

CD-ROM WARRANTY

This software is protected by both United States copyright law and international copyright treaty provision. You must treat this software just like a book. By saying "just like a book," McGraw-Hill means, for example, that this software may be used by any number of people and may be freely moved from one computer location to another, so long as there is no possibility of its being used at one location or on one computer while it also is being used at another. Just as a book cannot be read by two different people in two different places at the same time, neither can the software be used by two different people in two different places at the same time (unless, of course, McGraw-Hill's copyright is being violated).

LIMITED WARRANTY

McGraw-Hill takes great care to provide you with top-quality software, thoroughly checked to prevent virus infections. McGraw-Hill warrants the physical CD-ROM contained herein to be free of defects in materials and workmanship for a period of sixty days from the purchase date. If McGraw-Hill receives written notification within the warranty period of defects in materials or workmanship, and such notification is determined by McGraw-Hill to be correct, McGraw-Hill will replace the defective CD-ROM. Send requests to:

McGraw-Hill
Customer Services
P.O. Box 545
Blacklick, OH 43004-0545

The entire and exclusive liability and remedy for breach of this Limited Warranty shall be limited to replacement of a defective CD-ROM and shall not include or extend to any claim for or right to cover any other damages, including but not limited to, loss of profit, data, or use of the software, or special, incidental, or consequential damages or other similar claims, even if McGraw-Hill has been specifically advised of the possibility of such damages. In no event will McGraw-Hill's liability for any damages to you or any other person ever exceed the lower of suggested list price or actual price paid for the license to use the software, regardless of any form of the claim.

McGRAW-HILL, SPECIFICALLY DISCLAIMS ALL OTHER WARRANTIES, EXPRESS OR IMPLIED, INCLUDING, BUT NOT LIMITED TO, ANY IMPLIED WARRANTY OF MERCHANTABILITY OR FITNESS FOR A PARTICULAR PURPOSE.

Specifically, McGraw-Hill makes no representation or warranty that the software is fit for any particular purpose and any implied warranty of merchantability is limited to the sixty-day duration of the Limited Warranty covering the physical CD-ROM only (and not the software) and is otherwise expressly and specifically disclaimed.

This limited warranty gives you specific legal rights; you may have others which may vary from state to state. Some states do not allow the exclusion of incidental or consequential damages, or the limitation on how long an implied warranty lasts, so some of the above may not apply to you.